Making Sense of Construction Improvement

Making Sense of Construction Improvement

Stuart D. Green
School of Construction Management and Engineering
University of Reading

⟨W⟩WILEY-BLACKWELL

A John Wiley & Sons, Ltd., Publication

This edition first published 2011
© 2011 Stuart D. Green

Blackwell Publishing was acquired by John Wiley & Sons in February 2007.
Blackwell's publishing program has been merged with Wiley's global Scientific,
Technical and Medical business to form Wiley-Blackwell.

Registered office
John Wiley & Sons, Ltd, The Atrium, Southern Gate, Chichester, West Sussex,
PO19 8SQ, UK

Editorial offices
9600 Garsington Road, Oxford, OX4 2DQ, UK
The Atrium, Southern Gate, Chichester, West Sussex, PO19 8SQ, UK
2121 State Avenue, Ames, Iowa 50014-8300, USA

For details of our global editorial offices, for customer services and for information
about how to apply for permission to reuse the copyright material in this book
please see our website at www.wiley.com/wiley-blackwell.

The right of the author to be identified as the author of this work has been asserted
in accordance with the UK Copyright, Designs and Patents Act 1988.

Library of Congress Cataloging-in-Publication Data
Green, Stuart, 1958–
Making sense of construction improvement / Stuart D. Green.
    p.   cm.
  Includes bibliographical references and index.
  ISBN 978-1-4051-3046-2 (pbk.)
1. Construction industry–Management.   2. Economic history.
3. Building–Superintendence.   I. Title.
  HD9715.A2G698 2011
  690.068–dc22

                                                    2010051095

A catalogue record for this book is available from the British Library.

This book is published in the following electronic formats: ePDF 9781444341072;
Wiley Online Library 9781444341102; ePub 9781444341089; Mobi 9781444341096

Set in 9.5/12.5pt Palatino by SPi Publisher Services, Pondicherry, India
Printed and bound in Malaysia by Vivar Printing Sdn Bhd

1   2011

Contents

Preface

This book is about construction improvement. More specifically, it is about making sense of the construction industry debate. It is based on 30 years' experience working in, or interacting with, the UK construction sector. In many respects, the book is an account of a personal journey. My engagement with construction firms commenced in 1979 following my graduation from the University of Birmingham with a degree in civil engineering. Following several years' experience with a national contracting firm, I returned to academia in 1985 to study for an MSc in Construction Management at Heriot-Watt University in Edinburgh. Thereafter, I was fortunate enough to stumble into a research career in the School of Construction Management and Engineering at the University of Reading. Initially, I looked at construction management as most engineers tend to look at it. My primary interest was in developing tools and techniques which could improve construction management practice. This remains a worthwhile aim for many, but it is not the primary focus of this book.

In contrast to other textbooks, the aim of the current book is not to offer advice on how to manage construction projects more effectively. What the book does attempt to do is to understand the forces which have shaped the construction sector improvement agenda over time. The book was initially predicated on a sense of unease with the way fashionable improvement recipes such as business process re-engineering (BPR), partnering and lean construction were promoted in the construction sector. Particular motivation was derived from the publication of *Rethinking Construction* (Egan, 1998). Perhaps it was the iconoclast within me which caused me to question what it seemed everybody else was taking for granted. Senior colleagues at the time even warned me that to go against the grain of the accepted orthodoxy would have an adverse impact on my career. But it rapidly became apparent that many industry leaders shared my reservations about the so-called 'Egan agenda', even if they were not at the time willing to say so in public. There is of course an argument that I have myself become unhealthily preoccupied with *Rethinking Construction*, and am hence guilty of the same obsession as which I accuse others. But it remains the case that the construction improvement continues to be linked back to the Egan report with little recognition of what had gone before. It is almost as if what happened prior to 1998 to shape the current industry is of no importance. Recognition is occasionally given to the contribution of *Constructing the Team* (Latham, 1994), but more often than not the role of Latham is increasingly reduced to 'paving the way for Egan'.

Recurring arguments

The motivation to write the current book was also derived from the disinclination of many sources on construction management to place the advice offered in the broader context of industry change. The overriding tendency is to point towards the latest new management panacea, and then to advocate its application to the construction sector as an essential component of 'best practice'. Such recipes are invariably sourced from other industrial sectors and all too often are promoted as generic instrumental solutions that are universally applicable (Bresnen and Marshall, 2001). There is a repetitive pattern to the arguments mobilised that is nothing if not predicable. The tendency is for the advocates of any new improvement recipe to claim – usually with great rhetorical flourish – that it has an established theoretical base and an outstanding track record in other sectors. The case would then be made for its application to construction, with particular emphasis on the 'barriers' which need to be overcome. A commonly cited barrier is the construction sector's fragmented structure, which is invariably seen to present significant challenges to the implementation of the latest management thinking. But an even greater challenge is presented by the construction sector's supposedly ingrained 'regressive attitudes'; otherwise presented as its 'adversarial culture'. In other words, construction practitioners are out-of-touch with the latest management thinking; they are rooted in an old-dated paradigm; they are old-fashioned; they are generally somewhat ill-informed and backward (see Fernie *et al.*, 2006). Such an interpretation categorically does not accord with the author's personal experience. The people who work in construction are the same people with whom we all socialise every weekend. They have the same strengths and the same weaknesses; they have varying levels of education and they possess the same diversity of political opinion as can be found within the broader society within which they are embedded. From the outset, we should therefore be wary of interpretations which rest on the assumption that the people who work in the construction sector are inherently different from those who work in other sectors. We should be even more wary of claims that the construction sector has an 'adversarial culture' which can be changed simply through the adoption of advanced management techniques.

Management fashions

It will be argued throughout the current book that many of the improvement recipes which are directed at the construction sector owe more to management fashion than to scientific endeavour. While the notion of 'fashion' is undoubtedly pejorative, this does not mean that management fashions can be dismissed as irrelevant to 'real world' issues. The way managerial discourses are mobilised and diffused impacts directly upon the material reality of the construction sector. This is because they provide the narratives which practitioners use to shape their working lives. But such narratives are only persuasive in the first place if they provide practitioners with the means of making sense of the changes they observe happening around them. The contention is therefore that fashionable management improvement recipes often reflect and reinforce pre-existing dynamics of change within the sector. Rather than viewing each improvement recipe as a new panacea for the industry's problems, the aim is to understand why such recipes become fashionable in the first place. Hence the book searches for a broader understanding of the construction improvement debate which strays beyond the boundaries of narrowly-defined instrumental rationality. The overriding message is the need to understand construction management improvement recipes in the context of ongoing sectoral structural change. Of particular importance is the need to understand how the debates of the present are shaped inexorably by the events of the past.

Precisely what constitutes 'good management' is further shaped and conditioned by the prevailing political consensus. Discourses of management only become fashionable if they concur with the mood of time. For example, BPR became popular during early 1990s because it reflected and reinforced the prevailing belief that organisations were over-sized and needed to be slimmed down. But in truth, downsizing and outsourcing were already part of the corporate landscape prior to the ideas of Hammer and Champy (1993). BPR was persuasive because it provided middle managers with the means of making sense of what they observed happening around them. Even more importantly, it provided them with a means of bolstering their own self-identities in a threatening and increasingly uncertain environment. Simply put, it was far better to become an advocate of BPR than to be a victim. All other things being equal, middle managers who are open to management fashions have a propensity to remain in gainful employment for longer than those who habitually dismiss such ideas as short-lived fads. It tends to be much better to be part of the 'groovy gang' who are seen to be advocating enlightened management ideas, rather to be perceived as a 'dinosaur' standing in the way of progress.

The ebb and flow of management fashions also points towards the highly transient nature of 'best practice'. The very idea of best practice is suggestive of the 'one best way' promoted by the advocates of scientific management in the early 1900s (Taylor, 1911). To have any expectation of finding optimal ways of working irrespective of context is to deny the influential ideas of contingency theory (Burns and Stalker; Lawrence and Lorsh, 1967) In short, notions of best practice are inevitably time and context specific. Hence universally accepted definitions of 'best practice' will forever remain elusive. In the construction sector, continuously shifting policy objectives render any fixed interpretation of best practice especially challenging.

Even within relatively localised contexts, the existence of multiple interest groups further undermines the idea of that there can be a single model of best practice which serves the needs the all interest groups. 'Best practice for whom?' is the question which must be continually asked. Throughout the course of this book it will become apparent that arguments in support of 'best practice' are invariably rhetorical devices which are mobilised by different interest groups in a continuous competition for power and influence. This again points towards the need for the infrastructure of industry improvement to be an essential part of any analysis. The position adopted throughout the current book is that best practice is a concept which is socially negotiated and continuously contested over time. Indeed, the very idea of best practice is best understood as a management fashion.

An unfolding story

In seeking to understand the context within which construction improvement is enacted, it is clearly also necessary to understand the structural characteristics of the construction sector. Unfortunately, management improvement recipes are invariably considered entirely separately from economic analyses of industry change (and *vice versa*). This has more to do with the institutional separation of management scholars from economists than with anything which occurs in practice. Following Pettigrew (1997), management practice is not only *shaped* by the structural characteristics of a particular industry sector, but it is also actively engaged in *shaping* those same structural characteristics. Popularised construction improvement recipes frequently act to inform action, which in turn acts to create the constraints within which practitioners operate. This immediately conjures up the idea of a continuously unfolding interaction between industry structure and management practice. One cannot be considered in isolation of the other.

Hence attention is given to the way in which events have unfolded over time, and the language within which the debate has been conducted. Acknowledgement is given to the active role played by key individuals at different points in time. But attention is also afforded to the evolving infrastructure of construction improvement within which the debates about construction improvement are situated. At the same time, it is clear that debates within the construction sector at any particular point in time cannot be understood in isolation from the prevailing socio-political consensus. It has therefore been necessary to chart the way in which the discourse of construction improvement has been influenced continuously by political shifts within government. General election results have frequently had a profound subsequent impact on the construction improvement debate. Governments of course tend to change on very specific dates immediately following general elections, whereas political discourses are in a state of continuous flux.

At any one point in time there are several counter discourses competing for attention. Likewise there are invariably several differing discourses of industry improvement plying for the attention of practising managers. In pointing to the relationship between management discourses and the broader socio-political consensus it is important to avoid any argument of causality. Political and management discourses are to a large extent mutually constituted, and both to no small extent rely on counter discourses against which they can be positioned. Discourses can further be seen to comprise a complex web of institutions and policy initiatives with direct material consequences, as well as rhetorical components. Hence it is possible to perceive structural change in the form of outsourcing and downsizing as part of the same discourse as the justifying narrative of BPR.

Political shifts are in turn frequently influenced by global events, coverage of which is limited to those which are considered essential for the purposes of telling the story. The essential problem with construction improvement is that it is forever chasing a moving target. Even if the multitude of interest groups within the sector were to focus their attention effectively upon a single set of shared objectives, it likely to be rapidly replaced by some other set of imperatives as a result of a shift in the broader policy environment. The construction sector has long-since been central to government policy objectives in housing, health, education and transport. Any shift in policy in any of these areas inevitably has implications for the construction sector. More recently, construction has found itself at the centre of society's aspirations for social, economic and environmental sustainability, and hence prone to any shift in emphasis between these three often conflicting agendas.

Arguments beyond efficiency

In light of the construction sector's sensitivity to subtle shifts in government policy, it is striking how much of the improvement debate remains focused on narrow issues of cost efficiency. To focus on improving efficiency is of course no bad thing. Few people in the construction sector (or elsewhere) argue in favour of inefficiency. But to focus on efficiency alone – to the detriment of all other considerations – is undoubtedly a bad thing. Yet it is not cost efficiency that the sector's senior managers spend the majority of their time worrying about. They tend to be much more concerned with continuity of workload, and in consequence most successful companies have learned to expand and contract rapidly in the face of fluctuating levels of demand. Contracting firms tend to privilege structural flexibility over narrowly-defined efficiency; the risks of being inefficient can always be delegated to subcontractors, who can in turn delegate them to their own sub-subcontractors.

It is possible for individual firms to become very efficient while at the same going out of business. The defining issue here is often an inability to develop new responses to changing circumstances. For example, there is little point in becoming very efficient at providing a service for which there is a diminishing demand. A rather more sophisticated storyline would seek to balance the need for efficiency in the operation of current activities while at the same time developing effective responses to emerging opportunities. Strangely, the construction improvement agenda rarely engages with the real day-to-day challenges faced by managers in the construction sector. It is much easier to advocate supposedly generic instrumental solutions with track records of success in other sectors.

There is also a recurring assumption that if individual firms can be made to be more efficient the benefits will aggregate to the sector as a whole. Unfortunately, there is little evidence to support this belief. Such arguments tend to take it entirely for granted that there is a common understanding about what the construction sector is *for*. In truth, such a consensus has never existed, and never will. Different interest groups will always articulate different objectives for the construction sector. Clients struggle consistently to find common ground with the aspirations of general contracting firms, which in turn rarely have shared interests with specialist suppliers. Even clients can hardly be characterised as a single coherent group with a shared set of interests, and public sector clients frequently utilise procurement as means of achieving broader government policy objectives. The latter tendency was less evident during the 1980s when the 'enterprise culture' was dominant, but was very much in evidence under the New Labour government of Tony Blair. Contracting firms are nothing if not resourceful, and responding to ever changing sets of requirements has become an undoubted strength of the sector.

Definition of construction sector

It must be conceded out the outset that the construction sector is not a homogeneous entity. The author is well-aware of the continuing debate about boundary definition (e.g. Pearce, 2003) and the contention that the construction industry is not a single industry, but several separate sub-industries each with its own distinguishing characteristics (Ive and Gruneberg, 2000). The most notable distinction lies between building and civil engineering. Yet even here there is considerable overlap in the use of materials, labour and plant, and the activities of many contractors span across both building and civil engineering (Hillebrandt, 1984).

Pearce (2003) differentiates between narrow and broad definitions of the construction sector. The narrow definition limits itself to on-site assembly together with repair and maintenance activities carried out by contractors. This interpretation concurs with the definition of construction enshrined within Division 45 of the Standard Industrial Classification (SIC) developed by the Office for National Statistics (2003). This narrow interpretation excludes those involved in professional services such as surveying, architecture and engineering. It also excludes those who involved in self-build and those who are directly employed by organisations whose main business is something other than construction. Further confusion has been caused as construction companies have also become increasingly diversified in terms of their operations. Some have even had cause to re-list themselves on the Stock Exchange as service companies.

Reports such as Emmerson (1962) and Banwell (1964) tended to refer to the 'construction industries' in the plural, whereas Latham (1994) and Egan (1998) both refer to the 'construction industry' as if it were a single homogeneous entity. The modern practice is to refer to the construction sector. It will be seen that the construction improvement agenda up until, and including *Rethinking Construction* (Egan, 1998) was commonly limited to Pearce's (2003) narrow definition. However, more recently there has been a tendency within the improvement debate to include professional services within the boundary of the construction sector, and even to some extent the construction materials suppliers.

The position adopted in the current book is that the boundary which is drawn around the 'construction sector' is not a matter of fact, but is something which is continuously renegotiated. The book therefore does not set out to confine itself to any particular false boundary, but rather to follow the coverage of the construction improvement debate in all of its glorious imprecision. It must also be recognised that even the idea of a 'national' construction sector is increasingly challenged by globalisation. The cutting edge of the modern UK construction sector undoubtedly bears little resemblance to the 'construction industry' of popular perception. The leading design and engineering consultancy firms increasingly compete highly effectively across global markets at the forefront of the knowledge economy. Even relatively modest local construction projects increasingly rely on globally distributed networks of value-adding services. The construction improvement debate has yet to engage with this end of the construction business. But we must also remind ourselves that construction still requires labour to work on site in conditions which are invariably much less than perfect. In the final analysis, construction is a people business. Even if the narrow definition is adopted, the UK construction sector employs more than 2 million people and comprises approximately 8.5% to GDP. If the broader definition is adopted to include the whole construction value chain (i.e. architecture, engineering and construction products) the respective figures rise to 3 million and 10% (LEK, 2009). In short, construction matters.

Timeline

To tell the story of how the debate about construction improvement has evolved over time, it is clearly necessary to select a starting point. In many respects, the choice of when to start the description is somewhat arbitrary.

It is always possible to argue that the selection of a particular date excludes important historical precedents. What is clear is the debate about construction improvement is not new; humankind has debated how to improve the building process ever since we gave up on cave dwelling. A case could be made for beginning the story with the development of the general contracting system during the industrial revolution. Arguments could also be made that the modern structure of the construction sector owes much to the craft guilds which were established during the fifteenth and sixteenth centuries. However, the present book is not intended to be a work of history, but to offer to contextualised account of the contemporary construction improvement debate. The concern is with the modern construction sector, and its relationship with the modern economy. For this reason the story commences in the immediate aftermath of the Second World War.

Structure of the book

Chapter 1 charts the various influences on the construction improvement debate from the aftermath of the Second World War through to the election of Margaret Thatcher in 1979. Attention is given to the industrial unrest of the 1970s and the collapse of the post-war social consensus. Chapter 2 charts the rise of what has become known as the 'enterprise culture' in the United Kingdom, both in terms of its constituent policies and its impact on management practice. Chapter 3 addresses the material manifestation of the enterprise culture in the construction sector, giving particular attention to the diversification of client demand and the emergence of the hollowed-out contracting firm.

Chapter 4 analyses the way in which the construction improvement agenda took shape during the course of the 1980s and 90s. Consideration extends to the extent to which the improvement agenda both reflected and reinforced the discourse of enterprise. Particular attention is given to *Constructing the Team* (Latham, 1994) and the resultant infrastructure of industry improvement. Chapter 5 is dedicated entirely to a critical review of *Rethinking Construction* (Egan, 1998), with due acknowledgement to the context within which it was produced. As with the preceding chapter, consideration is extended to the organisational infrastructure subsequently tasked with implementation. It will be argued that *Rethinking Construction* deserves special consideration because of its pivotal position in the modern improvement debate. Chapter 6 breaks from the chronological storyline to challenge the way the improvement debate is too often limited to narrow issues of machine efficiency. Attention is given to a range of alternative metaphorical perspectives which can be used to understand organisations. Particular attention is given to understanding client organisations.

Chapter 7 directs critical attention at two of the most popular management improvement recipes of the 1990s: business process re-engineering (BPR) and partnering. Both are critiqued in terms of their vagueness of definition and also in terms of their underlying metaphors. It is argued that both are best understood in the context of structural change in the form of privatisation, downsizing and outsourcing. Chapter 8 directs similar critical attention at lean construction, another improvement recipe advocated by *Rethinking Construction*. In common with BPR and partnering, lean construction is also found to suffer from an essential definitional elusiveness. The discourse of lean thinking is further argued to have been popular because of the way it reflected and reinforced ways of organising which had already been adopted.

Chapter 9 extends the discussion to the shifting policy objectives of the New Labour government from the election of Tony Blair in 1997 through to the uncontested succession of Gordon Brown in 2007. Particular attention is given to the policy objective of combining enterprise with aspirations of social democracy and its implications for the improvement debate in the construction sector. Coverage includes a review of *Modernising Construction* (NAO, 2001) and *Accelerating Change* (Strategic Forum, 2002). Chapter 10 bring the discussion up to date with a review of the current 'legacy of dilemmas'. Attention is given the changing infrastructure of construction improvement together the disconnected agendas which characterise the current debate. Consideration is also given to the improvement discourses which flourished during the New Labour years, especially those relating to the quality of design and the notion of a value-creating built environment industry. Coverage is concluded with a consideration of health and safety in the construction sector and a brief review of the Wolstenholme (2009) report, *Never Waste a Good Crisis*. A final postscript addresses the likely implications for construction improvement of the advent of the coalition government led by David Cameron which was formed in Spring 2010.

Acknowledgements

In writing a book such as this I have undoubtedly been influenced by many of the colleagues with whom I have worked at the University of Reading. Brian Atkin, John Bennett and Roger Flanagan in no small way acted as role models for how academics should behave in a university environment. But perhaps I have learned most from the research fellows with whom I have been fortunate to work since I joined the University of Reading in 1987. Their ideas inevitably over time became my ideas, and in return I can only hope that some of my ideas also became theirs. The following are especially deserving of mention: Scott Fernie, Chris Harty, Chung-Chin Kao, Graeme Larsen, Roine Leiringer, Susan May and Steph Weller. Several of these have since progressed to bigger and better things; some will undoubtedly go on to write better books than this one. But in citing the contributions of others, it is also necessary to make it clear that the responsibility for any faults and misconceptions lie clearly with the author. Notwithstanding the contributions of colleagues, my main thanks must lie with my family for staying with me as I have struggled towards completion.

For the benefit of readers who have followed my work over the years, it should also be acknowledged that sections of the text are in part based on previous work. For example, the discussion in Chapter 6 on organisational metaphors comprises an extension of ideas previously published in Green (1996) (which in turn drew heavily from the ideas of Morgan, 2006). The discussion of business process re-engineering in Chapter 8 likewise adapts material from Green (1998) and the section on partnering in the same chapter similarly updates material previously published as Green (1999a). Finally, the discussion of lean construction in Chapter 9 draws from polemics previously published as Green (1999b; 1999c) and from empirical research previously presented in Green and May (2005).

References

Banwell, Sir Harold (1964) *The Placing and Management of Contracts for Building and Civil Engineering Work*. HMSO, London.

Bresnen, M. and Marshall, N. (2001) Understanding the diffusion and application of new management ideas, *Engineering, Construction and Architectural Management*, **8** (5/6), 335–345.

Burns, T. and Stalker, G.M. (1961) *The Management of Innovation*, Tavistock, London.

Egan, Sir John. (1998) Rethinking Construction. Report of the Construction Task Force to the Deputy Prime Minister, John Prescott, on the scope for improving the quality and efficiency of UK construction. Department of the Environment, Transport and the Regions, London.

Emmerson, Sir Harold (1962) *Survey of Problems Before the Construction Industries*. HMSO, London.

Fernie, S., Leiringer, R. and Thorpe, T. (2006) Change in construction: a critical perspective. *Building Research and Information*, **34**(2), 91–103.

Green, S.D. (1996) A metaphorical analysis of client organisations and the briefing process, *Construction Management and Economics*, **14**(2), 155–164.

Green, S.D. (1998) The technocratic totalitarianism of construction process improvement: a critical perspective, *Engineering, Construction and Architectural Management*, **5**(4), 376–386.

Green, S.D. and May, S.C. (2005) Lean construction: arenas of enactment, models of diffusion and the meaning of 'leanness', *Building Research & Information*, **33**(6), 498–511.

Green, S.D. (1999a) Partnering: the propaganda of corporatism?, *Journal of Construction Procurement*, **5**(2), 177–186.

Green, S.D. (1999b) The missing arguments of lean construction, *Construction Management and Economics*, **17**(2), 133–137.

Green, S.D. (1999c) The dark side of lean construction: exploitation and ideology, in *Proc. of the 7th Conference of the International Group for Lean Construction*

(IGLC-7), (ed. I. D. Tommelein), University of California, Berkeley, USA, pp. 21–32.

Green, S.D. and May, S.C. (2005) Lean construction: arenas of enactment, models of diffusion and the meaning of 'leanness', *Building Research & Information*, **33**(6), 498–511.

Hammer, M. and Champy, J. (1993) *Re-engineering the Corporation*, Harper Collins, London.

Hillebrandt, P.M. (1984) *Analysis of the British Construction Industry*, Macmillan, London.

Ive, G.L. and Gruneberg, S.L. (2000) *The Economics of the Modern Construction Sector*, Macmillan, Basingstoke.

Latham, Sir Michael (1994) *Constructing the Team*. Final report of the Government/industry review of procurement and contractual arrangements in the UK construction industry. HMSO, London.

Lawrence, P.R. and Lorsh, J.W. (1967) *Organization and Environment*, Harvard Press, Cambridge, Mass.

LEK (2009) *Construction in the UK Economy: the Benefits of Investment*, L.E.K. Consulting, London.

NAO (2001) Modernising Construction. Report by the Comptrollor and Auditor General of the National Audit Office, The Stationery Office, London.

Office for National Statistics (2003) *UK Standard Industrial Classification of Economic Activities 2003*, Stationery Office, London.

Pearce, D. (2003) *The social and economic value of construction: the construction industry's contribution to sustainable development*. nCRISP, Davis Langdon Consultancy, London.

Pettigrew, A.M. (1997) What is processual analysis? *Scandinavian Journal of Management*, **13**(4), 1–31.

Strategic Forum (2002) *Accelerating Change*, Rethinking Construction, London.

Taylor, F.W. (1911) *Principles of Scientific Management*, Harper & Row, New York.

Wolstenholme, A. (2009) *Never Waste a Good Crisis*, Constructing Excellence in the Built Environment, London.

1 Construction in the Age of the Planned Economy

1.1 Introduction

This chapter begins with the election of Clement Attlee as Prime Minister on 5 July 1945 in the final weeks of the Second World War. This was the first Labour administration to be elected with a sizable majority and to its full term of office. Britain at the time was essentially bankrupt and was faced with a vast legacy of challenges as it faced up to the demobilisation of its armed forces. Housing was a top government priority together with the modernisation of British industry. The Atlee government is further credited with creating the British Welfare State, the legacy of which has lasted until the present day. Atlee's government was followed by a series of Conservative governments which did little to unpick the broad political consensus which characterised the immediate post-war era. The Labour Party returned to power in 1964 under the leadership of Harold Wilson with a renewed commitment to a planned economy.

The post-war years were marked by Britain's dramatic decline as a global power. Increasing exposure to global competition exposed the long-term structural weaknesses of British industry which was hampered by decades of under-investment. Nevertheless, the domestic standard of living improved dramatically as the worst excesses of poverty were eliminated. However, by the late 1960s the post-war social consensus was starting to crack and Britain's industrial decline was exacerbated by widespread industrial unrest. In the early 1970s strikes became part of the fabric of industrial life as militant trade unionism became more widespread. Although some industrial action was undoubtedly driven by an overtly political agenda, much was prompted by justifiable grievances and poor management. The national building strike of 1972 left a legacy of bitterness within the construction sector and was undoubtedly a turning point in the industry's development, with significant long-term implications for patterns of employment. Things were never quite the same after the national building strike as they had been before.

Making Sense of Construction Improvement, First Edition. Stuart D. Green.
© 2011 Stuart D. Green. Published 2011 by Blackwell Publishing Ltd.

The early 1960s saw two important government-commissioned reports which will be referred back to on numerous occasions through this book. The Emmerson report of 1962 presented on a 'survey of problems before the construction industries' whereas the Banwell report of 1964 was concerned primarily with the placing and management of contracts. Many of the recommendations of these two reports are as valid today as they were when they were published. It is certainly possible to trace subsequent exhortations in favour of partnering and teamwork back to these early reports. However, the Emmerson (1962) and Banwell (1964) reports are also indicative of a world which longer exists. They were written in an era of relative stability which was characterised by a widespread faith in government intervention. At the time it was widely accepted across the political spectrum that there was a 'mutually of responsibility' for construction sector development which extended across government and industry. There was much talk of demand management in order to ensure the steady and predictable flow of work which was seen to be essential if the private construction sector was to invest in the improvement of efficiency. However, the world was changing, and the carefully-crafted construction improvement debate of the 1960s was soon to be overtaken by events.

1.2   An inherited legacy

1.2.1   The task of nation building

The election of 1945 was undoubtedly a pivotal turning point in Britain's modern political history. Britain's war-time leader, Winston Churchill, was voted out as surplus to requirements as the electorate put its faith in a new beginning. Middle Britain was perhaps unsure whether or not it really wanted the socialism that was on offer, but it was sure that it did not want a return to the unemployment and social insecurity of the 1930s. The incoming government was committed to an extensive programme of nationalisation, a significant expansion of social welfare and the achievement of full employment. Jobs, housing and social security were seen as a just return for the sacrifices made during the war. The policy agenda was in no small way shaped by the experience of massive government intervention during the war. Centralised state planning had been instrumental in defeating fascism and overcoming the malaise of the Great Depression. The organisation of resources necessary for a successful war effort had been deemed to be beyond the means of the private sector with the result that large swathes of the British economy had been under *de facto* government control. The interventionist philosophy prevailed beyond the war by means of the Beveridge settlement in the United Kingdom and Roosevelt's New Deal in the United States. The post-war social consensus on both sides of the Atlantic derived its theoretical support from Keynesian economics, which questioned the validity of the classical economic doctrine of *laissez-faire* and emphasised the need for government intervention.

The task of the newly elected government was seen in terms of nation building. The Atlee administration is best remembered for laying the foundations of the welfare state. But it was also committed to the modernisation of British industry, which had been starved of meaningful investment for decades. Long-term economic advantage had been sacrificed in favour of short-term military imperatives. The legacy of victory included obsolete capital equipment, an exhausted labour force and massive damage to the housing stock (Tomlinson, 1997). But perhaps the biggest legacy was a huge balance of payments deficit. The problems faced by the building industry were a microcosm of the problems faced by the British economy at large. It was outdated, inefficient, under-financed and hugely fragmented. Yet at the same time the building industry was of central importance to the government's agenda across a range of policy areas. Education, health and, most importantly, housing all depended upon an efficient building industry

1.2.2   *Homes for heroes*

The expression 'homes fit for heroes' was originally coined in the wake of World War I. it was the Housing Act 1919 which first required local authorities to provide housing with the help of central-government subsidies. The housing shortage returned to the top of the political agenda following World War II. Britain at the time was not only struggling to recover from the economic cost of the war, but was also struggling to come to terms with its loss of status as a global power. In rapid succession, former colonies were granted independence throughout the 1950s, with the consequent loss of protected markets for British manufactured products. Britain remained heavily indebted to the USA as a result of the wartime lend-lease scheme, and in consequence lacked the capital to modernise either its infrastructure or its manufacturing base. Relative to its major global competitors, Britain was undoubtedly in decline. But in absolute terms, the economy enjoyed steady growth and society became increasingly affluent. The late 1950s saw the beginning of the consumer revolution as Britain slowly shrugged off the austerity of the immediate post-war period. The building industry was of central importance to the government's economic policy, and there was widespread concern among policy makers with regard to its fitness to perform. In terms of industry structure, little had changed since the recommendations of the Simon report in 1944.

The Labour government of the immediate post-war period was especially obsessed with housing. Half a million hones had been destroyed or rendered uninhabitable by the Luftwaffe. Overall, a staggering 25% of the housing stock had been damaged during the course of the war (Marr, 2007). Urban landscapes throughout Britain were blighted by derelict and uninhabitable houses. The post-victory demobilisation of the armed forces added a sudden and unprecedented demand for family accommodation which was in short supply. Many major cities still exhibited the social legacy of the industrial revolution. The aspiration was for 'homes for heroes'; the reality was semi-derelict urban areas and dilapidated slums inherited from a previous era. Inner-city areas were characterised by outdated terraced housing which frequently lacked adequate sanitation. Illegal squatting was widespread as homeless families took action into their own hands. Something had to be done.

Output expanded rapidly from 1945 with a view to alleviating the housing shortage. In the immediate post-war years relief came in the form of prefabricated factory assembled housing units, often put together in hastily converted aircraft factories. From the late 1950s and throughout the 1960s emphasis shifted to high-rise tower blocks. Local authority output peaked in 1953 at 198 000 units, although an increasingly vibrant private sector contributed to an overall peak of 352 000 units in 1968. This was a level of housing output which has yet to be surpassed, and in all likelihood never will be. In accordance with the dominant ethos of the day, the commitment to public housing prevailed throughout the 1960s. Modernity had arrived, and Victorian terraces were replaced by symbols of a new age, complete with fitted kitchens and up-to-date bathrooms. Thousands of almost identical tower blocks were built throughout Britain, transforming the urban landscape and severely disrupting social cohesion. Too often it was assumed that entire communities could be transferred to high-rise accommodation, but problems soon became apparent. Lifts were easily vandalised leaving families stranded in mid air. Shops and schools were located too far away. In the haste to build high and build quickly, the quality of construction often suffered. Too many flats had condensation problems from the outset. Insufficient consideration was given to sound insulation such that occupants were afforded little privacy. One problem family in a single block could rapidly make life unbearable for all. Local authorities were encouraged to build high by central government. Corruption flourished as builders shared the available subsidies with dishonest local officials. Quality was widely compromised as the optimism of the 1960s gave way to the realisation that the old-fashioned slums of the 1940s had been replaced by the modern slums of the 1970s. This was not the construction sector's finest hour.

1.2.3 Ronan point

The problems with quality of high-rise construction were typified by the Ronan Point disaster in London's East End in 1968. A gas explosion on the eighteenth floor of a newly-constructed high-rise residential block in Newham caused a chain reaction resulting in the collapse of an entire corner of the block. The lift had stopped working and families had to evacuate in their nightclothes via the staircases. The building had been occupied for just two months. It had been built by Taylor Woodrow Anglian under contract from Newham Council. Eleven residents were injured and four were killed. The block had been constructed using Large Panel System (LPS) building whereby prefabricated cladding sections were constructed in a factory environment and then bolted into place on site. The consequent public inquiry concluded that the Ronan Point block had been structurally unsound, and the building regulations were later changed to ensure that concrete slabs on large panel systems could withstand explosions. In the case of Ronan Point, the issue was primarily about the lack of restraint ties to withstand an explosive force; it was not about poor workmanship. But poor workmanship and lack of adequate on-site supervision were undoubtedly widespread in the rush to build homes as high and as quickly as possible.

The Ronan Point block was quickly rebuilt, but its metaphorical shadow hangs over the UK construction sector to this day. The incident led to a major backlash against high-rise residential blocks as a solution to the post-war housing shortage. The emphasis thereafter shifted back to low-rise housing with a consequent drop in the rate of output. In a microcosm of the broader trend, Ronan Point itself was knocked down in 1986 and replaced with terraced housing. The high-rise developments of the 1960s tarnished the reputation of prefabricated building systems which in popular perception became (rather unfairly) synonymous with low quality.

There is of course no inevitable link between the use of prefabricated building components and poor quality. But such a link had been created in the public's perception. The Ronan Point disaster also left a professional bias against system building which remains largely intact. The real quality problems associated with the high-rise experience of the 1960s owed more to inadequate research prior to widespread adoption of new construction techniques than to prefabrication *per se*. Insufficient attention was given to on-site supervision and responsibility for the quality of the finished product was dissipated between too many parties.

1.2.4 The Poulson scandal

The credibility of high-rise tower blocks was further dented by a public perception of widespread corruption. The extent of corruption was personified by the Poulson scandal which came to public light in 1972. John Poulson was an architect who was found to have been involved in the extensive bribery of public officials at all levels in return for the award of building contracts. Poulson was tried for bribery and corruption in 1973 at Leeds Crown Court. He was found to have given away more than £500 000 in gifts in order to win contracts and was sentenced to five years in prison, subsequently extended by a further two years. Also convicted was T. Dan Smith, who had been the Labour leader of Newcastle Council in the 1960s. Smith was typical of the 'local authority barons' who prevailed at the time. In addition to being leader of the council, Smith also chaired the planning, finance and housing committees. It was Smith who resided over the demolition of the slums of Scotswood and their subsequent replacement by high-rise tower blocks. He was a powerful figure in the North East of England and was exceedingly well-connected within the Labour Party (Osler, 2002). By 1965 he had been appointed chairman of the Northern Economic Planning Council. Poulson meanwhile had built a huge architectural and planning practice on the back of the boom in high-rise construction. He had also been hugely successful in winning work from the newly nationalised industries within which he also cultivated powerful contacts. Poulson's practice eventually went bankrupt in 1972 revealing a range of illegal payments to companies owned by Smith. Poulson's network of contacts included several major building contractors and extended throughout the political establishment of the day. The list of those who had accepted his favours included several MPs, police officers, health authorities and civil servants. In addition to Smith and Poulson others to be jailed included Andrew Cunningham, leader of Durham Council, and George Pottinger, a senior civil servant with the Scottish Office. Other notable casualties included Reginald Maudling, the Conservative Home Secretary at the time. Although Maudling was not convicted of any wrongdoing, he was forced to resign on the announcement of the police investigation as a result of his strong connections with Poulson, including the receipt of 'gifts'.

The Poulson scandal was undoubtedly one of the biggest British corruption scandals of modern times, and had Poulson not spread his favours so evenly across the political parties the fall-out would have been much greater. The legacy of the Poulson affair continues to hang over the UK construction policy agenda, especially in respect of procedures for the award of public-sector contracts. Subsequent notable cases of public sector involvement in fraud include the Labour-run Doncaster Council in the 1990s, which resulted in the conviction of no less than 21 councillors. Even more notorious was the 'homes for votes' scandal involving Dame Shirley Porter, the leader of Conservative-run Westminster City Council in the late 1980s. We all like to think that British society is relatively free of corruption, but there is little room for complacency.

1.2.5 *Prefabrication discredited*

The case for prefabrication was to suffer a further blow as a result of a *World in Action* television exposé of prefabricated timber-frame houses broadcast in 1983. The programme focused on the quality problems with respect to a small group of timber-framed houses constructed in the West of England. The reported problems included a lack of fit between the prefabricated timber frame and the conventionally constructed foundations. Related concerns were expressed about the dwellings' lack of water-tightness, and hence their tendency towards subsequent decay and deterioration. The programme received massive publicity and severely dented the market for timber-framed housing. But in truth, the *World in Action* programme was based on investigative journalism of the worst kind. The extrapolation from a small sample to the industry at large was at best ill-informed and unrepresentative. A subsequent investigation by the Building Research Establishment (BRE) surveyed over 400 timber-framed dwellings and found no evidence in support of the problems predicted by the *World in Action* team (Ross, 2002). But the damage to the reputation of prefabricated timber frames had already been done. The exposé of timber-framed housing reawakened memories of Ronan Point and together they discredited the very idea of prefabrication. They also served to sharpen the public's perception that the building industry was a cowboy industry which consistently offered low quality. The industry was of course much less than perfect, but the problems with quality were undoubtedly grossly exaggerated. But unfortunately mud sticks, and it took the image of prefabricated construction decades to recover before being subsequently reinvented under the label of 'modern methods of construction' (MMC).

1.2.6 *The shadow of nationalisation*

Leaving housing aside, and returning to our account of the post-war building industry, for several decades after the Second World War nationalisation of the British building industry was held by many to be a serious possibility. Throughout the late 1940s and 1950s nationalisation was almost synonymous with modernisation. The coal industry had been nationalised in 1947 putting an end to years of cut-throat competition between private mine owners. The National Coal Board had taken possession of around 800 mines and became responsible for 750 000 employees. The electricity supply industry was also bought under public ownership. Nationalisation of the railways followed in 1948, together with the nationalisation of long-distance road haulage industry. Both the railways and road transport fell within the remit of the newly-created British Transport Commission. A further major nationalisation was that of the steel and iron industry, which was taken into state control in 1948. There was little opposition to the nationalisation of the coal industry and the railways as both were unprofitable. However, the nationalisation of the steel and iron industry and the road haulage sector had both been bitterly contested. But it is notable that when the Conservatives returned to power in 1951 they chose only to return road haulage to the private sector; the three major nationalised industries were retained under public ownership in line with the broad political consensus of the day. There were of course traditionalists within the Conservative Party who continued to equate nationalisation with Stalin's industrial policy in the Soviet Union. It must also be conceded that many within the socialist movement of the time were open admirers of the Soviet

Union. Mainstream opinion fell between these two extremes in favour of a mixed economy. But the nationalisation of large sectors of the British economy was widely accepted as an essential prerequisite of modernisation.

1.2.7 Modernisation stalled

The coal industry and the railways had undoubtedly suffered from decades of chronic under-investment prior to being nationalised. Both also suffered from fragmentation of ownership and damaging cut-throat competition. Much of British industry had of course been under *de facto* state control during the Second World War. Hence the transition to nationalisation was not as radical a change as it may now appear. Few would deny that nationalisation improved working conditions in the coal industry and the railways. For example, the National Coal Board introduced paid holidays, sick pay and rest homes for injured miners. Safety also became an increasingly important issue within all the major nationalised industries, and worker fatalities were no longer accepted as an inevitable risk. But nationalisation did not bring the hoped-for harmony in industrial relations. Miners and railway workers continued to agitate for a fairer share of the productivity gains. The coal industry was especially strike prone, with national stoppages in 1972 and 1974. The miners famously played a significant part in the downfall of the Conservative Prime Minister, Edward Heath. British Rail was somewhat less strike prone, with no national stoppage until the 1980s. So nationalisation did not mean the end of industrial unrest, but strikes were much less common than subsequent myth would have it.

The nationalised industries were at the forefront in the application of modern management techniques of the time. Centralising planning was very much in vogue and the principles of scientific management (Taylor, 1911) were implemented widely in the form of bonus schemes. On a more strategic level, the Operational Research Executive was especially influential within the National Coal Board and is still held in high esteem among operational research (OR) practitioners. The OR profession had cut its teeth on the organisation of the Atlantic conveys during World War II. It was therefore a relatively easy transition to state-controlled enterprises such as the National Coal Board and British Railways. It was some time later that OR techniques began to be applied within private sector companies. In the 1970s major contractors such as Laing and Wimpey maintained OR departments with a view to improving productivity. It is also notable that many of the early construction management textbooks were dominated by OR techniques such as critical path analysis, linear programming and statistical sampling (e.g. Barrie and Paulson, 1991; Pilcher, 1992). In today's climate it is easy to forget that these techniques were initially piloted primarily within the public sector. The discipline of project management was built similarly upon quantitative techniques such as the so-called Program

Evaluation and Review Technique (PERT) initially developed by the US Navy for the Polaris programme in the 1960s.

It follows from the above that if the UK building industry had been nationalised in the 1950s it might have made much faster progress in the implementation of modern management methods. Wholesale nationalisation of the building industry was never quite adopted as Labour Party policy. But there were numerous supporters of nationalisation within the Labour Party who continued to argue the case throughout the 1950s and 60s. Even as late as 1978 official Labour Party policy was advocating a measure of public ownership through the establishment of a national building corporation, and through the development and extension of local authority direct labour organisations (DLOs). The threat of building industry nationalisation was seen to be sufficient by many employers within the industry to sustain the Campaign Against Building Industry Nationalisation (CABIN) into the late 1970s. During the mid-1980s it was by no means unusual to see posters on building site huts advocating the importance of 'saying no to building nationalisation'. Such posters of course tended to disappear quickly at the hands of UCATT members. But what cannot be denied is that the proportion of the building industry under public control in the form of local authority Direct Labour Organisations (DLOs) continued to decline throughout the 1960s. The rate of decline was to increase significantly throughout the 1980s with a continued steady decline thereafter (see Chapter 2).

1.3   Improving construction

1.3.1   Survey of problems before the construction industry

A useful insight into the construction sector of the early 1960s is provided by the Emmerson (1962) report. The report was commissioned the previous autumn by Lord Hope, then Minister of Works. The report was commissioned in the context of increasing demand and an ongoing concern about construction sector efficiency (Moodley and Preece, 2003). It was the first major report on the construction sector since the end of the Second World War and provides an invaluable insight into the construction sector policy debate of the early 1960s. The title of the report was *Survey of Problems before the Construction Industry*, although Emmerson did note that the industry had recovered well from the disruption of the Second World War. He further noted that the industry had to date been flexible in meeting demand and had avoided large industrial disputes. Emmerson's general argument was that the problems that the industry faced were not of its own making. The report excluded explicitly Direct Labour Organisations (DLOs), presumably on the basis that these were exempt from the problems of inefficiency which prevailed within the private sector. This assumption was of course entirely consistent with the prevailing ethos of the day.

Sir Harold Emmerson was a civil servant who spoke with the tone and authority of the establishment which prevailed at the time. It is notable that he considered construction sector efficiency to be a cause worth pursuing in the national interest, rather than a worthwhile end in itself. Emmerson was sympathetic to the much repeated argument that efficiency in the construction sector was dependent upon a steady and expanding construction programme for some years ahead. The repetitive 'boom and bust' cycle was cited as being preventive of any investment by industry in its own productivity. Emmerson therefore saw the adoption by government of a policy to alleviate fluctuations in demand to be vital. He emphasised the need for a long-term approach by government; this was seen to be essential in instilling the industry with the confidence in its own future. Throughout the 1960s there was of course a much greater faith in the benefits that could be achieved through active government intervention. By the mid-1970s this confidence had been largely replaced by a rhetorical commitment to the mechanisms of the free market. The collapse of the post-war social consensus in the mid-1970s rendered the possibility of an interventionist approach on the part of government unthinkable. However, it notable that *laissez faire* seemed to apply to some sectors more than others. For example, the government often intervened in the aerospace sector throughout the 1980s in ways that were unthinkable for construction.

1.3.2 Relations and contracts

Emmerson (1962) was especially clear that efficiency was dependent upon effective relations between the building owner, the professions and the contractor. He saw the private sector client as being dependent primarily upon the advice of the architect. The concept of the 'expert client' was seemingly limited at the time to the public sector. Emmerson placed great emphasis on the need for adequate numbers of quality professionals. He expressed particular concern that many architects worked within small practices, which were seen to lack the necessary skills of organisation, cost control and office management. He therefore advocated consolidation of architectural firms into larger practices. Emmerson was also concerned about the professional restrictions placed upon architects and surveyors, who were barred at the time from working for contracting firms. He saw this as a false separation which was detrimental to closer relations between professionals and building firms.

Despite the longstanding recommendations of Simon (1944) in favour of selective tendering, the Emmerson report reveals that open tendering was still widespread in the construction industries of the early 1960s. It was argued that open tendering worked against the interests of firms who sought to compete on the basis of quality. The contention that clients do not achieve best value through lowest-price tendering was much later to become an acclaimed plank of the Egan agenda. Emmerson further argued against the increasing trend within the private sector towards the use of nominated sub-contractors. This was in contrast to practice within the public sector which tended to leave the choice of sub-contractor to the main contractor. The latter policy was judged to be preferable as it afforded maximum control to the main contractor. Emmerson, it seems, was striving for an 'integrated team' who were subservient to the main contractor. Nevertheless, he recognised that some highly specialised elements of construction were best conducted by nominated sub-contractors.

Emmerson also identified the growing use of 'package deals' through which building contractors took responsibility for design and construction. This was seen to enable better planning thereby resulting in greater cost efficiency. Subsequent concerns about the design quality which could be achieved through design-and-build contracts were not yet seemingly on the agenda at the time. Emmerson further noted the increasing occurrence of serial contracting whereby contractors tendered for one job with the expectation that the successful candidate would be appointed for a continuing programme (Moodley and Preece, 2003). Such methods were seen to require a greater degree of collaboration between contractors and design teams. Emmerson suggested the need for a further study in this area which also engaged with the local authorities. This recommendation led directly to the subsequent Banwell (1964) report (see below).

1.3.3 Training and employment

For many involved in today's construction sector it seems that the apprenticeship system has always been in crisis. However, in the early 1960s such structures were still working effectively. Emmerson (1962) offered a few minor comments, including the recommendation that there should be a fresh look at the optimum length of the various craft apprenticeships. He also observed that building techniques were changing and these necessitated the recognition of new craft skills. Particular mention was made of the differing skills that were required in connection with the use of new materials, prefabrication and industrialised building methods. Emmerson further commented on the increasing number of on-site disagreements which were rooted in demarcation disputes between the different crafts.

Of further note is the way Emmerson recorded his satisfaction with the implementation of recommendations by the national Joint Council for the Building Industry and the Civil Engineering Conciliation Board relating to holiday pay, payment for lost time due to inclement weather and a 'guaranteed minimum weekly pay packet'. These measures had apparently led to a substantial reduction in casual employment, something of which Emmerson clearly approved. The corollary of this was that the larger contracting firms increased their reliance on permanent labour, but many were nevertheless still suffering from an unacceptably high level of turnover of labour. Yet Emmerson was clearly not satisfied and asserted that a great deal more could be done to improve working conditions in the industry. He was especially keen that a high proportion of operatives within the building industry should be directly employed. Emmerson of course was writing in an age when the techniques of marketing were still in their infancy, and spin doctors were yet to be invented.

1.3.4 The placing and management of contracts

The ink was barely dry on the Emmerson report before Sir Harold Banwell was commissioned to produce a follow-up. The Banwell committee was established in October 1962 as a direct result of the Emmerson's recommendation that there should be an independent inquiry into the placing and management of contracts. The Banwell report was published in 1964 with the uninspiring (but predictable) title of *The Placing and Management of Contracts for Building and Civil Engineering Work*. The Banwell committee had been given a remit broad by the Minister for Public Works, Geoffrey Rippon, with the following terms of reference:

> 'To consider the practices adopted for the placing and management of contracts for building and civil engineering works; and to make recommendations with a view to promoting efficiency and economy.' (Banwell, 1965; v)

From the outset Banwell strove to overcome the limitations of Emmerson by consulting as widely as possible. The methodology comprised a widely distributed questionnaire survey together with extensive consultation with interested parties. The Banwell committee met on no less than 30 occasions and collected evidence from some 119 organisations and individuals. The extent of consultation was therefore impressive, but it also reflected the way in which government-appointed committees operated in the early 1960s. Geoffrey Rippon was many years later to find a degree of fame in his role in the downfall of Margaret Thatcher.

Hardcastle *et al.* (2003) offer a useful overview of the Banwell report and provide an especially pertinent description of the employment environment which prevailed at the time:

> '[j]ob stability was … a major factor at the time, in that nationalised industries were the norm rather than the exception that they are today. In a very real sense workers expected a job for life rather than the significant turnover of workers seen today.'

Expectations in the 1960s were indeed very different from those of today, and it is perhaps in the area of employment stability where the greatest contrast lies. In the 1960s, a significant proportion of operatives were employed by public-sector Direct Labour Organisations (DLOs). At their peak, local authority DLOs employed 200 000 building workers (Harvey, 2003). This was indeed construction before the age of enterprise. But the privately-owned contractors of the 1960s were markedly different from those of today in that they typically retained large directly-employed workforces. Modern contracting firms routinely delegate concerns about labour productivity to their subcontractors, but this was much less the case in the early 1960s. At the time, the productivity of the employed work-force was a major concern for main contractors. This caused several of the largest contracting firms to establish their own OR departments with a view to implementing 'modern management techniques'. The stable employment environment also encouraged investment in training and provided the context for the apprenticeship system which flourished during the 1960s.

The Banwell report was especially prescient of more recent government reports in the way in which it discriminated between the industry's 'world-class leading edge' (to use Egan's terminology) and a much larger rump which was seen to be under-performing. Banwell was especially effusive about the 'progressive members' of the industry which he saw to be 'lively and full of new ideas, willing to experiment and not afraid to change their practices and procedures'. These were perhaps the forerunners of the later much-heralded Movement for Innovation (see Chapter 5). Banwell went on to compare the attitudes of the industry's progressive members with those of the broader majority, who were found to lack the same level of urgency, enthusiasm and sense of purpose. Indeed, the majority of the industry was found to adhere to 'inflexible contractual and professional conventions', which in Banwell's view were no longer fit for purpose.

1.3.5 The team in design and construction

Banwell was noticeably fond of invoking the team metaphor when describing the parties involved in a construction contract. The committee was especially concerned that construction teams too often failed to invest sufficiently in up-front planning. This was seen to lead to subsequent changes during the construction stage which resulted in variation orders and additional costs. The advice to clients was that they should bring in specialist consultants at the earliest stage. Banwell was especially keen that the main contractor should also be bought into the team prior to design completion. To all extent and purposes this was a call for 'early contractor involvement' (ECI) which was to be introduced much later by the Highways Agency to much critical acclaim. Indeed, as one reads through the Banwell report it becomes apparent that many recommendations echo those of more recent reports. In the words of Hardcastle *et al.* (2003):

> '...if the Banwell committee were to redraft their report using contemporary phraseology, then it is likely that they would emphasise

the fact that "buildability" is enhanced by an "integrated team" consisting "of multi-skilled, multifunctional professionals".'

It would be easy to conclude that nothing much had changed since the early 1960s. Certainly the construction industries were not changed by action being taken on Banwell's recommendations. The changes which did happen could not have been reasonably foreseen in 1964, and they unfolded in direct contrast to Banwell's recommendations. One of the most notable changes was the complete disintegration of the stable employment environment which prevailed during the 1960s. But the advocates of industry improvement were still calling for integration in 2010 just as their predecessors had done in 1964. The interesting point is that they hardly seem to have noticed the radical changes which took place in the intervening period.

1.3.6   Contract procedures

The Banwell Committee was equally clear on the need to integrate subcontractors into the construction team. It was recommended that the same standards of fairness should apply to the selection of subcontractors as were applied to the selection of the main contractor. The report further echoed Emmerson's concerns about the inefficiencies of open competitive tendering. Local authorities were especially criticised for their rigid adherence to outmoded procedures and for their reluctance to move towards selective tendering or negotiated contracts. Pre-tender administration procedures were further found to be widely inadequate. On the positive side, the Banwell Committee encouraged serial tendering as a fruitful way forward, especially in terms of the advantages of improved continuity of employment. The emphasis on the merits of serial tendering is suggestive of later thinking along the lines of 'strategic partnering' (see Chapter 7). Of particular interest is the way that local authorities were seen to be stuck in the past along with the regressive majority within the industry. The 1960s would seem to have been characterised by the same mutuality of distrust which was to be alluded to subsequently by Latham (1994) (see Chapter 4). The Banwell report further recommended that there should be a common form of contract for all construction work in England, Scotland and Wales. The report is especially dated by Banwell's insistence that there should always be an accurate and comprehensive bill of quantities to militate against claims. This call was entirely consistent with the emphasis on the need for greater up-front planning. Firm price contracts were seen to be paramount, and the emphasis lay with the basic administrative procedures and the need to complete the necessary paperwork. This was at least good news for the embryonic quantity surveying profession. The focus on bureaucratic procedures was in no small way a reflection of the stable economic environment which prevailed at the time. Inflation was not yet a major concern.

Finally, it is worth noting the emphasis Banwell gave to payment and retention mechanisms and the importance of cash flow to contracting firms. A recurring concern was that inefficient administrative procedures should not be allowed to delay contractually due payments. There was a recognition that clients could not expect to receive good service if they failed to live up to their obligations for prompt payment. And the importance of prompt payment was seen to apply equally to subcontractors. This was a theme to which Latham was to return 30 years later. Many other things may have changed in the interim, but cash flow was to remain the life blood of the industry.

1.4   Planning for stability and predictability

1.4.1   *The changing context*

Ive (2003) cites *The Public Client and the Construction Industries* as a classic case of a report left behind by 'changes in the macroeconomic and political-economic context'. Big words indeed, but these were big changes. The Emmerson (1962) and the Banwell (1964) reports can reasonably be regarded as two reports in the same series. But there was a third report which ploughed a very similar furrow. The Wood report *The Public Client and the Construction Industries* was published in 1975 and to all extent and purposes comprised the third instalment. The Working Party was formed in December 1971 and it set about its work in the same diligent fashion as had characterised the Banwell Committee. But the important difference was that between 1971 and 1975 the stable economic context which had been taken for granted by Emmerson and Banwell was rocked by a series of broader external events.

The story commences with the election of Edward Heath as Tory Prime Minister in 1970. Heath was a staunchly pro-business politician who believed that a radical approach was required if the British economy was to be successful. He was committed to free-market economics and was convinced of the benefits of 'small' government. Influential voices within the Conservative Party were for the first time advocating the wholesale privatisation of the previously nationalised industries. Upon taking power the Heath government immediately cut public expenditure, reduced income tax and abolished the Prices and Incomes Board through which the government had sought to manage the economy throughout the 1960s. The industrial relations climate deteriorated rapidly. Heath further refused to intervene in a damaging dock strike and sought to curtail the power of the trade unions through legislation. Plans were also put into place to test the efficiency of public sector services against private sector comparators. However, in the face of extensive trade union opposition such ambitions were quickly shelved. The winter of 1971/2 saw the first national miners' strike since the 1920s. Inflation was rocketing and unemployment exceeded one million for the first time since the Great Depression of the 1930s.

In the face of extreme pressure from all directions, the Heath government implemented a stark U-turn and abandoned its commitment to free-market economics and returned to the interventionalist policies which had prevailed throughout the 1960s. The miners' wage demands were met and large subsidies were granted to struggling private sector businesses such as Rolls-Royce and Upper Clyde Shipbuilders. Heath strove to implement a fresh tripartite approach to consensual economics involving government, the employers and the trade unions. A statutory Pay Board was established with a view to setting pay levels over a three-year period. But the political and economic climate was to get much worse before it got better. Of particular note was the first ever national building strike in 1972 (see below). The sense of national crisis was exacerbated by the global economic shock instigated by OPEC's oil embargo following Israel's victory in the Yom Kippur War of 1973. Buoyed by their victory the previous year, the executive of the National Union of Miners (NUM) voted for a second national strike in pursuit of higher pay. These were the days before commercial North Sea oil production and Britain was at the time heavily dependent upon coal. The strike was immediately effective and resulted in phased power cuts and the implementation of a national three-day week. The government found itself subsumed into crisis once again, the U-turn having been widely interpreted as a sign of weakness. In February 1974, Heath called a general election on the issue of 'who governs Britain?' The electorate responded with a resounding 'not you'; although they were clearly not totally convinced by Harold Wilson's Labour Party either. The election resulted in a hung parliament with Wilson taking office in the midst of a deep and damaging recession. He quickly settled with the miners and proclaimed the 'Social Contract', offering the trade unions a privileged place in government in return for voluntary pay restraint. A second general election in October 1974 secured Wilson a narrow workable majority and Edward Heath was thus confined to history.

The 1970s were also difficult times for the construction sector which, as always, suffered even more severely than other parts of the economy. By 1974, the steady growth in output enjoyed throughout the 1960s was a distant memory (see Figure 1.1). The years of declining output caused many contractors to embark upon a trend of shedding direct employees, which was to continue right through until the mid-1990s.

1.4.2 Ignoring reality

In amongst all the political tumult, Wood continued to concentrate on the job in hand. The image which comes to mind is that of Nero playing his fiddle as Rome burned. But Sir Kenneth Wood had been given a job to do

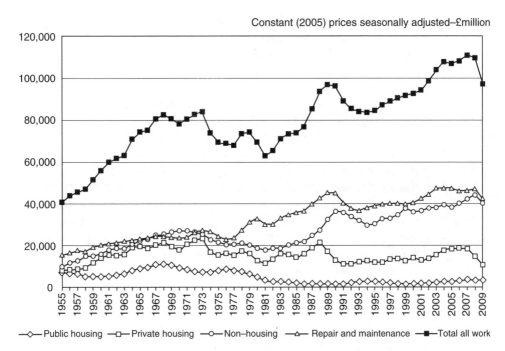

Constant (2005) prices seasonally adjusted–£million

—◇—Public housing —□—Private housing —○—Non–housing —△—Repair and maintenance —■—Total all work

Figure 1.1   Great Britain construction output by type of work (1955–2009) (*Source*: Office for National Statistics)

and he was determined to see it through. However, it should be noted that Wood only took over as chair of the working party in 1973, following the resignation of David Morrell. The recommendations of the Wood report were entirely consistent with the direction previously set by Emmerson (1962) and Banwell (1964). Significant attention was given to the need to reform public sector procurement procedures to better realise value for money. The report demonstrated that ten years after the Banwell report open competitive tendering was still widespread amongst public sector clients, and that this was still found to be damaging to the cause of industry efficiency. Wood reinforced previous exhortations in favour of 'alternative methods' such as two-stage tendering, negotiated contracts and serial contracting. However, it must further be noted that the cause of negotiated contracts had not been helped by the Poulson scandal of 1973, which involved the extensive bribery of public officials at all levels in return for the award of building contracts.

The Wood report further emphasised the importance of government departments and local authorities maintaining restricted lists of approved contractors. Support was also found for the recommendations of Banwell in respect of providing contractors with continuity of work. Associated advantages were seen to include accumulated learning and a better understanding of clients' attitudes and policies. The absence of continuity of work for individual contractors was seen to hamper training and to limit the incentive to innovate.

The Wood report was especially characterised by its plea for public sector agencies to manage their capital expenditure budgets actively for the purpose of industry development. A lack of stability and predictability in demand were held to be the greatest obstacles to efficiency and performance. It was recommended that spending authorities should be allowed to develop and adhere to rolling programmes of construction expenditure on a three-year time horizon. It was seen to be vital that such spending profiles were protected from fluctuating macroeconomic conditions if the construction industries were to be able to modernise. As Ive (2003) observes, such recommendations were 'dead in the water' at the outset. The government had already been seen to have been helpless in the face of global economic shocks. The political consensus in favour of macroeconomic management through tripartite institutions had already been largely discredited. Few had much faith in Harold Wilson's 'Social Contract' as a means of achieving economic stability. The last vestiges of such faith dissipated rapidly during the subsequent 'winter of discontent' in 1978/9 (see below).

1.4.3   Undermined by events

Taking the recommendations of Emmerson (1962), Banwell (1964) and Wood (1975) together, they are characterised by a common concern about the ability of the construction industries to deliver government policy objectives. This was especially true of the 1950s and 60s when there was a national need to build the required 'homes for heroes'. The reports are further marked by the assumption that 'expert clients' were limited largely to the public sector. There was a collective concern about the damaging effects of open tendering, and strong support for 'alternative' approaches such as design-and-build, negotiated contracts and serial contracting. The reports also complained bitterly about the adverse consequences of the prevailing lack of stability in demand. This was seen to undermine the industry's capacity to invest in its own efficiency. The reports were consistent notably in their support for direct employment and in their criticism of labour casualisation through the 'lump', which was seen to be highly damaging.

But perhaps the most striking commonality between the three reports was the assumed mutuality of responsibility for industry improvement between the industry itself and its clients. Emmerson took it entirely for granted that government had an important role in shaping performance improvement in the building industry. However, it was a vision of partnership between government departments and industry rather than one of control and regulation. Both Emmerson and Banwell placed great emphasis on the importance of the knowledge and expertise which lay within the works directorates which existed within a range of government departments. They also pre-empted modern storylines of knowledge management in recognising the benefits of sharing knowledge between the various heads of the departments responsible for major policy areas such as housing, schools and hospitals. All three reports were clear on the need for local authorities and other public bodies to be given 'strong guidance' on appropriate procedures. Attitudes and practices within many construction firms were seen to be outdated and regressive, but the same was also seen to be true for many within the public sector. Local authorities were singled out for especially sharp criticism. The Wood report went further than the other two reports in advocating that departments should plan three-year rolling programmes with a view to providing selected construction firms with continuity of work. To a greater or lesser extent, all three reports were shaped by the institutional mechanisms of the tripartite approach to industry development that prevailed throughout the 1950 and 60s. Collectively, they provide a fascinating insight into the state of the construction industries at the time. But the world within which they had been created was already starting to crumble by the time the Wood report was published in 1975. Interventionist approaches to industry improvement were going out of fashion as long-accepted notions of collectivism were to be suddenly cast aside in favour of market forces. Macroeconomic management was yesterday's news; the enterprise culture was about to be unleashed, with deep and lasting consequences for the construction sector.

1.5 Trouble and strife

Before describing the tenets of the enterprise culture and its manifestation within the construction sector it is necessary to backtrack a few years to flesh out a few important details. The legacy of the national building strike of 1972 was to subsequently loom large in the mindsets of building employers. Images of a strike-bound industry holding employers to ransom were to become central to the justification for a shift in employment practices. The entertainer Ricky Tomlinson was to play a small but iconic role in unfolding events. But previous disputes had also left their mark on both building employers and property developers.

1.5.1 Wild-cat strikes

It should initially be re-emphasised that the 1960s had been a relatively stable and successful period for the British economy. There had been relatively few days lost to strike action and inflation up until 1968–69 had been kept more-or-less under control. But it was towards the end of the decade when things stated to go awry. The 1960s was an era when trade unionism was strong. Closed shops were by no means unusual, and trade unions had the ability to initiate and enforce strikes. Throughout the 1960s days lost due to industrial action increased progressively. Especially disruptive were the so-called wild-cat strikes which were initiated by local trade union officials often in defiance of national trade union bureaucracy. In the construction sector, the most famous such strike was at the Barbican Development in London in 1967.

The dispute at the Barbican was characteristic of the tension that often existed between local shop stewards and office-based union bureaucrats. On the Barbican development many of the local shop stewards were (allegedly) communist party members who created mayhem through a succession of strikes in pursuit of better wages. A secondary, and entirely understandable, concern related to the 'lump' and the increasing occurrence of cut-and-run subcontractors. Such subcontractors had the unfortunate tendency to disappear suddenly leaving unpaid workers in their wake. The Barbican dispute also witnessed sporadic violence between the police and pickets which attracted extensive media headlines. The extent to which such disputes were politically motivated remains a moot point. The devious deeds of Trotskyite agitators provides one explanation; an alternative explanation would focus on such disputes being an extension of the 'crack' enjoyed by exuberant young men living away from home in the London of the 1960s. Violence, excessive drinking and wild behaviour have long-since been associated with transient building workers (cf. Joby, 1984). Indeed, it has even been argued that young men are attracted to casual building work by the promise of cash-in-hand payments coupled with minimal responsibilities. But violence, excessive drinking and wild behaviour have never been the sole preserve of working class men in the building industry. Similar behavioural traits within the Bullingdon Club, and other so-called Oxbridge 'dining clubs' do not seem to steer their members towards casual work in the building sector.

But irrespective of the degree of political motivation, it is easy to imagine how the Barbican dispute caused property developers and building employers to suffer numerous sleepless nights. The number of strikes grew steadily throughout the 1960s, with a particular peak at the end of the decade (see Figure 1.2). However, the number of days which the United Kingdom lost to industrial action during the 1960s and 70s were by no means out of line with our major European competitors. But – as the old cliché states – it is perception which counts. And as the 1970s unfolded, the perception grew that Britain was in crisis and that 'something had to be done'.

1.5.2 National building strike

Leaving the Barbican dispute to one side, it is the 1972 national building strike which deserves a greater place in our story. It has already been established that the construction sector of the early 1970s was very different from the construction sector of today. This was still in essence the industry that had emerged from the Second World War. The strike was led by the then newly-formed UCATT together with the GMWU and TGWU. Construction workers throughout the country went on strike demanding a minimum weekly wage of £30. But an important secondary objective was to seek the abolition of contractors' use of casual labour. The so-called 'lump' referred to the pool of casual labour which was hired on a daily basis and was

Thousand

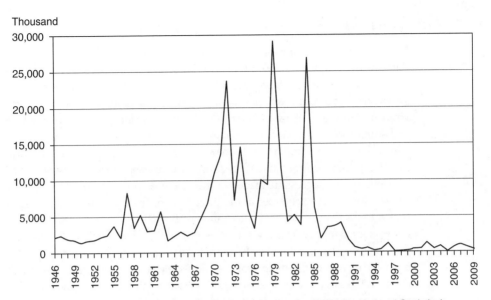

Figure 1.2   Number of working days lost (1946–2009) (*Source*: Office for National Statistics)

invariably paid in cash without any deduction for tax. The existence of the 'lump' inevitably undermined the machinery of collective bargaining. It was also widely seen as detrimental to health and safety on site. Members of the 'lump' of course did not tend to be unionised and its institutionalised use acted as a disincentive to investment in training. Nevertheless, in 1972 the majority of construction operatives were directly employed and the level of trade union membership was such that a coordinated national strike was possible. The strike resulting in a 13-week stoppage which affected numerous major construction sites; ultimately the employers were forced to the negotiating table and many of the strikers' aims were met. A number of strikers were investigated subsequently for alleged acts of vandalism and sabotage. The investigations resulted in two building workers, Ricky Tomlinson and Des Warren, being tried and convicted at Shrewsbury Crown Court for common law conspiracy as a result of their picketing activities. The so-called 'Shrewsbury Two' subsequently became a *cause célèbre* within the labour movement where there was a widely held view that the convictions were unsafe and politically motivated. Des Warren died tragically in 2004 without the pardon that activists had continued to campaign for. The cards of fate fell more kindly for Ricky Tomlinson, who forged a fresh career as a comedian/entertainer. Irrespective of whether the convictions of the 'Shrewsbury Two' were justified or not, the building strike of 1972 left a legacy of bitterness on both sides. And the 'lump' continued to operate in much the same way that it had previously. In fact, the percentage of self-employed operatives within the construction industry grew significantly from 1975 onwards, eventually reaching a peak in 1995 (see Chapter 2).

1.5.3   Crisis, what crisis?

To pick up the story from our previous account, you will recall that the Labour Party resumed power in 1974. Harold Wilson had achieved an overall majority in the House of Commons (at the second time of asking) and proceeded to enact the 'Social Contract' as an alternative to the failed policies of Heath. Unfortunately, any sense of optimism did not last long. The economy was still suffering from the oil price shock of 1973, unemployment continued to rise, inflation was rampant and the pound sank steadily against the US dollar. The 'Social Contract' did succeed in reducing the number of strikes, but was much less successful in delivering wage restraint. Denis Healey was Chancellor of the Exchequer at the time. One of the few policy instruments that were realistically available to him was the taxing of high earners to fund tax cuts for the poor. The Conservative opposition saw this as a direct disincentive to wealth creation; Healy preferred to blame low investment in industry, lack of commitment to training and outdated management practices (Marr, 2007). All three criticisms would seem to be uncontroversial in the case of the construction industries. Wilson then suddenly announced his retirement from politics as sitting Prime Minister. He was succeeded by James Callaghan who inherited an economy in a mess and was soon to be swamped by images of crisis. The pound was seemingly in free fall, inflation was rampant and unemployment continued to rise alarmingly.

The government had already introduced swingeing cuts in public expenditure but with seemingly little effect on the downward pressure the pound. It was against this background that Callaghan, and his Chancellor of the Exchequer Denis Healy, opened negotiations with the International Monetary Fund (IMF) for a huge bail-out loan. In return, the IMF negotiators demanded a further squeeze on public expenditure. IMF officials at the time were highly orthodox with little sympathy for left-leaning governments (Tiratsoo, 1997). The guiding belief was that too much welfare provision acted routinely to limit economic growth. The IMF insisted that the British economy should be deflated by a further £2.5 billion in return for a loan of US$3.9 billion. In the final analysis, the government only made use of half of the extended credit loan, all of which was repaid by the time the Labour government left office (Turner, 2008). Nevertheless, the enforced shift towards monetarist economic policy by the Labour government of 1976 sowed the seeds for a radical change in direction which was to have lasting structural implications for the British economy at large. And the construction sector was no exception.

1.5.4   Monetarism embraced

It is at this point appropriate to offer a brief explanation of monetarism, although the underpinning economic theory is at best arcane. Monetarism is perhaps best understood as a mixture of theoretical concepts, philosophical beliefs and economic policy prescriptions. Monetarists argue that the main cause of inflation is 'too much money chasing too few goods'. Hence tight control of the money supply is required to maintain price stability. Markets are seen to work best when left to their own devices; too much government intervention is held to be liable to do more harm than good in the long term. Monetarists advocate that *laissez faire* is invariably the best advice; small government is argued to be good government. Monetarists had been fighting a running battle with Keynesian economists throughout the 1960s and 70s. In the United Kingdom, the introduction of monetarist policy is usually linked with the Conservative government of Margaret Thatcher elected in 1979. But James Callaghan's Labour government had already taken several steps in this direction, even before the intervention of the IMF.

Governments of course are rarely consistent in the policies they follow. History tends to be more characterised by policy fudges than positions of theoretical purity. But our primary interest here is the way in which monetarism become accepted as the dominant economic doctrine. The renewed respectability of monetarist policy in the mid-1970s was an important catalyst in the initiation of the enterprise culture. The enterprise culture has since had a profound and lasting impact on construction sector best practice. The discourse of the enterprise culture continued to gather strength irrespective of fluctuations in economic policy. It is the discourse of the enterprise culture which provides the storylines for a succession of influential reports on the construction sector throughout the 1980s and 90s. It will be argued that the 1998 Egan Report *Rethinking Construction* represented the highpoint of the enterprise culture in the construction sector. It will be further argued that the enterprise culture has shaped the structural reality of the sector. Furthermore, the enterprise culture has progressively become central to the self-identity of managers – and operatives – throughout the construction sector. At the time of writing, monetarist economy policy has briefly given way to 'quantitative easing', but the enterprise culture continues to reign – modified but essentially unabated.

1.5.6 *The winter of discontent*

The IMF crisis of 1976 undoubtedly played an important role in shaping the legacy of the 1970s, but the subsequent 'winter of discontent' (1978–79) made a deeper and more lasting impression on the public consciousness. The overriding image, still remembered by many, was that of a country rendered inept and helpless by countless prolonged and bitter industrial disputes. Schools were closed, ports were blockaded, rubbish was piled up in the streets and the dead remained unburied. The overall image of crisis was undoubtedly exaggerated by the right-wing press who were at the time determined to undermine the credibility of the Callaghan government. But these were real events which left a lasting impression on the public consciousness. They were caused in part by the drastic squeeze on public spending in conjunction with tough pay constraints. These in turn were symptomatic of prolonged economic underperformance throughout the 1970s. There is also no denying that disputes were occasionally enflamed by irresponsible trade union officials who were actively encouraged by the 'loony left'. There was a widespread fear that the trade union movement harboured a number of Trotskyite militants who were at the time acting to undermine the machinery of democratic government. Yet on the whole fears about militant activists were grossly overstated. In the construction sector senior trade union officials often acted in conjunction with employers to constrain militant industrial action. But images are often more influential than facts, and the images generated during the winter of discontent would be mobilised repeatedly to discredit and marginalise the role of trade unions.

1.6 Summary

The primary purpose of this chapter has been to demonstrate that the debate about construction is not immune to political change or unaffected by external events. Any discussion about industry improvement during the 1970s was inevitably different from the discussions which prevailed during the 1950s. And the shift in the nature of the debate had much less to do with construction *per se*, and much more to do with the broader environment within which construction operates. In the wake of the Second World War the construction sector – in common with many other industry sectors – suffered from severe under-investment and was characterised by outdated management practices. It was highly fragmented and in severe need of modernisation. Factions within the Labour Party even flirted with nationalisation as a possible solution. While this was never quite a serious proposition, many within the industry felt it necessary to lobby against the possibility of nationalisation well into the 1980s.

Yet the construction sector played a huge and highly significant role in the modernisation of Britain's housing stock throughout the 1950s and 60s. Output peaked in 1968 at a remarkable 352 000 housing units in a single year. But the boom in housing output was accompanied by recurring problems with quality which by the 1970s were already becoming far too apparent. The local authority housing boom had also been accompanied by a succession of corruption scandals. The two exemplars of failure were provided by the Ronan Point disaster and the Poulson scandal. The legacy of these two events still overshadows the construction sector of today. The truth was that modernisation came too quickly, and modernisation was found wanting. The failures were due primarily to inadequate management and poor supervision. The housing sector thereafter retrenched into a reliance on traditional construction techniques which to this day remains highly conservative and heavily institutionalised. The poor reputation of prefabricated construction techniques continues to disincentivise investment in so-called 'modern methods of construction'.

The other big change described in this chapter relates to the collapse of the post-war Beverage settlement. The industrial unrest of the 1970s was a symptom of the failure of the social-democratic consensus which had shaped tripartite relations between government, industry and the trade unions since 1945. It is possible to point at a number of contributing causes. Weak government, poor management and radical trade unionism all played their part and any apportionment of blame is inevitably shaped by one's personal political leanings. The iconic events which figure consistently in any description of the troubles which prevailed in Britain during the 1970s were the intervention of the IMF and the subsequent 'winter of discontent' (1978–79). These events have since taken on mythical proportions in the way they are invoked repeatedly to justify the subsequent shift to the political right characterised by the election of Margaret Thatcher in 1979. However, is would be a mistake to give Margaret Thatcher too much credit (or blame) for the enactment of the enterprise culture. The reality was that Britain could not remain immune from the creeping onset of globalisation. In many respects, Thatcher just happened to be the political beneficiary of being in the right place at the right time.

References

Banwell, Sir Harold (1964) *The Placing and Management of Contracts for Building and Civil Engineering Work*. HMSO, London.

Barrie, D.S. and Paulson, B.C. (1991) *Construction Project Management*, 3rd edn., McGraw-Hill, New York.

Emmerson, Sir Harold (1962) *Survey of Problems before the Construction Industries*. HMSO, London.

Hardcastle, C., Kennedy, P. and Tookey, J. (2003) The Placing and Management of Contracts for Building and Civil Engineering Works: The Banwell Report (1964) in *Construction Reports 1944–98*, (eds. M. Murray and D. Langford), Blackwell, Oxford, pp. 55–68.

Ive, G. (2003) The public client and the construction industries, in *Construction Reports 1944–98*, (eds. M. Murray and D. Langford) Blackwell, Oxford, pp. 105–113.

Joby, R.S. (1984) *The Railwaymen*. Newton Abbot: David & Charles.

Latham, Sir Michael (1994) *Constructing the Team*. Final report of the Government/industry review of procurement and contractual arrangements in the UK construction industry. HMSO, London.

Marr, A. (2007) *A History of Modern Britain*, Macmillan, London.

Moodley, K. and Preece, C. 2003 in *Construction Reports 1944–98*, (eds M. Murray and D. Langford) Blackwell, Oxford, pp. 161–177.

Osler, D. (2002) *New Labour Plc: New Labour as a Party of Business*, Mainstream, Edinburgh.

Pilcher, R. (1992) *Principles of Construction Management*, 3rd edn. McGraw-Hill, London.

Ross, K. (2002) *Non-traditional Housing in the UK: A Brief Review*, BRE, Garston, Watford.

Simon, Sir Ernest (1944) *The Placing and Management of Contracts*. HMSO, London.

Taylor, F.W. (1911) *Principles of Scientific Management*, Harper & Row, New York.

Tiratsoo, N. (1997) 'You've never had it so bad': Britain in the 1970s, in *From Blitz to Blair: A New History of Britain since 1939* (ed. N. Tiratsoo), Weidenfeld & Nicholson, London, pp. 163–190.

Tomlinson, J. (1997) Reconstructing Britain: Labour in power 1945–1951, in *From Blitz to Blair: A New History of Britain since 1939* (ed. N. Tiratsoo), Weidenfeld & Nicholson, London, pp. 77–101.

Wood, Sir Kenneth. (1975) *The Public Client and the Construction Industries*, HMSO, London.

2 The Dawn of Enterprise

2.1 Introduction

The immediate beneficiary of the failure of the 'Social Contract' was Margaret Thatcher. She had taken over from Edward Heath as leader of the Conservative Party in 1975. In 1979 she was elected Prime Minister with a majority of 30. Her pitch to the electorate had been distinctly middle-ground with little suggestion of the radicalism with which she was later associated. Indeed, the Conservative Manifesto of 1979 was noticeably less strident than Heath's had been in 1973. Council-house sales were an important and popular policy, but there was little hint in the manifesto of the large-scale privatisations that were to come. The commitment of Thatcher and her immediate advisors to hard-line monetarism seemed at the time to pass the electorate by. The intricacies of economic policy are of course rarely of interest to the voting public. And in any case they had heard monetarist rhetoric before from the Heath government of 1973 and that had not translated into anything unusual. Heath of course subsequently implemented a dramatic U-turn in the face of widespread resistance to his initial monetarist leanings. This was a mistake which Thatcher was determined not to replicate.

Margaret Thatcher remains a very divisive figure in British politics. She is credited by many with the economic miracle whereby the UK economy broke free from restrictive practices and widespread over-manning. Those who adhere to this perspective recall all that was bad about the 1970s. Industry was unproductive, workers were lazy, trade unions were too powerful. The 'Social Contract' had been tried, and had failed. The contention is that when Thatcher came to power Britain was at its nadir; it was the 'sick man' of Europe. The debacle of the three-day week summed up years of weak government and industrial unrest. Britain had been forced to go 'cap in hand' to the IMF. Rubbish had piled up in the streets and the dead had remained unburied (Turner, 2008).The legend is that Mrs Thatcher intervened and re-energised the economy. She took on the militant trade unions and won. Enterprise was unleashed, and the flexible economy took over from the outdated and grimy industries of the 1970s.

Making Sense of Construction Improvement, First Edition. Stuart D. Green.
© 2011 Stuart D. Green. Published 2011 by Blackwell Publishing Ltd.

But there is another interpretation. The counter-perspective is that the problems of the 1970s were over-exaggerated by a right-wing press which was determined to oust the Labour Party from power. The number of days lost due to strike action in Britain is held by many to have been no greater than those endured by our major European partners/competitors. Many also point at the way British industry succeeded in maintaining productivity despite the three-day week, and had the Treasury done its sums correctly there would have been no need to negotiate a loan from the IMF. Furthermore, the counter-argument contends that the 'Social Contract' had very nearly worked, and would have done so had trade union leaders done more to keep their part of the bargain. Arguably, what was needed was a sticking plaster rather than radical surgery. Critics of Thatcher go on to blame her personally for decimating Britain's manu-facturing base and for undermining the nation's social cohesion. Certainly it cannot be denied that within two years of her election UK unemploy-ment had risen to three million. But had Argentina not chosen to invade the Falkland Islands in 1982, Thatcher would undoubtedly have been defeated by Michael Foot's Labour Party in 1983. History would then have unfolded very differently.

2.2   Uncertain beginnings

Margaret Thatcher from the outset saw herself very much as an outsider who offered a radical break from the established Conservative tradition (Jenkins, 2006). At the time of her election the economy was already plung-ing into recession. The immediate impact of the new government's policies was to make things worse. Clumsy attempts to implement a monetarist economic policy resulted in reduced public spending and increased inter-est rates. High interest rates coupled with high inflation hit manufacturing industry very badly. Output fell dramatically and unemployment soared from 1.2 million in 1979 to 3 million in 1982. The situation was not helped by an artificially strong pound bolstered by the prospect of North Sea Oil which was beginning to flow. The harsh economic climate resulted in the loss of almost 25% of Britain's manufacturing capacity. The government maintained that this was medicine which had to be taken; it was presented as a necessary and overdue process of weeding out the inefficient. Tough decisions which had been ducked consistently throughout the 1970s were finally being grappled with, or at least so it was claimed. Thatcher was clear that there was no benefit in bailing out 'lame ducks'. This was seen to be inflationary and counter-productive. But it must also be said that the rhetoric was much harsher than the reality. The government continued to support nationalised basket-cases such as Rolls-Royce and British Leyland. Nevertheless, the Prime Minister certainly had little appetite for managing demand for the benefit of the construction industry.

2.2.1 Flexibility in the marketplace

Adamson and Pollington (2006) describe the construction sector's 'Group of Eight' visit to the Prime Minister in the early 1980s. The Group of Eight was a grouping of the industry's three main professional institutions: Royal Institute of British Architects (RIBA), Royal Institution of Chartered Surveyors (RICS), Institution of Civil Engineers together with the Building Employers Confederation (BEC), the Federation of Civil Engineering Contractors (FCEC) and the Building Materials Producers (BMP). The Group of Eight was completed by two trades unions: the Transport and General Workers Union (TGWU) and the Union of Construction Allied Trades and Technicians (UCATT). The Group of Eight had been created with the purpose of presenting a united front to government. It was particularly concerned with lobbying for a greater degree of demand management in accordance with the spirit of the Wood report. When faced with such a representation Thatcher must have immediately recalled the failed 'tripartism' of the Heath era. She apparently quickly lost patience and 'sent the group away with a flea in their collective ear':

> 'It was made quite clear that if the UK construction industry was incapable of performing in a modern de-regulated economy, the Government, and the public sector generally, would obtain its construction requirements from overseas sources'. (Adamson and Pollington, 2006; 10)

The times were indeed a changing. The Group of Eight was quickly sidelined and was replaced in time by the Construction Industry Council (CIC) and the Construction Industry Employers' Council (CIEC). It is easy to imagine the message which would have been picked up by construction company chief executives:

> 'Blimey, we can forget about notions of a planned economy, we're going to have to face the endless stop-go demand cycles on our own. Competitiveness from now on is going to depend upon our ability to be flexible in response to changing market conditions.'

The severity of the recession of the early 1980s would have served to concentrate their thinking even further. The immediate challenge was survival. In such circumstances, investment in training would certainly have been viewed as expendable; likewise any vestiges of research and development activity. In short, contracting firms would have concluded very quickly that they could not afford to carry any excess 'fat'; 'leanness' and 'agility' were soon to become the favoured organisational metaphors.

2.2.2 *Treasury control and the incubation of the target culture*

Other than its commitment to vaguely defined monetarist policies, Thatcher's first administration initially gave very few indicators of the radical agenda which was to follow. However, there was from the outset a determination to curb local authority spending. This was achieved initially through the imposition of ever-stricter caps on expenditure. Local authorities were for the first time given specific spending targets, and through such mechanisms the Treasury, slowly but surely, started the process of taking public expenditure under its direct control. The grandiose spending sprees by local authority barons which characterised the 1960s were well and truly over. Increasing centralisation, coupled with a culture of target-setting and performance monitoring, became established as part of the lexicon of government (Jenkins, 2006). This was a trend which in later years was to reach ridiculous proportions and sat ill-at-ease with the government's rhetoric of free enterprise and small government. It was also a trend which was to shape the construction sector's improvement agenda. The price that was paid during the early 1980s was a decline in local democracy and the incubation of a mechanistic culture of performance management.

The achieved reductions in local authority spending were undoubtedly real, but they were unfortunately largely counter-acted by the need to provide unemployment benefit for 3 million people. In contrast to her popular image, Thatcher would not entertain radical cuts to the safety net provided by the benefit system. Hence, despite the clampdown on local authorities, public expenditure actually increased from 41% to 44% over the course of her first period of office (Jenkins, 2006). The important change was the shift in public expenditure from local authorities to central government.

2.2.3 *The origins of privatisation*

The first period of Thatcher's government also saw a succession of legislation aimed at reducing the power of the trade unions. Each step was hesitant and largely uncontroversial, but the accumulated effect was significant. Of particular note was that unions were now required to hold a ballot ahead of strike action, rather than relying on the traditional showing of hands. The only notable privatisation of the early 1980s was that of Britoil, although even here the government was careful to retain a 'golden voting share'. The state also sold its shares in Amersham, British Petroleum and British Aerospace. But there was as yet little indication of the mass privatisations which were to follow.

Privatisation in the 1980s was still a long way from accepted common sense, and its origins and supposed merits have been carefully researched by Bishop and Kay (1988). Perhaps the most intriguing argument is that privatisation was an 'accidental policy' originally devised as a means of reducing the power of public sector trade unions. Bishop and Kay describe how it grew almost surreptitiously from being a little-used mechanism of labour market reform into a core macroeconomic policy. Even more intriguingly, they contend that the headline privatisation of British Telecom grew from a conversation about how to keep the necessary new investment out of the public sector borrowing requirement (PSBR). Privatisation was apparently the conclusion they arrived at rather than being a predetermined policy priority. This of course is not the story which is presented in subsequent political memoirs (e.g. Thatcher, 1993; Lawson, 1992), but it is not unreasonable to suppose that such sources are inevitably prone to a degree of *post hoc* rationalisation. Irrespective of its origins, the policy of privatisation was soon to blossom to such an extent that it became synonymous with what was to become known as Thatcherism. One of the reasons that privatisation was to gather so much momentum was that it served the interests of several different agendas. Certainly those who were against nationalisation could hardly fail to support its reversal. At the same time, it suited the purposes of those who were concerned about the power of public sector trade unions. It also matched with the espoused monetarist requirement to reduce PSBR and with the vision to widen share ownership. It was to be advocated finally primarily as a means of improving efficiency through the introduction of market competition. But during Thatcher's first term the justifying arguments were still under development.

2.2.4 Right to buy

If privatisation were an accidental policy, this categorically was not the case with the 'right to buy' which was extended to council house tenants. The 'right to buy' was arguably one of the most radical policies of Thatcher's first term, and it was implemented against the wishes of local authorities. The sale of council homes was central to the vision of a 'property-owning democracy' which the 1979 Conservative government sought to progress from the outset. Almost one million council homes were sold in the period 1979–83, invariably at less than their market value. The Thatcher years saw a dramatic shift in the balance of housing tenure, with a sharp decline in the percentage of homes rented from local authorities with a corresponding increase in the number of owner occupiers (see Figure 2.1).

Inevitably it was the better quality houses in the most desirable locations which were more likely to be purchased by their resident tenants. Hence local councils were left with a housing stock which was on average inferior, and thereby became more expensive to maintain in proportion to revenue receipts. Local authorities would have been less adverse to the sale of council houses had they been allowed to invest the proceeds in the construction of new houses. But the rules specified that the revenue generated from sales had to be added to reserves rather than being reinvested. Local authority investment in new housing had been in sharp decline since the mid-1960s,

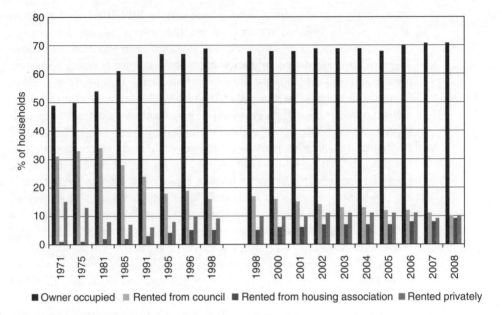

Figure 2.1   Housing tenure in the UK (1971–2008) (*Source*: *General Lifestyle Survey 2008*, Office for National Statistics, London)

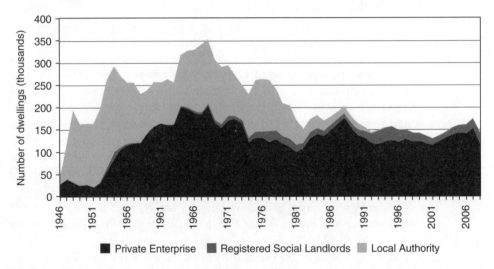

Figure 2.2   House building: permanent dwellings completed, by tenure, England, historical calendar year series (*Source*: *Housing and Planning Statistics 2009*, Department for Communities and Local Government, London)

and by the early 1990s had almost ceased (see Figure 2.2). The concerns about quality pertaining to public housing which had prevailed throughout the 1960s and 70s had been solved at a stroke; the state simply withdrew from its role as a provider of mass housing.

2.2.5 *Demise of the DLOs*

At their peak in the inter-war period, local authority Direct Labour Organisations (DLOs) employed 21% of the total building labour force (Ball, 1988). The DLOs had been created originally at the end of the nineteenth century to offset private sector problems of profiteering through collusion, price-rings, labour casualisation and inadequate training provision (Langford, 1982). Throughout the late 1970s the Conservative Party had campaigned to open up the DLOs to private sector competition. The underlying rationale was made explicit in the 1979 Conservative Party Manifesto:

> '[t]he reduction of waste, bureaucracy and over-government will yield substantial savings....local direct labour schemes waste an estimated £400 million a year.'

Following their electoral victory in 1979, the Conservative government acted quickly to introduce legislation which made significant changes to the way DLOs were organised, and the way in which they were required to account for their activities (Kirkham and Loft, 2000). The legislation required important changes to accounting practices with a view to emphasising economic efficiency. In short, the DLOs were required increasingly to operate like private companies. Their broader contributions in terms of the provision of training and local employment were no longer to stand in the way of narrowly defined economic efficiency. The legislation relating to DLOs in the early 1980s was the forerunner of the subsequent 1988 Local Government Act, which introduced 'market testing' into the delivery of public services in the form of compulsory competitive tendering (CCT). Local authority construction DLOs had been singled out to pilot an approach which was later to be applied to all local authority direct service organisations (DSOs). The construction DLOs also suffered disproportionately from the enforced withdrawal of local authorities from the provision of council houses. The long-term impact of the enacted legislation on the numbers DLOs was to be dramatic. By the mid 1970s they still employed in excess of 200 000 building workers, but by the late 1990s their role in new construction was virtually over (Harvey, 2003) (see Figure 2.3). As an aside, it is worth noting that the DLOs had consistently offered better employment conditions than those which prevailed in the private sector. They also invested more in training and proportionally supported a greater number of apprentices. The immediate effect of the legislation was to force them to adopt the employment practices which prevailed in the private sector. The gradual demise of the DLOs was of course inevitable once they had been opened up to market competitive. This may well have been positive in terms of economic efficiency, but it had long-term adverse effects on labour casualisation and training.

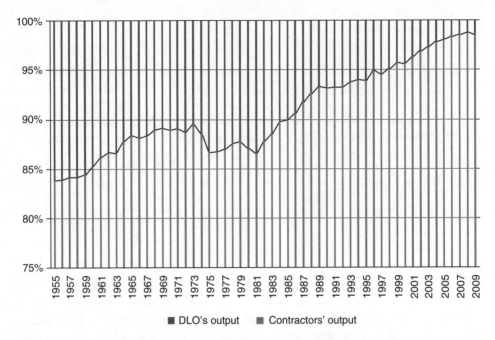

Figure 2.3 Great Britain construction output by contractors and DLOs (1955–2009) (*Sources*: *Construction Statistics Annual 2010*, Office for National Statistics, London; *Digest of Data for the Construction Industry 1996*, Department of the Environment, London)

As an aside, it is interesting to note that the reasons cited by Langford (1982) for the initial establishment of the DLOs – collusion, price rings, labour casualisation and inadequate training provision – all remain highly topical concerns. But strangely nobody has yet advocated the reinstatement of the DLOs as a solution to the ongoing problems of fragmentation.

2.2.6 *Providence intervenes*

But, in truth Thatcher's first term of office was very nearly a short-lived failure. Despite her determination to avoid Heath-style U-turns, she surrendered very quickly in 1981 in the face of a threatened coal strike. She came under intense pressure throughout 1981 to change direction and to re-inflate the economy in accordance with the doctrines of Keynesian (Jenkins, 2006). Her response was to cut public spending even further, and to deflate the economy by a further £7 billion. The screams of pain from the construction sector can easily be imagined. Several of the cabinet 'wets' were close to open revolt and Thatcher's survival as Prime Minister was under threat. Interest rates at the time stood at 16% and the economy was deep in recession. The pressure was eased somewhat in 1982 by a modest and mild economic recovery. Inflation had been reduced to 4.6%, but the consensus of opinion at the time was that Thatcher was heading for defeat in the general election of 1983. Fate intervened in the unlikely form of General Galtiari, President of Argentina.

Argentina had invaded the far-flung Falkland Islands in the spring of 1982. A naval task force was cobbled together and set sail for the South Atlantic amidst much nationalistic jingoism. In military terms, this was an extremely risky adventure. But luck frequently favours the adventurous, and so it was with the Falklands Task Force. Troops were landed successfully on the Falkland Islands (*Las Malvinas*) and the reluctant Argentine conscripts were duly evicted. Britain suddenly felt good about itself and Margaret Thatcher was the immediate political beneficiary. Despite the slight economic recovery of 1982, output was still less than it had been in 1979 and unemployment remained stubbornly high. The government's espoused commitment to monetarism had been close to shambolic from the outset, and the benefits of harsh medicine remained stubbornly unproven. It was the feel-good factor generated by the Falklands War which propelled Thatcher to electoral victory in 1983 with an improved majority of 144. Some credit for Thatcher's electoral victory should also be ceded to the Labour Party, which had allowed itself to become increasingly divorced from mainstream opinion. This was personified in the leadership of Michael Foot. The problem was not with Foot's ability, nor with the conviction of his politics. The problem was that Foot failed to deal with the militant tendency within the Labour Party, and in consequence the Labour Party was no longer seen as a safe pair of hands. Politics was also increasingly conducted in the harsh glare of the media, and sound-bites were becoming more important than well-thought out policy positions. This was a world for which the duffle-coated Foot was decidedly ill-equipped.

2.3   Enterprise in formation

2.3.1   *Milestone privatisations*

Thatcher undoubtedly derived considerable confidence from her victory in the 1983 general election. But equally important had been her victory over the 'wets' within the Conservative Party. Thatcher's second term started apace and began to define the parameters of what was to become subsequently known as 'Thatcherism'. Privatisation was suddenly top of the agenda and the 'accidental policy' was soon to be embraced as a central component of policy. Despite the initial uncertainties, the official policy rhetoric emphasised increasingly that privatisation would inevitably secure efficiency increases as a result of exposure to market competition. This was a belief which was subsequently proven to be true, at least in terms of narrow definitions of economic efficiency. But the critics often argue that the resultant increases in efficiency were derived primarily from the downtrading of employment conditions rather than any private sector innovation.

However, there is a broader point to be made: not all privatisations resulted in exposure to market competition. This was to be especially true of the privatisations during Thatcher's third term; inefficient public sector bureaucracies were all too often replaced by inefficient private sector monopolies. But the worst examples of this were yet to come. Headline privatisations during Thatcher's second term of office included British Airways, British Gas, British Shipbuilders, British Leyland, British Telecom and Rolls- Royce.

The first privatisation to go through was British Telecom in November 1984. This was rapidly followed by British Gas. The public sensed the possibility of windfall profits and both floatations were heavily over-subscribed. The sale of British Telecom raised £3.9 billion, and British Gas raised £5.4 billion. The figures involved illustrate a further obvious advantage to privatisation: it raised government revenue which could be used to balance the books. Most purchasers quickly cashed in their profits, thereby undermining the objective of increasing wider share ownership. But if privatisation was to be used as a means of generating revenue, the most important challenge was to ensure that the assets were sold for a fair price. In consequence, more care was taken with subsequent privatisations to ensure that the shares were not sold too cheaply. But the under-pricing of British Telecom and British Gas suggests that revenue generation was, at least initially, a side benefit rather than the prime motivation. Perhaps one of the biggest failures of the Thatcher era was that the proceeds of privatisation were not used to invest in industry modernisation, or in the nation's decaying infrastructure (Hirst, 1997). Most of the funds raised were used to boost current spending. The same was also true of the revenues provided by North Sea Oil.

2.3.2   Confronting the 'enemy within'

The epochal event of Thatcher's second term of office was undoubtedly the 1984–85 Miners' Strike. This was industrial conflict at its starkest, and it took on all the trappings of medieval warfare. Thatcher had already backed down to the miners in 1981, and the memory of Heath's humiliation at the hands of the miners in 1972 doubtlessly fuelled her determination. The National Union of Mineworkers (NUM) was led by the egotistical Arthur Scargill who was consistently demonised by the right-wing press. In truth, he was an easy target and his rabble-rousing rhetoric did little to endear him to public opinion. The government had been preparing for a showdown with the miners since 1981 and coal stocks had been built up at the power stations in anticipation. Trade union legislation had since outlawed secondary picketing and had ended trade union immunity for damages. Unions were further required to hold a secret ballot prior to going on strike, rather than relying on the traditional showing of hands at mass meetings. Britain was also slightly less reliant on coal as a result of a number of power stations having been converted to burn oil or gas. It was against this background that Ian McGregor, National Coal Board (NCB) chairman, announced of list of uneconomic pits for closure.

The response from Arthur Scargill was predictable. He called for a national strike but failed to implement the secret ballot which was now a legal requirement. The NUM strongholds saw high turnouts in support of Scargill. Trade unions such as the NUM were at the time held in great esteem in traditional working class communities. They had played a central role in the creation of their members' self-identities. Great emphasis was placed on loyalty and solidarity. When the union said 'out' the members obeyed. But in some parts of the country, namely Nottinghamshire, this did not happen. The Nottinghamshire miners continued working and formed their own union: the Union of Democratic Miners (UDM). The strike saw medieval-style pitched battles between pickets and the police on a daily basis. It was the first time ever that the police force had been organised on a national level in order to mobilise a coordinated response to illegal flying pickets. Intimidation and violence were common place, and the police were no less guilty in this respect than the striking miners. Thatcher referred to the need to defeat the 'enemy within', the implication being that she had already defeated the enemy without in the guise of General Galtieri. This was a scandalous comparison which the miners ill-deserved. Dirty tricks and illegal tactics were mobilised by both sides. Scargill was even (falsely) accused of seeking funds from Colonel Gaddafi of Libya. The decisive factor was that sufficient coal continued to reach the power stations to enable them to produce electricity throughout the winter of 1984. Substantial stockpiles had of course been built up at the power stations in anticipation of a showdown with the miners. Many miners' families suffered considerable hardship, and striking miners were ultimately torn between a loyalty to the NUM and a loyalty to their families. Slowly but surely, amidst much acrimony, the miners trickled back to work. The strike finally collapsed in March 1985.

The striking miners marched back to work to the accompaniment of their colliery brass bands, with union banners fluttering proudly in the breeze. Thatcher had won. The legacy of bitterness within the mining communities lived on for decades, and the police force was never quite viewed in the same way again. But at least the NCB had been brought under managerial control. Management had won the right to manage, at a price.

2.3.3 Docklands transformed

The following year saw a further bitter and prolonged strike centred upon Wapping in the East End of London. The newspaper industry was long overdue for modernisation and suffered from anarchic working practices. Wild-cat strikes were common place. Eddie Shah had been the first newspaper magnate to mobilise the new trade union legislation in the cause of modernisation, and was quickly followed by Rupert

Murdoch's News International. Murdoch had built and clandestinely equipped a new production plant in Wapping. In response to industrial action on Fleet Street, production was switched unceremoniously to Wapping and dismissal notices were served to the striking Feet Street operatives. The Wapping plant saw picketing on a similar scale to that which had occurred during the miners' strike; and it was met with the same heavy-handed approach to policing. Violent scenes from Wapping became a standing item on the evening news broadcasts. The dispute finally collapsed in February 1987, by which time thousands of workers had survived for over a year with no pay.

Looking beyond Wapping, it is easy to imagine how building employers would have followed both the print workers dispute – and that of the miners – extremely closely. But in truth, by the mid-1980s they were probably losing relatively little sleep over the threat of large-scale industrial action in the construction sector. Nevertheless, the government's industrial relations legislation would have been warmly received and would have helped ease the exaggerated memories of the wild-cat strikes on the Barbican project and the 1972 national building strike. But in truth, the construction sector was already well on the way towards solving its industrial relations 'problem' through its own devices. Construction firms were simply withdrawing from the onerous responsibility of employing people (see Chapter 3). The times were indeed a changing.

The Wapping dispute coincided with the redevelopment of London's Docklands, under the guiding hand of the London Dockland's Development Corporation (LDDC) which had been established in 1981. Much to the chagrin of the local boroughs, the LDDC had been granted full planning powers; another example of local democracy being overridden by centralised government transferring power to non-accountable quangos. The LDDC presided over several iconic construction projects. The Canary Wharf development, with its striking tower at One Canada Square, was completed in 1991. It remains emblematic of the Thatcher years and the lifestyles enjoyed by the much-derided yuppies of the time. The Canary Wharf development was also much criticised for its failure to provide jobs for local people. Other iconic construction projects in Docklands included the development of London's City Airport which was opened in 1987 and the Docklands Light Railway, which also opened in 1987.

2.3.4 Broadgate breaks the mould

Demand for office space in Canary Wharf had been driven by the financial deregulation of the City of London. The so-called 'Big Bang' happened on 27 October 1986. Deals once confined to the trading floor of the Stock Exchange were now allowed to take place by computer screen and telephone. The Stock Exchange's monopoly on share dealing had ended. Immediately prior to 'Big Bang' Britain had been exporting £2 billion of financial services per year. This expanded almost overnight to £24 billion per year (Marr, 2007) and revitalised London's position as financial capital of Europe. The booming trade in financial services also counter-balanced the balance of trade deficit in manufactured goods, which had slipped into deficit for the first time in 1983 (Hirst, 1997). Financial services hence became of central importance to the nation's economy.

One of the immediate physical outcomes of the financial deregulation was to shift the centre of gravity of the City of London eastwards. The Canary Wharf development was preceded in this respect by the £800 million-plus Broadgate development adjacent to Liverpool Street Station. The Broadgate scheme was designed specifically to facilitate real-time financial trading between London, New York and Tokyo. The development spanned 29 acres and set new standards for large-scale urban commercial development.

The initial phases of Broadgate were completed in 1992. The vision had been set by a new breed of sophisticated developer in the form of Rosehaugh Stanhope. Under the leadership of Stuart Lipton and Godfrey Bradman, Rosehaugh Stanhope set precedents that have been followed by large commercial developments ever since. Many within the construction industry were at the time looking across the North Atlantic for new ideas. Reported construction times for office developments in the United States were significantly faster than those being achieved within the United Kingdom (Flanagan *et al.*, 1986). Rosehaugh Stanhope took this one stage further by engaging American designers and construction managers in the hope of forcing change on the UK construction sector.

Developments such as Broadgate not only set new standards for architecture, they also set new standards for fast construction (Gray, 2008). The term 'fast-track' construction was widely adopted on major projects as a direct result of what was achieved at Broadgate. The client at Broadgate was further notable for pioneering the use of 'construction management' in the United Kingdom, to the consternation of the major contractors at the time (see discussion in Chapter 4). Central to the Broadgate model were flexible open floor plans unencumbered by disruptive columns. The development also made extensive use of off-site prefabrication, and was especially notable for its pioneering use of prefabricated toilet pods. Following innovations that were developed initially in the United States, completely fitted-out and serviced pods were manufactured off-site and plugged into position as a sealed unit. The commitment to prefabrication at Broadgate resulted not only in fast, efficient construction, but also in a quality finished product. It also enabled assembly to take place in non-unionized factory locations in the North of England where wage levels were much lower than those which prevailed in London.

London in the late 1980s was therefore not only an exciting place for those who worked in financial services; it was also an exciting place to be for those who worked in construction. Peter Rogers was at the time the construction director of Stanhope – he was later destined to become chairman of the

Strategic Forum (see Chapter 9). Rogers' reaction to an advertisement by one of the Broadgate trade contractors for a 'claims surveyor' has since passed into legend. The response was swift and brutal; firms which employed claims surveyors were unceremoniously declared *persona non grata*. Developers such as Stanhope undoubtedly enjoyed a significant degree of power in the marketplace, especially in London. But Broadgate was categorically not representative of the industry at large. For most contractors, claims surveyors remained an essential part of the adopted business model.

2.3.5 Labour mobility

Returning once again to the bigger political picture, the gleaming new developments in London were not replicated across the post-industrial landscapes of the North. The London-based boom in financial services served to amplify the fact that Britain in the mid-1980s was a divided nation. The recession of 1979–82 which had decimated so much of the nation's manufacturing base was concentrated inordinately in the North. Where once there had been industry there was now post-industrial wasteland. The differences between the North and the South were exacerbated by the miners' strike. The affluent inhabitants of the leafy suburbs of the Home Counties looked to Mrs Thatcher to protect them against the likes of Arthur Scargill. In contrast, the populace of the post-industrial North felt their traditional way of life was under attack. The difference was felt as starkly in the construction sector as anywhere. Construction clients in the North of England were few and far between. Clients in Scotland were even rarer. There was a joke doing the rounds at the time about Liverpool's rush hour which was said to take place on a Friday afternoon at London's Euston Station. London had always known regional migrant workers, but the 1980s saw a renewed flood of commuting Northerners bent on securing a share of the capital's prosperity. They stayed in London during the week and returned to their northern homes at the weekend. They had taken Norman Tebbitt's (the then Employment Secretary) 1981 advice to 'get on their bikes' in search of work. In a previous life the commuting Northerners would have been trade union members; but they were now entrepreneurs contributing to the new enterprise economy. And going on strike was not on option because they were increasingly self-employed.

2.4 Enterprise unleashed

2.4.1 Privatisations gather pace

Thatcher was emboldened by her third electoral victory in 1987 and began to shrug off much of the intuitive caution which had characterised her first two terms of office. Large parts of the country undoubtedly remained set against Margaret Thatcher. But large parts of the country continued to support her. The split was largely along regional lines, but the polarisation also manifest itself in terms of increasing inequality. The increase in UK wage inequality in the 1980s was matched only by the United States (Machin, 1996). There was also a marked growth in part-time working and temporary employment, with a widespread reduction in employment protection. Thatcher's third term departed rapidly from the monetarist rhetoric which had prevailed previously. The Chancellor, Nigel Lawson, was allowed to initiate a consumerist boom that quickly ran out of control. Any pretence of monetarist economic policy was abandoned in favour of the ultimate of give-way budgets. The 1988 budget reduced the top rate of tax from 60% to 40% and set the standard rate at 25%. Prudence this was not. Inflation started to rise, reaching 9.5% in 1990. The balance of payments also went from bad to worse, with a deficit of over £20 billion in 1989.

The privatisation agenda also picked up pace with British Steel being followed rapidly by the water industry and electricity distribution. The assumption that the privatised companies would become more efficient as a result of the stimulus of market competition was entirely consistent with the spirit of the times. Certainly nobody in the cabinet was brave enough to point out the potentially damaging effects of cut-throat competition. But many of the privatised companies were not exposed to harsh winds of competition. This was especially the case for the subsequently privatised utility companies which enjoyed quasi-monopolistic positions. In these cases regulation became the norm rather than the espoused aspiration of competition. Hence a host of new public regulatory bodies were created with unlikely sounding names such as Ofcom, Ofgas, Oftel and Ofwat. The regulatory bodies implemented regimes based on detailed targets and were given the power to invoke penalties when the targets were not achieved. The privatised utility companies went on to become important clients of the construction sector, and their changing procurement policies became subject to continuous analysis. Many more privatisations were to follow. By the 1992 election, 46 publically-owned businesses had been privatised, involving the transfer of 900 000 people (Marr, 2007).

A further privatisation with important implications for the construction sector was that of British Airports Authority, which in 1987 became BAA plc. This was one of the more bizarre privatisations in that BAA enjoyed a considerable monopoly over Britain's airports. It owned London's three main airports, including the world's busiest international terminal at Heathrow. Collectively these three airports handled 92% of passengers entering the capital. BAA also took control of Glasgow, Edinburgh, Aberdeen and Southampton airports giving it control of in excess of 60% of flights to and from the United Kingdom, including around 80% of those in Scotland. Precisely how the creation of such a huge quasi-monopolistic business was supposed to aid the cause of competition remains unclear. British Airways had been privatised in 1984, and if its customers became dissatisfied with its services they could conceivably transfer their business to another airline. But this was not the case for the customers of BAA, especially if you were unfortunate enough to live in London or Scotland. BAA became the epitome of the privatised regulated business whereby market competition was replaced by an all too familiar regime of targets and performance monitoring.

BAA was of course a hugely important client of the construction sector, and was destined to play a major role in the 'best practice' debate. However, it continued to display many of the characteristics of public sector bureaucracy while at the same time adopting the rhetoric of modern management techniques. BAA had been subject to much public criticism before its privatisation, and this continued unabated afterwards. But it was a client that construction firms liked to keep sweet because of its inherent stability and continued investment in its infrastructure. It was also a client which was to go on to pilot the use of framework agreements (at least for while). Banwell, Emmerson and Wood would no doubt have approved of BAA, which seemed to echo the mutuality of responsibility which they had advocated so strongly. Privatisation of course enabled BAA to improve its own technical efficiency despite the misgivings of its many customers. At least this was assumption which prevailed throughout the 1990s. Unfortunately it was not true. Parker (1999) has analysed technical efficiency within BAA both before and after privatisation and concluded there was no discernable difference.

2.4.2 Creeping centralisation

Despite all the rhetoric about 'small government', the 1980s saw increased centralised control over a range of policy areas. In 1988 the national curriculum was imposed on state schools together with a centralised testing regime. Extensive creeping institutional centralisation also characterised ongoing reforms within the National Health Service (NHS). Jenkins (2006) opines that the degree of centralisation matched nothing seen outside of the Soviet bloc. The reforms also demanded indicators of demonstrable success, such that the NHS was soon awash with targets relating to waiting lists, referrals, appointments, mortality rates and anything else to which a number could be attached (Jenkins, 2006). These reforms formed part of a 20-year trend whereby power was consistently taken way from democratically-elected local politicians and placed in the hands of centrally appointed quangos. 1988 also saw responsibility for low cost housing shift from local authorities to Housing Action Trusts – again with important implications for construction. Mrs Thatcher really did not like those local authorities. Charitable housing associations were also favoured over local authorities. Despite their charitable status, housing associations received 90% of their funding from central government via the Housing Corporation. The budget of the Housing Corporation grew from £50 million in 1979 to £1 billion by the time Thatcher left office in 1990 (Jenkins, 2006). The Housing Corporation was funding three times as many new houses as were local authorities. But the government's primary housing policy aim was to encourage home ownership, and the primary policy instrument was mortgage interest tax relief. The 1980s were get-rich-quick time for many in the housing market. Borrowing money to buy houses with the intention of doing them up and selling them on for a profit became a national pastime. The nation's wealth was gambled on bricks and mortar with a sizable army of *ad hoc* builders involved in cosmetic renovation. In consequence, the grey economy grew exponentially; at least until the crash came.

The consumerist housing boom of the late 1980s very quickly turned to recession. The boom-bust cycle imposed itself with ruthless brutality. But the construction sector had by now given up any expectation of a planned economy and had in turn reorganised itself on the model of structural flexibility. The legions of workers from the North, along with other large tracts of the construction workforce, cost nothing to shed because they had never been employed. But the growth of self-employment in the construction sector is a topic which deserves particular attention (see Chapter 3).

2.4.3   Poll tax debacle

The credit-fuelled inflationary expansion of the late 1980s did much to dent Thatcher's credibility for economic management, although the Chancellor Nigel Lawson was arguably primarily to blame in respect of the 1988 give-way budget. Many within the Conservative Party were increasingly disgruntled with Thatcher's autocratic style, and her deep unpopularity across large swathes of Wales, Northern England and Scotland caused many to doubt the possibility of a fourth electoral victory. But the issue which really brought about the end of Margaret Thatcher's political career was the poll tax debacle. The Thatcher years had seen a long succession of measures aimed at preventing local authorities from spending money. They had also seen numerous measures to limit their revenue raising powers through the imposition of central government control. The perceived difficulty was that elected left-wing councils were able to implement high spending policies which a relatively small number of local ratepayers had to pay for. The idea was to make local councils much more accountable by making all voters pay an equal amount. This was described officially as the 'community charge', but was rapidly dubbed by the media as the 'poll tax'. Although allowance was made for the unemployed and low paid, the poll tax was essentially regressive in that it bore little relation to peoples' ability to pay. The poll tax had been originally implemented in Scotland, to much local opposition and resentment. The warnings were not heeded, and the eve of the poll tax's introduction in England and Wales saw a massive demonstration in Trafalgar Square which deteriorated into a full-blown riot. The events shook the cabinet as the government's ratings slumped in the national opinion polls. The subsequent political manoeuvrings are beyond the scope of this book, but suffice to say that Margaret Thatcher was forced to stand down and was replaced by the comparatively grey John Major.

2.4.4 Enterprise re-invigorated

When John Major entered Downing Street in 1990 the nation heaved a sigh of collective relief. Even Thatcher's most ardent supporters had grown tired of her autocratic style. Major's elevation to the office of Prime Minister succeeded in boosting the Conservative Party's ratings in the opinion polls overnight. The stridency of Thatcher was suddenly replaced by images of warm beer and cricket. Major was further buoyed by Britain's role in the 1990 Gulf War which resulted in Saddam Hussein being ousted from Kuwait. Although Major was very much Thatcher's preferred successor, he acted quickly to distance himself from his mentor through the abolition of the poll tax which was replaced to general relief by the council tax. Major also inherited the poisoned chalice of having to negotiate the Maastrict Treaty on European political unity. Playing to a large gallery of baleful Conservative Party Euro-sceptics, Major demonstrated that he was no soft touch and famously negotiated an opt-out from the Treaty's Social Chapter. Many advocates of the enterprise culture perceived the Social Chapter as a regressive step back towards the corporatist structures of the 1970s. The opt-out was to set the United Kingdom's labour market on a very different trajectory from the rest of Europe. Employment protection was categorically rejected in favour of the 'flexible economy'. The enterprise culture therefore remained firmly in place while its critics pointed towards Britain's apparent destiny as Europe's low-wage, off-shore production platform.

To almost everyone's surprise, John Major went on to win the general election of 1992. This represented a fourth consecutive victory for the Conservative Party, a remarkable achievement given the almost continuous upheaval and discord. But those who were expecting a relatively quiet life under the unassuming John Major were to be disappointed. The enterprise culture continued unabashed, and was reinvigorated by Major's election as Prime Minister. Hirst (1997) characterises the Major years as 'Thatcherism on autopilot'. The commitment to privatise British Rail had been made prior to the 1992 election, and at the time there was every reason to assume that the Major government would not get the opportunity to honour its pledge. Certainly had the Labour Party under the leadership of Neil Kinnock been elected in 1992 the privatisation of the railways would not have happened. Thatcher had previously considered the railways as 'a privatisation too far'; but the Major government rushed in where others had feared to tread.

It is perhaps notable that the Major government made the commitment to privatise the railways without expecting to be given the opportunity to put it into practice. This would do much to explain the lack of thought and confusion which characterised this most disastrous of all privatisations. Marr (2007) describes the eventual solution as the 'Complete Horlicks' option. In 1993 British Rail was broken up and privatised. Ownership of the track was separated from a series of franchises which were created for operating trains. The privatisation was allegedly done in the cause of promoting competition, but given that different routes were allocated to different operators it was difficult to see how this resulted in increased competition. The intricate details of the rail privatisation are beyond the scope of this book, but suffice to say it was heralded by many as an unmitigated disaster. The privatisation cost the taxpayer billions with little visible improvement in performance. The end result was a heavily regulated, and predominantly centrally planned, system that was dominated by targets and performance indicators. This was privatisation for the sake of it; enterprise it was not.

Major also embarked upon a programme of pit closures in the coal industry as a precursor to privatising what remained of this once proud industry. Many considered this a major betrayal of the strike-breaking Union of Democratic Mineworkers (UDM), but the enterprise culture seemingly held no truck with honouring previous debts. Change continued at a remorseless pace. The NHS was reorganised along the lines of an 'internal market' and the influence of local authorities over state education was reduced in preference to the creation of Ofsted, which became one more quango to add to the ever-expanding list. Parents were now expected to operate as 'informed consumers' of the education system and to select the schools on the basis of league tables and publically available reports. The reality was that the best state schools were impossible to access unless one was fortunate enough to live in the pre-assigned catchment area. Progressively, this began to distort local property markets such that housing within the most desirable catchment areas was valued more highly than housing within less attractive catchment areas.

Further market reforms were also introduced into the NHS whereby doctors 'purchased' the required services from hospitals. Such reforms afforded increased control to the Treasury and established what subsequently became known as the 'cult of audit'. The end result of the long-running processes of centralisation is that central government progressively became responsible for activities which had previously fallen within the remit of regional health authorities. Localised decision making and accountability was replaced by a highly bureaucratic infrastructure of target setting and performance monitoring. Here lay one of the central paradoxes of the enterprise culture: despite the ongoing emphasis on individual responsibility a significant number of managers within the public sector found their personal responsibility being continuously eroded. Reforms of the health and education sectors are of course always of interest to the construction sector because they provide a degree of workload continuity. The notion of 'performance measurement' was subsequently to have a significant impact on the construction sector improvement debate.

However, the advent of the Private Finance Initiative (PFI) was to have much more immediate impact on the structure of the construction market. The evolution of PFI and its implications for the construction sector will be described further in the next chapter, where attention will also be given to other manifestations of the enterprise culture in construction.

2.5   The enterprise culture

2.5.1   The antecedents of enterprise

This chapter has so far described some of the epochal events which characterised the period 1979–97 together with the associated shift in the prevailing political consensus. Numerous specific references have been made to the implications of the adopted policies for the construction sector. But at this point it is appropriate to depart from the preceding factual description to focus on the characteristics of what had become known as the 'enterprise culture'. In Britain, it was Thatcherism which provided the immediate political context for the theory and practice of the enterprise culture, but the two terms should not be taken as synonymous (Keat, 1991). As has already been observed, government policy during the Thatcher years was characterised by several unfolding paradoxes and inconsistencies. State control of the economy increased rather than decreased, and the rhetoric of the free market frequently resulted in regulated private-sector monopolies. The enterprise culture also derived much of its legitimacy from the advent of the 'New Right' in the United States. In essence, the enterprise culture was a cultural phenomenon.

Although the antecedents of the enterprise culture can be traced back as far as we can find evidence, the expression first entered the political lexicon in the United Kingdom around the time of the 1987 general election. Of particular note was a series of speeches and articles by the unelected Lord Young of Graffham who had been appointed Secretary of State for Employment in 1985 (see, e.g. Young, 1986; Young, 1987). As an ultra-loyalist to the cause, he was chosen to run Thatcher's 1987 election campaign and was subsequently promoted to Secretary of State for Trade and Industry. The enterprise culture was first and foremost a rhetorical slogan aimed at winning the 1987 election. It was further presented as an attempt to change the prevailing political culture in Britain. Sources often refer to the *rhetoric* of the enterprise culture, focusing on the written and spoken words in which the ideas were communicated. However, by the early 1990s authors were referring to the *discourse* of the enterprise culture to include not only the associated rhetoric, but also underpinning ideas and associated policy initiatives (Keat, 1991; du Gay and Salaman, 1992). Even more broadly, the discourse of the enterprise culture can be taken to include a complex web of ideas, linguistic expressions, policies, social institutions and associated material practices (Green *et al.*, 2008). In this sense, this chapter in its entirety can be seen to be about the discourse of the enterprise culture. And the remainder of the book can be seen to be about the way the discourse of the enterprise culture has played out over time in the construction sector.

2.5.2   Commercial enterprise

It is appropriate initially to outline the main parameters of the enterprise culture as articulated originally by its proponents in the late 1980s. To start with the obvious, the essential building block of the enterprise culture was a belief in the wealth creation capabilities of enterprise capitalism. Capitalism of course comes in many forms, and is contextualised in different places in different ways (Berger, 1987). In the context of 1980s Britain, the advocated belief in enterprise capitalism was positioned against the perceived (and real) failings of the 1970s. Enterprise capitalism was seen to create wealth, and wealth creation was perceived to be the ultimate measure of success. The desirable ideal was therefore to have privately-owned enterprises operating in a free market. Successful firms succeeded, and unsuccessful firms failed. This was seen to be the 'natural order' of competition. The role of government was to keep out of the way. Firms' survival and growth depended upon their 'leanness' and 'fitness' in dealing with the rigours of the marketplace (Legge, 1995). Certainly there was little perceived merit in subsidising 'lame ducks' for the purposes of maintaining full employment. It followed that government intervention in the market was an intrinsically 'bad thing'. The guiding espoused philosophy was that the market 'must take its course'. In its ideal form, the free market should of course be an unregulated market (by definition). And certainly the Thatcher administration did much to reduce the 'regulatory burden' which had been placed on industry, although even here the rhetoric was much more strident than the reality. It should be emphasised that although the enterprise culture drew from the lexicon of economics, it was not in itself an economic policy. Over the course of the 1980s the word 'enterprise' became elevated to a cultural status (Heelas and Morris, 1990). The enterprise culture was about much more than economics; it was nothing less than an attempt to re-engineer the culture of the nation.

As has already been noted, the rhetoric of the enterprise culture was mobilised progressively (albeit retrospectively) throughout the early 1980s to justify the policy of privatisation. The belief was that state-owned industries would inevitably operate more efficiently once exposed to the cold wind of competition. The paradox here of course was that several state-owned industries and public utilities subsequently became regulated private sector monopolies. The prime example was perhaps BAA plc, which was later destined to play a significant role in the construction industry improvement debate throughout the 1990s. Few at the time were sensitive to the paradoxical position of BAA, although the Competition Commission did eventually catch with them following their take-over in 2006 by Spanish infrastructure company *Grupo Ferrovial.*

The enterprise culture further provided the retrospective justification for the changes to the operating status of local authority DLOs in the early 1980s and the associated introduction of 'market testing'. This was later expanded to compulsory competitive tendering (CCT) for the majority of public sector services. The de-regulation of financial services and removal of currency exchange controls also became constituent parts of the enterprise culture. The most important manifestation of the enterprise culture in the construction sector was the incentivisation of self-employment, but this deserves separate consideration.

2.5.3   *Individual enterprise*

The discussion so far has focused on enterprise in the sense of the 'commercial enterprise'. But the word 'enterprise' can also be understood as a verb in the sense of the 'enterprising individual' (Fairclough, 1991; Keat, 1991). From the outset, the enterprise culture gave emphasis to a highly individualistic form of capitalism. Guest (1990) describes how both the Thatcher government in the United Kingdom and the Reagan administration in the United States emphasised the virtues of 'rugged entrepreneurial individualism'. The notion of enterprise capitalism is therefore very different from the collectivist ideas that dominated during the 1960s and 70s. The idea of enterprise is also positioned against hierarchical command-and-control corporate cultures. Enterprising individuals are supposed to take risks; they are supposed to challenge the status quo. Above all else, they are supposed to innovate. Enterprising individuals are governed primarily by self-interest; they expect to receive financial rewards for the efforts. There is also an assumption that they are prepared to take control of their own careers and to take personal responsibility for their personal development (Heelas and Morris, 1990). Such characteristics were certainly encouraged throughout the 1980s, and became synonymous with the so-called 'yuppie' (i.e. young, upwardly-mobile professional). Yuppies, stereotypically, lived in gleaming new apartments in London Docklands and worked in Canary Wharf. In the 1980s, the phenomenon was largely limited to the South East of England. It would be some years before yuppification took hold in urban pockets in the North.

But the enterprise culture was not directed solely at yuppies; enterprise was for everyone, not just professionals. The advocates of enterprise exhorted society at large to make maximum use of their talents in the cause of wealth creation. It should be recalled that many skilled workers voted for Thatcher in 1979 because of their concern about 'eroded differentials' during the years of the 'Social Contract'. The stereotype here was that of 'Essex man', which is frequently mobilised to explain Thatcher's electoral successes. The clichéd 'Essex Man' can be characterised as a working-class, and hence slightly coarser, version of the yuppie. In popular imagination,

Essex Man would drive a white van and would sport tattoos on both arms. He would also decorate his house with England flags during international football tournaments. Throughout the 1970s Essex Man would have been a union member and would have consistently voted Labour. But somewhere along the line he embraced the enterprise culture, and started to vote for Margaret Thatcher. The stereotype is strangely silent on the voting preferences of 'Essex Woman'.

2.5.4 The cult of the customer

Enterprise reforms aimed at extending and intensifying the role of market forces mean inevitably that 'consumers' become increasingly influential (Heelas and Morris, 1990). This translated to an increasing differentiation of demand, with radical implications for the way producers were organised. Previously predominant forms of mass production had been positioned against mass consumption (any colour you like as long as it's black). However, the exercise of consumer choice led to differentiated forms of consumption thereby requiring flexible production patterns. The 1980s also saw a number of other external pressures for change, including: increased competition from foreign industry, greater quality consciousness within the consumer population; rapidly changing product markets, deregulation and readily available new technologies (Fuller and Smith, 1991). Many such similar lists have been generated subsequently to promote the need for change in the construction sector. But radical organisational change in response to the above pressures was the norm throughout the 1980s, and the restructuring of the construction sector was in many ways following the same script. Such trends were by no means limited to the United Kingdom, which in many respects was following the pattern of change already established in the United States. Justifying narratives came in a variety of forms, but the common denominator was the iconic status of the customer.

The onset of consumerism and the 'cult of the customer' were by no means limited to the private sector. Previously passive recipients of state services such as housing, education and health were similarly jolted out their apathy and were expected to exercise 'informed consumer choice'. Such thinking later informed a host of wide-ranging public sector reforms throughout the 1990s. There was no place in the enterprise culture for passive consumers of public services who simply accepted what they were given. And there was of course nothing better than a 'short, sharp shock' to jolt 'unenterprising' people out of their apathy – or at least this was the advocated storyline. Passengers on the railways were redefined as customers, and patients within the National Health Service were similarly re-conceptualised. Students within universities were encouraged to see themselves as customers of educational services who paid fees in exchange for knowledge. Even the unemployed were encouraged to regard themselves as enterprising consumers of the welfare system.

But the introduction of the term 'customer' into the day-to-day language of managers and workers was about much more than cosmetic window dressing. The promotion of the enterprise culture throughout the 1980s translated directly into extensive organisational restructuring in both the United Kingdom and United States (du Gay and Salaman, 1992). Previously dominant bureaucratic forms of organising were replaced progressively by a form of organising based on market relations. Individuals within organisations were required to re-position themselves continuously in terms of the imperative of the 'sovereign customer' (Keat, 1991). The necessity to meet the needs of the sovereign customer did not only apply to relations between the organisation and its external clients, it also permeated the internal organisation of firms. Internal departments became tasked with meeting the demands of internal customers. All organisational units (and individuals) therefore became 'enterprises' within an internal marketplace. Of particular note is the way in which relational transactions were reduced to commercial transactions between suppliers and customers. Du Gay and Salaman's (1992) diagnosis is especially notable for the way they argue against dichotomising between external pressures and internal strategic responses. They contend that these should be conceptualised as if they were part of the same phenomenon, which they label as the discourse of the enterprise culture.

The discourse of the enterprise culture promoted the 'short sharp shock' as something which was cleansing; an essential cold shower in the cause of improved competitiveness. The empathy of the past whereby individuals could rely on the certainty of the next pay cheque is eradicated. Stability and certainty were replaced by uncertainty and the chaos of the marketplace. Firms in the construction sector had received the message very early: notions of 'mutually of responsibility' between the construction sector and its clients were gone. Individual workers picked up the same message. There was no place anymore for the 'salary man'; they were essentially on their own. They became individual profit centres tasked with 'adding value'. Even if they were not self-employed, they were expected to act as if they were self-employed. Boundaries were formed where previously there had been none; every individual was tasked with 'staying close to the customer' (Peters and Waterman, 1984). When faced with the chaos of the marketplace, continuous business improvement became the new default position. Innovation was an essential requirement; only through continuous innovation could individual enterprises remain in touch. Much of this is now accepted as common sense; in the late 1980s it was heralded as a significant cultural shift.

2.5.5   Enterprise and the prevailing management orthodoxy

The enterprise culture was to have a profound and long-lasting impact on management theory and practice. Many of the iconic management improvement recipes which came to the fore during the 1980s were dependent upon the redefinition of relationships which was central to the cult of the customer (du Gay and Salaman, 1992). Total Quality Management (TQM) and Just-in-time (JIT) systems both depended upon an organisational model which gave primacy to supplier and customer relationships. Such recipes both reflected and reinforced the enterprise culture and can only meaningfully be understood within the broad context of industry restructuring. TQM hinges upon a definition of quality in terms of conforming to customer requirements. But this applies to internal customers as well external customers. Each producer is thereby tasked with meeting the specified quality requirements. The model of course assumes that customers are able to pre-determine their precise requirements. It further assumes that customers speak with one voice and are consistent in their specification of require-ments over time. In the context of the construction sector there is a danger that this interpretation of quality oversimplifies the role of the construction client, but this is an issue which is best explored separately (see Chapter 6). The idea that the paying customer has sole responsibility for defining the quality of buildings also had significant implications for the architecture of the public built environment. The homogeneous monotony of Britain's high streets is a direct reflection of the enterprise culture and its longevity. But to see the very worse of British architecture it is only necessary to travel through city centres on the railway to view the endless functional conform-ity of industrial/retail units designed through design-and-build architec-ture. Conformity with client requirements is close to perfection; but the quality of the end product is nothing short of depressing. Quality and inno-vation can be strange bedfellows in the world of the enterprise culture. The problem was to be addressed partially by the subsequent creation of the Commission for Architecture in the Built Environment (CABE) and the associated promotion of design quality indicators (DQIs) (see further dis-cussion in Chapter 10). The seemingly endless creation of quangos and various forms of key performance indicators (KPIs) to measure things which the market ignores has been an important part of the story ever since the enterprise culture came to prominence. These are the mechanisms through which the damaging side effects of the enterprise culture are man-aged and their adverse effects are limited. If all else fails, at least the exist-ence of KPIs gives the impression that 'something is being done', even if their material impact is limited. This will be a recurring theme throughout the remainder of this book.

    The concept of Just-in-Time (JIT) management can also be understood as a constituent part of the enterprise culture. JIT rests on a production proc-ess model which involves a series of organisational units in a sequence of interdependent operations. TQM dictates that each unit delivers its output to its immediate customer in accordance with specified requirements. JIT emphasises that the work should be done only when required (Sayer, 1986). JIT improves efficiency because there is no longer any necessity to keep stocks. The timing of the entire production process is hence dictated by the required delivery date to the external client. And the delivery date for the output from each unit involved in the production process is similarly dictated by its immediate internal customer. Control of the pace of work is therefore no longer predetermined by centralised bureaucratic control, but is now delegated to a series of internal customers. Such ideas were later to become inseparable from 'lean production' and considerable debate was devoted to the extent to which such methods were applicable in construction. This debate will be addressed in some detail in Chapter 8, but for present purposes it is sufficient to note that the advocated methods were constituent parts of the discourse of the enterprise culture. Thus, they should be considered in combination with other essential parts of enterprise culture.

The underlying contention is that management ideas reflect and reinforce the broader discourse of the prevailing socio-political consensus. Management recipes such as TQM and JIT can certainly be understood to be constituent parts of the enterprise culture. But, as has already been noted, it would be a mistake to see the discourse of the enterprise culture as something which was homogeneous or fixed in terms of its content. The enterprise culture was in no small way defined in opposition to the prevailing counter-discourses. And the enterprise culture itself contained a variety of sub-strands that were competing continuously for prominence. Discourses such as the enterprise culture are constantly renegotiated and their flexibility in this respect is central to their longevity. But there is little doubt that the 'modern management techniques' of the enterprise culture displaced progressively the management ideas which had prevailed throughout the 1970s.

It was perhaps during the late 1950s that the optimisation algorithms of operational research and the narrow machine metaphors of scientific management represented the epitome of modernity. In the prevailing conditions of stability and certainty such techniques had much going for them, but their credibility was seriously challenged by the turbulent environment of the 1970s. The emphasis among the *cognoscenti* switched progressively to 'systems thinking' and the need to adapt to the increasingly dynamic business environment (cf. Berrien, 1976; Burns and Stalker, 1961; Rice, 1963). Lawrence and Lorsch (1967) were especially influential in emphasising the interface between organisations and their broader environment. Kast and Rosenzweig (1985; first published 1970) pulled much of this thinking together to focus on the relationship between businesses and the societal environment within which they operate:

> '[o]rganizations are subsystems of a broader suprasystem - the environment. They have identifiable but permeable boundaries that separate them from their environment. They receive inputs across these boundaries, transform them, and return outputs. As society becomes more and more complex and dynamic, organisations need to devote increasing attention to environmental forces.'

Operational managers in the construction sector were understandably resistant to this particular brand of 'systems-babble', despite its increasing popularity within academic circles. But implicit within the advocated storyline was the assumption that managers were not expected to confine their attention to narrowly construed notions of profit maximisation. Such arguments are of course currently popular once again under the guise of corporate social responsibility (CSR) (Green, 2009). However, in the 1980s the notion that managers should do anything other than meet the requirements of the customer was rapidly marginalised by the advent of the enterprise culture. Satisfying the demands of the customer was accorded an iconic status in the search for effectiveness and profitability. Satisfying the demand of diverse societal stakeholders was 'old thinking' and constituted inefficiency and waste.

And yet stakeholders were to creep back slowly into fashion again along with CSR. The original development of stakeholder theory dates back to the 1970s (Freeman and Reed, 1983; Freeman, 1984). In essence, stakeholders were seen to be groups or individuals who have a stake in the activities of the corporation, i.e. they provide support and in some way the corporation is seen to be responsible for safeguarding their interests. However, there was always an important distinction between the broad and narrow definitions of stakeholders (Freeman and Reed, 1983). In the broad sense, stakeholders were seen as 'any identifiable group or individual who can affect the achievement of an organisation's objectives or who is affected by the achievement of an organisation's objectives'. In the narrow sense, stakeholders were defined as those groups or individuals that the organisation depends upon for its continued survival. In common with systems theory, the wider definition of stakeholders was quickly undermined by the cult of the customer; to expend resources on satisfying the aspirations of anyone other than the 'customer' very soon became synonymous with waste. In the construction sector, this line of thinking reached its highpoint with the 1998 Egan report *Rethinking Construction* (see Chapter 5).

Yet even the narrow interpretation of stakeholders struggled to survive in the face of the enterprise culture's relentless emphasis on short-term economic imperatives. The subsequent articulation of partnering (see Chapter 7) could in part be read as an attempt to remind managers that any short-term exploitation of stakeholders may not be in the organisation's long-term interests. But some stakeholders were always more expendable than others, and in consequence partnering was always unlikely to be applied consistently. Given the broader project to establish an 'enterprise culture', subsequent localised calls for a culture shift towards partnering cannot be understood in isolation.

2.5.6   *Dissenting voices*

The enterprise culture attracted many influential disciples during the 1980s, but it also attracted many critics. Indeed, it could be argued that the active promotion of the enterprise culture depended upon the identification of suitable sets of 'fall guys' who could be cast as the defenders of the previously dominant 'dependency culture'. Critics were further liable to be labelled as the 'enemy within'; a jibe which was directed especially at trade union members. Old-style paternalistic Tories (i.e. the 'wets') were also frequently cast into the role of the 'other'. Privileged members of the nation's professional classes were similarly allocated bit parts as the protectors of vested interests and barriers to progress. The 'enemy within' was indeed a broad church, and trade unionists therefore frequently found themselves in good company.

In terms of substantive arguments against the enterprise culture, the most obvious was the argument that the enterprise culture promoted a society based on greed. The counter-argument promoted by the advocates of enterprise emphasised the so-called 'trickle-down effect', whereby the mechanisms of wealth creation inevitably benefit the poor. And for those who were left behind, the Thatcher government was consistent in emphasising the importance of the 'safety net' of the welfare system. This of course was scant consolation for those who found themselves stranded within deprived post-industrial communities where job opportunities were few and far between. The danger here was that they would seek to be enterprising in the 'wrong way' and turn to quick profits through crime.

A further critique of the enterprise culture related to quality of life. Those seeking to live their lives in accordance with the principles of the enterprise culture operated as if they were small, self-contained businesses. The critics of enterprise argued that they hence became obsessed with only a small part of what life has to offer. Fear of failure therefore looms large, incubating stress at work. The 1980s saw a boom in the occurrence of chronic fatigue syndrome (so called 'yuppie flu') and other associated ailments. Much discussion therefore concentrated around the 'human cost' of competition and the cultural poverty of the 'loadsamoney' society (see Chapter 3). The ruthless conceptualisation of individuals as profit-maximising units also de-emphasised the importance of ethical behaviour and eroded long-established notions of social responsibility. The moral and societal implications of the enterprise culture are of course beyond the scope of this book. But the criticisms articulated against the enterprise culture *per se* were to be repeated frequently in response to the associated managerial recipes. This was to become as true within construction as within any other sector. That the criticisms now seem dated, and largely irrelevant, is indicative of how persuasive the discourse of the enterprise culture has become.

Perhaps the most prescient criticism of the enterprise culture related to its lack of focus on sustainability. The enterprise culture infamously did not recognise the existence of 'society', and hence accorded no importance to social sustainability. Environmental concerns were not at the time high on the political agenda, but the 'green' lobby was to become increasingly mainstream by the turn of the millennium. As we shall see, the increasing importance accorded to sustainability resulted in several strange contortions in the construction sector policy agenda. Even more strikingly, the enterprise culture proved to be not even sustainable in economic terms. The iconic status of enterprise and the guiding doctrine of monetarism were severely dented by the global credit crunch of 2007–2009. At the time of writing, the future is difficult to predict, but a modified and regulated form of enterprise seems likely to emerge. The task of marrying enterprise with sustainability primarily rests with our political leadership, but the resulting policies will inevitably present new challenges for the construction sector.

2.6 Summary

It is of course easy to demonise Margaret Thatcher for the negative consequences of the enterprise culture. But it must be remembered that she was supported by a sustained electoral coalition that enabled her to win three successive electoral victories. Thatcher made herself electable by appealing not just to the property-owning middle classes, but also to skilled workers who felt disgruntled by the erosion of differentials bought about by the 'Social Contract'. These included skilled tradesmen in the construction sector many of whom voted for Thatcher in consecutive elections. There are numerous supporters of Thatcher who continue to argue that she imposed a much needed modernisation on a structurally misaligned economy. It must further be recalled that Keynesianism had already been abandoned by the preceding Labour government of James Callaghan at the bidding of the International Monetary Fund (Gray, 1998). The winds of change were already blowing, and Thatcher's policies were in no small way driven by external forces. The shift to the political right was in part precipitated by the onset of globalisation, driven by time-space compression due to technological change (cf. Harvey, 1989). Ultimately, the managed economy of the post-war period would have expired irrespective of the political opportunism of Margaret Thatcher.

The Thatcher years were also instrumental in the transformation of the British Labour Party, which progressively abandoned its socialist heritage in favour of 'New Labour' managerialism. The epochal event was the 1984 Miners' Strike and the defeat of the National Union of Mineworkers (NUM) which changed the face of British industrial relations forever. It is especially notable that the New Labour government elected in 1997 failed to repeal any of the industrial relations legislation their predecessors had so roundly condemned. Indeed, the New Labour government embraced the discourse of the enterprise culture and perhaps even extended it to new heights.

But it would be a mistake to characterise the enterprise culture as something which was ever homogeneous. Neither was it a project which was ever finished; the enterprise culture has been in continuous evolution ever since the constituent ideas were first articulated. Although privatisation became widely recognised as a core component of the enterprise culture, this was by no means clear from the outset. Indeed, it was only during Thatcher's third term of office that the enterprise culture took on any degree of coherence, and even then paradoxes and contradictions continued to abound. The elusive nature of the enterprise culture did not of course prevent it from being pursued even more enthusiastically by the subsequent government of John Major. From 1990 onwards, privatisation was seemingly pursued for its own sake. Whether or not privatisation resulted in increased competition ceased to be important; but in truth the justification of increased competition always was a badge of convenience rather than conviction.

The precise definition of the enterprise culture may well be difficult to pin down, but few would argue that it has not had a significant and lasting impact on the UK economy and society at large. Indeed, it can be argued that it was the very elusive nature of the enterprise culture which accounted for its longevity. It is also important to understand that the notion of enterprise was mobilised *against* other pre-existing discourses. Indeed, there is a school of thought that postulates that it is only ever possible to understand discourses in conjunction with the counter-discourses to which they are opposed. Throughout the 1980s and beyond the discourse of the enterprise culture was consistently positioned against the perceived pre-existing 'dependency culture' whereby individuals supposedly looked to the state for free hand-outs. It was also positioned against the collectivist institutions which characterised the social contract of the 1970s. The elusive nature of the enterprise culture is further illustrated by the way in which it has been continuously renegotiated and repositioned over time.

What cannot be denied is that the rhetoric of the enterprise culture became highly persuasive among opinion makers during the late 1980s, and that its influence continued to prevail throughout the 1990s. It had an especially strong influence on management practice such that many of the constituent ideas became accepted 'common sense'. It will be argued in the following chapter that the enterprise culture had a particularly profound and lasting influence on the construction sector. It will further be contended that the construction improvement agenda from the late 1980s onwards reflected and reinforced the discourse of the enterprise culture. The improvement agenda also suffered from the same paradoxes and inconsistencies.

References

Adamson, D.M. and Pollington, T. (2006) *Change in the Construction Industry: An Account of the UK Construction Industry Reform Movement 1993–2003*, Routledge, Abingdon.

Ball, M. (1988) *Rebuilding Construction: Economic Change and the British Construction Industry*, Routledge, London.

Berger, P. (1987) *The Capitalist Revolution*, Wildwood House, Aldershot.

Berrien, F.K (1976) A general systems approach to organizations, in M.D. Duneet (ed.) *Handbook of Industrial and Organizational Psychology*, Rand McNally, Chicago.

Bishop, M. and Kay, J.A. (1988), *Does Privatisation Work? Lessons from the UK*, Centre for Business Strategy, London Business School.

Burns, T. and Stalker, G.M. (1961) *The Management of Innovation*, Tavistock Publications, London.

du Gay, P. and Salaman, G. (1992) The cult(ure) of the customer. *Journal of Management Studies*, **29**, 615–633.

Fairclough (1991) What we might mean by enterprise, in: Keat, R and Abercrombie, N. (eds), *Enterprise Culture*, Routledge, London.

Fuller, L. and Smith, V. (1991) Consumers' reports: management by customers in a changing economy, *Work, Employment and Society*, **5**(1), 1–16.

Flanagan, R., Norman, G., Ireland, V. and Ormerod, R. (1986) *A Fresh Look at the UK and US Building Industries*. Building Employers Confederation, London.

Freeman, R.E. (1984) *Strategic Management: a Stakeholder Approach*, Pitman, Boston.

Freeman, R.E. and Reed, D.L. (1983) Stockholders and stakeholders: a new perspective on corporate governance, *California Management Review*, **25**, 88–106.

Gray, C. (2008) The role of the professional client in leading change: a case study of Stanhope PLC, in *Clients Driving Innovation*, (eds P. Brandon and S-L. Lu), Wiley-Blackwell, pp. 234–240.

Gray, J. (1998) *False Dawn: The Delusions of Global Capitalism*, Granta, London.

Green, S.D. (2009) The evolution of corporate social responsibility in construction: defining the parameters, in *Corporate Social Responsibility in Construction*, (eds M. Murray and A. Dainty), Taylor & Francis, Abingdon, pp. 24–53.

Green, S.D., Harty, C. F., Elmualim, A.A., Larsen, G. and Kao, C.C. (2008) On the discourse of construction competitiveness, *Building Research & Information*, **36**(5), 426–435.

Guest, D.E. (1990) Human resource management and the American dream, *Journal of Management Studies*, **27**(4), 378–397.

Harvey, D. (1989) *The Condition of Postmodernity: An Inquiry into the Origins of Cultural Change*, Blackwell, Oxford.

Harvey, M. (2003) Privatization, fragmentation and inflexible flexibilization in the UK construction industry, in G. Bosch and P. Philips (eds): *Building Chaos: A International Comparison of Deregulation in the Construction Industry*, Routledge, London, pp. 188–209.

Heelas, P. and Morris, P. (1990) Enterprise culture: its values and value. In: Heelas, P. and Morris, P. *The Values of the Enterprise Culture*, Routledge, London, 1–25.

Hirst, P. (1997) Miracle or mirage?: The Thatcher years 1979–1997, *in* Tiratsoo, N. (ed) *From Blitz to Blair: A New History of Britain Since 1939*, Weidenfied & Nicholson, London, pp. 191–217.

Jenkins, S. (2006) *Thatcher and Sons: A Revolution in Three Acts*, Allen Lane, London.

Kast, F.E. and Rosenzweig, J.E. (1985) *Organization and Management: a Systems and Contingency Approach*, 4th edn., McGraw-Hill, New York.

Keat, R. (1991) Introduction: starship Britain or universal enterprise?, in R. Keat and N. Abercombie (eds), *Enterprise Culture*, Routledge, London, 1–17.

Kirkham, L.M. and Loft, A. (2000) Accounting and the governance of the public sector: the case of local authority direct labour organisations in the UK in the 1970s, paper presented at IPA Conference, London.

Langford, D.A. (1982) *Direct Labour Organizations in the Construction Industry*. Gower, Aldershot.

Lawrence, P.R. and Lorsch, J.W. (1967) *Organization and Environment*, Harvard Press, Cambridge, Mass.

Lawson, N. (1992) *The View from Number 11*, Bantam Press, London.

Legge, K. (1995) *Human Resource Management: Rhetorics and Realities*, Macmillan, Basingstoke.

Marr, A. (2007) *A History of Modern Britain*, Macmillan, London.

Machin, S. (1996) Wage inequality in the UK, *Oxford Review of Economic Policy*, **12**(1), 47–64.

Parker, D. (1999) The performance of BAA before and after privatisation, *Journal of Transport Economics and Policy*, **33**(2), 133–46.

Peters, T.J. and Waterman, R.H. (1982) *In Search of Excellence: Lessons from America's Best-Run Companies*, Harper & Row, New York.

Rice, A.K. (1963) *The Enterprise and its Environment*, Tavistock Publications, London.

Sayer, A. (1986) New developments in manufacturing: the just-in-time system, *Capital and Class*, **10**(3), 43–72

Thatcher, M. (1993) *The Downing Street Years*, HarperCollins, London.

Turner, A.W. (2008) *Crisis? What Crisis?: Britain in the 1970s*, Aurum Press, London.

Young, Lord (1987) Spreading the enterprise culture, *The Director*, **41**(2), 40–43.

Young, Lord (1986) Enterprise – the road to jobs, *London Business School Journal*, **11**(1), 21–27.

3 Leanness and Agility in Construction

3.1 Introduction

As highlighted in the preceding chapter, the UK economy saw extensive restructuring throughout the 1980s and early 1990s. It has already been argued that the enterprise culture provided a retrospective narrative for justifying changes which had already been initiated. But the discourse of enterprise also served to accelerate the pace of reform. Of particular note was the way in which the processes of privatisation and outsourcing were extended progressively throughout the economy. This was especially true following John Major's re-election in 1992 when the policy of privatisation operated largely on auto-pilot. What cannot be denied is that this process of restructuring had far-reaching implications for the construction sector. Many clients which had previously been part of the public sector had been privatised in the expectation that exposure to market forces would drive efficiency improvement. The shift to the private sector inevitably changed the way in which these clients engaged with the construction sector. Even those clients which remained within the public sector outsourced many in-house capabilities in accordance with compulsory competitive tendering and other associated policy initiatives which encouraged the contracting out of services. Faced with the introduction of yardstick competition, many such clients retrenched into the very worst kind of adversarial contractual practices. These were the same practices which were to be subsequently so lamented by the Latham (1994) report (see Chapter 4).

The restructuring of the construction sector's client base led inevitably to the diversification of client demand. Different clients wanted different things and hence developed different approaches to construction procurement. At the same time, competitive pressures forced construction firms to reorganise themselves in accordance with the principles of leanness and agility. Flexibility in the marketplace became the new dominant doctrine as the construction sector became dominated increasingly by the hollowed-out firm. Progressively, main contractors removed themselves from the physical work of construction, preferring to concentrate on management

Making Sense of Construction Improvement, First Edition. Stuart D. Green.
© 2011 Stuart D. Green. Published 2011 by Blackwell Publishing Ltd.

and coordination functions. This was in turn was accompanied by a dramatic reduction in the number of directly employed construction workers and a corresponding increase in the number of self-employed operatives. Such changes cannot easily be explained on the basis of cause-and-effect relationships. The underlying causes were in part economic and in part sociological. They are best understood as a seamless flux of interconnecting changes which collectively characterise the material manifestation of the enterprise culture in the construction sector.

The broad sweep of the privatisation programme was described in the preceding chapter. The present chapter commences with a discussion of the factors which shaped the emergence of the hollowed-out construction firm during the 1980s. Attention is thereafter is given to the issue of labour market casualisation and the government incentivisation of self-employment. The chapter is concluded by a further discussion of the restructuring of the construction sector's client base. Particular attention is given to the privatisation of the Property Services Agency (PSA) and to the reorganisation of highways management and maintenance. Coverage also extends to the privatisation of the Building Research Establishment (BRE) and to the initial introduction the Private Finance Initiative (PFI) in 1992.

3.2   Towards the hollowed-out firm

3.2.1   The growth of subcontracting

During the late 1970s and throughout the 1980s the demand for construction saw significant diversification. Of particular significance was the demise of the state as the provider of mass housing, coupled with the government's retreat from any pretence of demand management. Also of note was the emergence of new types of private-sector client organisation who were much more challenging in the demands they made of the construction sector. Developers such as Stanhope provided one example of a much more demanding client, but there were many more. The onset of the enterprise culture saw a significant restructuring of the industry's client base, which was in turn mirrored by extensive structural changes within the contracting sector (Ball, 1988). One of the most important changes was a significantly increased reliance on subcontracting.

The growth in subcontracting comprised part of a radical industry restructuring which took place over a 25-year period commencing in the early 1970s. Harvey (2003) uses the term 'flexibilization' to describe the process of rapid structural change which was seen to include the downsizing of firms and the associated outsourcing of functions. Harvey's diagnosis finds considerable support in the official statistics which describe the changing structure of the contracting sector. In 1971, 83% of firms employed less than 13 direct employees and accounted for less than 16% of industry

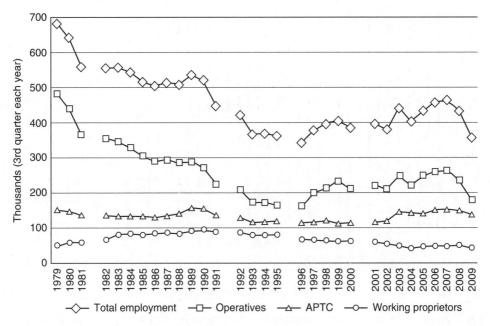

Figure 3.1   Main trades' employment by types (1979–2009) (*Sources: Housing and Construction Statistics 1980–1998; Construction Statistics Annual 2010*, Office for National Statistics, London)

output. At the same time the 0.1% of firms which employed more than 1200 employees accounted for 24% of the total value of output (Harvey, 2003). The subsequent 25 years saw a radical degree of restructuring such that by 1997 the number of workers employed by small firms had doubled, and the number of operatives employed by large firms had halved. This is indeed a dramatic structural shift in the construction sector's employment patterns. The net effect was to shift a large number of employees from large firms to smaller firms.

The decline in the number of operatives directly employed by contractors declined dramatically from 1979 through to 1995 (see Figure 3.1). Especially stark was the declining number of directly employed operatives in proportion to the number of administrative, professional, technical and clerical (APTC) employees. The evidence in support of the hollowing-out of contracting firms could hardly be more striking. One of the most striking proxies for the growth in subcontracting lies in the increase in the number of listed private contractors who rely on specialist trades in comparison to those who rely on main trades. In 1971, the percentage of contracting firms listed under specialist trades was only slightly in excess of 50%; by 2008 it was edging ever closer to 80% (see Figure 3.2).

However, while the figures above undoubtedly paint a vivid picture of structural change within the sector, they do not necessarily indicate the decline of large contractors in terms of their influence in the market place.

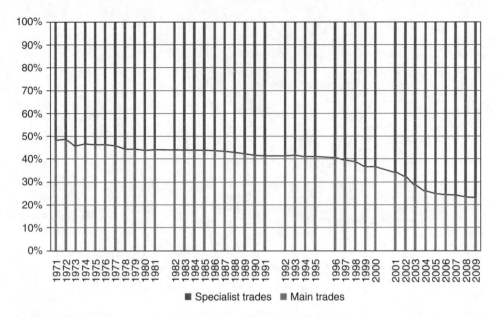

Figure 3.2   Private contractors: percentage of firms by trade (1971–2009) (*Sources*: *Housing and Construction Statistics 1979–1998*; *Construction Statistics Annuals 2000–2010*, Office for National Statistics, London)

What they do point towards is a significant and sustained growth in sub-contracting. As Ball (1988) pointed out at the time, the large contractors' declining share of total output sat ill-at-ease with their company accounts which showed expanding workloads throughout the 1980s. The explanation for this apparent paradox lies in the significant and sustained growth in subcontracting.

As a caveat to the above, an increased reliance on subcontracting was not unique to the construction sector. Many industry sectors in the 1980s experienced a significant growth in outsourcing as firms strove to reduce their fixed overhead costs in the face of a more competitive trading environment. What is perhaps unique to the construction sector is that subcontracting occurs at the point of production such that employees of different subcontractors routinely work side-by-side. This frequently results in operatives performing very similar jobs in the same location while being engaged on very different terms and conditions of employment.

3.2.2   The rationale for subcontracting

There are of course many good reasons for subcontracting, most of which have been well rehearsed by Hillebrandt (1984). Subcontracting offers an obvious advantage in that contractors can bring in specialist trades when required. Hence the main contractor does not have to worry about providing them with continuity of work, thereby minimising non-productive time and fixed labour costs. Subcontracting is also held to reduce the contractors' organisational and control costs, and perhaps more importantly shifts the responsibility for working capital to the subcontractor. An additional benefit of subcontracting is that contractors benefit from a secondary competitive tendering process which takes place much closer to the commencement of work.

It is important at the outset to distinguish between subcontracting which involves both materials and labour, and subcontracting which involves labour only. In the first case, the main contractor will subdivide the work into specialist packages which are then put out to tender on a 'supply-and-fix' basis. All other things being equal, the main contractor would then place the subcontract with the lowest bidder. M&E work is typically subcontracted in this way in its entirety. This avoids the need for the main contractor to develop the necessary skills in-house, and it similarly avoids the need for the main contractor's estimators to develop the skills to price such work. Structural steelwork and roofing are also typically subcontracted on a supply-and-fix basis. Indeed, such elements have always been subcontracted by main contractors, and there is little evidence to suggest that this changed over the course of the 1980s. However, the advent of increasingly sophisticated heating, ventilation and air conditioning (HVAC) systems meant that a higher proportion of a building's value was subcontracted. Cladding systems were likewise becoming much more specialised, and trends towards prefabrication and modularisation required the increased use of off-site factory environments. Buildings also required increasingly specialised control systems. Hence the construction sector became characterised by an increasing diversity of specialist niche markets, which in turn fed the growth in subcontracting.

It follows that the growth in subcontracting in the construction sector during the 1980s can in part be attributed to client diversification, greater technological complexity and a generally increasingly competitive environment. But these factors only go some way towards explaining why main contractors were subcontracting core trades such as brickwork and joinery which had previously been performed by directly-employed labour. In some cases, the main contractor continued to provide the materials, but delegated responsibility for on-site production to labour-only subcontractors. In other cases core trades such as brickwork and joinery were subcontracted in their entirety, but the first-tier subcontractor would in turn subcontract with labour-only subcontractors. It should be added that the extent to which such core trades were subcontracted had long been subject to regional differences. For example, in London such core traditional trades had been routinely subcontracted for some time. But the tendency to subcontract core trades undoubtedly became much more widespread throughout the 1980s. Groundwork and concreting work were also increasingly subcontracted on a labour-only basis. Even on civil engineering projects, shuttering joinery which had once been performed by directly-employed joiners was now being routinely subcontracted.

While larger firms increasingly relied on subcontractors, many small firms who had previously tendered for work from clients responded by reinventing themselves as specialist subcontractors. Given market pressure to reduce overheads, it increasingly made sense for such firms to specialise in terms of a single construction activity (or trade). This immediately reduced the required level of investment in specialist plant and equipment; it also simplified recruitment and training in that they were no longer required to carry multiple trades. In many respects, the decision to operate as a specialist subcontractor can be seen as a risk management strategy in terms of reducing a firm's exposure to fluctuations in workload; the demand for individual trades is frequently more stable than the demand for individual building types (Ive and Gruneberg, 2000).

3.2.3 Fragmentation in action

The logic in favour of subcontracting applies not only to main contractors, it also applies to firms who are themselves subcontractors. A subcontractor which has secured a supply-and-fix contract frequently finds it advantageous to sublet the installation work on a labour-only basis. Such logic can accumulate to encourage a multi-tiered system of subcontracts. The end result is that construction work has become increasingly organised through an extended vertical chain of subcontracting whereby each subcontractor in turn parcels out ever-smaller packages of work. The logic of subcontracting continues to apply until the value of the work package is too small to justify the transaction cost. On major construction sites, five tiers of subcontract are by no means out of the ordinary. The end result is that the main contractor's responsibilities for employment conditions, training and productivity are passed progressively down the contract chain. Ultimately, such responsibilities all too often end up resting with 'fly-by-night' firms who are least well placed to deal with them.

A further benefit of an increased reliance on subcontracting lay in the way in which it fragmented the industrial relations context, thereby reducing the impact of any industrial action. Given that throughout the post-war period construction was the most strike-prone industry after coal mining (Evans and Lewis, 1989), the risk of financial loss due to industrial action was by no means insignificant. Even more salient is that the vast majority of strikes in the construction sector were unofficial. The only notable exceptions were the national disputes of 1963–64 and 1972. In contrast to the coal mining industry, the battle lines in construction were not so starkly drawn between the employers and the trade unions. The real problem faced by employers was not the trade unions, but unofficial strike action which fell outside of union influence. Casting the NUM as the 'enemy within' had already stretched public credibility to the limit; UCATT would have been the most toothless enemy ever. The trade union movement within the construction sector has always suffered from fragmentation, and UCATT and TGWU were too busy competing with each other to be considered a threat to anyone. And in the face of 'wild-cat' strikes such as those which marred the Barbican project, trade unions leaders were frequently as impotent as the employers. Nevertheless, an increased reliance on subcontracting provided an additional benefit in that the risk of industrial action was shifted to subcontractors. If a subcontractor's workforce went on strike the subcontract could be terminated. A replacement subcontractor could then be found with a more compliant workforce. Such is the reality of risk management in the construction sector.

3.2.4 Antagonistic relations

An increased reliance on subcontracting enabled main contractors to remove themselves from the physical activity of construction. Rather than being directly involved in production, they increasingly focused their attention on the management and coordination of subcontractors. The shift towards subcontracting also enabled them to operate on the basis of much 'leaner' estimating departments. Once invited to tender, the main contractor would plan the overall sequence of work and identify any risks. But beyond this, the primary function of the main contractor was to subdivide the project into discrete work packages and then distribute the documentation to subcontractors for pricing. The lowest prices for each work package would be added together to form the total cost of the project. The main contractor would add on an allowance for overheads and profit and the tender would be submitted. As already discussed, in an ultra-competitive market the contractor may omit the mark-up for profit with a view to undercutting the competition.

But once the contract is secured, the emphasis routinely switches to regaining the profit margin through the aggressive pursuit of claims. In times of recession, contractors may occasionally even submit tenders at below cost for the purposes of maintaining turnover. This in turn would encourage an even more adversarial approach in their dealings with clients. In general, subcontractor relations with main contractors are aptly described as being 'subordinate, dependent and antagonistic' (Evans and Lewis, 1989). Contractors would seek to exert control over subcontractors by exploiting their financial vulnerability. It would be common practice to seek to impose onerous contractual conditions heavily stacked in the favour of the main contractor. The general intention would be to pay subcontractors as late as possible, while seeking every possible excuse to withhold certain 'retention monies'. Some subcontractors were of course more easily exploited than others; it depended primarily on the ease with which the skills on offer could be sourced from elsewhere. Such pressures inevitably encouraged subcontractors to reduce their own costs as much as possible, with the consequence that many would in turn subcontract to labour-only subcontractors. In times of recession, labour-only subcontractors would be particularly vulnerable to contractual abuse because their capabilities could be so easily replicated. Even in relatively buoyant markets, labour-only subcontractors have limited market power because the barriers to entry are so low. In contrast, specialist subcontractors whose skills were not so easily replicated would tend to have a significantly greater degree of negotiating power. Indeed, leading M&E subcontractors are frequently larger, and more capitally intensive, than many main contractors.

3.2.5 *Competitive pressures*

A particularly invidious practice on the part of main contractors is to engage in 'Dutch auctions' following the award of the main contract. Unscrupulous contractors would instigate such actions in order to secure their profit level. This is often referred to euphemistically as 'contract trading'. From the point of view of the main contractor, the process is primarily about risk management; the task is to source the subcontractor on the basis of the lowest possible price subject to an evaluation of the risk of default. But in the case of labour-only subcontractors, the risk of default is close to minimum. Such subcontractors are invariably paid in arrears, and even if they did default they could easily be replaced within a matter of days. It would not be unheard of for the replacement labour-only subcontractor to arrive onsite with the same labour as had been engaged by the previous labour-only subcontractor. Such is the degree of informality which characterises the construction sector at this level.

The bottom line is that for many contractors (and subcontractors) it is all too often simply a matter of accepting the lowest price on offer. This exerts an immediate downward pressure on wages and employment conditions, such that bad employers secure a marginal advantage over good employers. Labour sourced through labour-only subcontractors provides a saving of 20–30% over direct labour costs (Evans and Lewis, 1989). These savings result directly from the casualisation of the workforce and the associated downgrading of employment conditions. In consequence, contractors and subcontractors who employ their own workforce struggle to compete.

It should be noted that the downward pressure described above is not dependent on the main contractor instigating the 'Dutch auction'. The process of reverse bidding is often instigated by those subcontractors who declined the opportunity to price the work in the first instance. They would simply make a phone call to the main contractor and offer to complete the subcontract for less than the lowest price currently on the table. When faced with this conundrum, main contractors often consider that the ethical response is to go back to the original subcontractor and give them the opportunity to match the new price. Hence there is a regressive competitive dynamic which undermines the competitive position of those firms which are resistant to such practices. Even when the main contractor honours the original tender, the processes of reverse bidding may still take place at lower levels of the contractual chain. It is important to emphasise that such processes are not confined to the 'cowboy builders' of popular imagination. These are the bidding processes which often characterise the way work is organised on major construction sites presided over by contractors which are household names. Such practices are no less prevalent in the construction sector of today than they were in 1995; this is how the industry operates.

3.2.6 · Lean and mean in the marketplace

The net result of the processes described above was the emergence of the 'hollowed-out' firm whereby no one is employed other than a tight core of managerial personnel. Such trends were driven by the adoption of a competitive strategy based on structural flexibility, i.e. the ability to expand and contract in response to fluctuations in demand (Winch, 1998). Any increase in subcontracting clearly aids flexibility in that it reduces the number of workers who are entitled to statutory redundancy payments. And once numerical flexibility had become established as the dominant model of competitive strategy, most contractors had little alternative other than to follow the pack. Institutional theorists like to refer to 'isomorphic pressures' to conform to the structures and strategies adopted by the majority. Firms which retained a large directly-employed workforce risked becoming uncompetitive. In a volatile market, contractors which carried too many fixed costs were insufficiently agile. Leanness became a central prerequisite to survival. The increased reliance on subcontracting enabled firms to avoid the need to bear the cost of redundancies; it also enabled them to abrogate their responsibilities in training and human resource development (Harvey, 2003). In consequence, the large monolithic contractors of the 1970s largely disappeared from the landscape, together with the relatively stable employment and training regimes which they previously provided. Especially stark is the structural shift which took place from 1979 through to the early 1990s (see Figure 3.3).

The competitive dynamics described above accentuated the problems of fragmentation and contributed directly to labour casualisation. Subcontracting is not the problem in itself; the difficulty is that the employment context dissolves into a casualised 'twilight zone' of opaque and localised arrangements. Winch (1998) is clear that labour-only subcontracting is not the same as self-employment; but there is frequently a significant overlap between the two. Indeed, an entirely self-employed workforce is the logical end point of subcontracting whereby each individual operative becomes a self-employed subcontractor. Such a situation would of course be ridiculous, and would be the antithesis of a sustainable industry. But the reality is that by the mid-1990s parts of the construction sector were seemingly

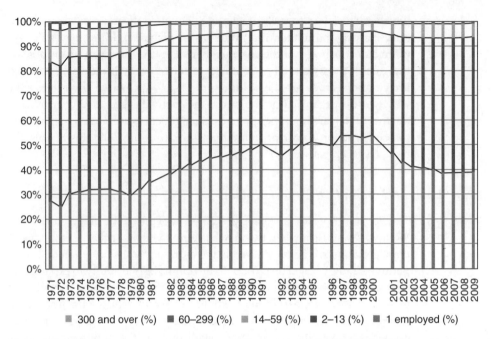

■ 300 and over (%)   ■ 60–299 (%)   ■ 14–59 (%)   ■ 2–13 (%)   ■ 1 employed (%)

Figure 3.3   Private contractors: percentage of firms by size (1971–2009) (*Sources: Housing and Construction Statistics 1979–1998*; *Construction Statistics Annuals 2000–2010*, Office for National Statistics, London)

moving irrevocably in this direction. The difficulty is that labour-only contractors frequently rely on labour supplied by labour agencies, which vary from legitimate businesses on the one hand to dodgy gang masters on the other. Such practices become even more impenetrable once migrant workers are added into the mix. Different gang masters frequently source labour from different geographic localities. In the 1980s the localities in question were cities such as Liverpool and Newcastle. In subsequent years they could equally be localities in rural Poland (cf. Langford and Agapiou, 2007). In either case, an outrageous situation was allowed to develop whereby operatives are reimbursed on the basis of where they come from, rather than on the basis of the job they do. This was the price to be paid for leanness and agility in the marketplace. Such trends developed despite the expressed preferences of the Emmerson (1962) report in favour of direct employment. It should be emphasised once again that the situation described is by no means limited to the grey economy of workers who serve the domestic sector. In all too many cases this is the situation which prevails on the nation's major construction sites.

The increasing fragmentation of the construction sector shadowed the differentiation of client demand. The demise of the state as a mass producer of housing removed significant stability from the marketplace and challenged the viability of the monolithic contractors which benefited from the 1960s public housing boom. The newly privatised utility companies likewise had little interest in providing continuity of work for the construction sector. Indeed, the entire philosophy of demand management advocated by Wood (1975) had slipped decidedly out of fashion. An increased reliance on subcontracting was a logical response to the changing environment. Those who were best attuned with the shifting competitive dynamics of the market place had read the runes as early as the mid-1970s. And for those who remained unconvinced, the recessions of the early 1980s and early 1990s provided salutary lessons on the benefits of leanness. Hence the dominant recipe of numerical flexibility established leanness and agility as favoured metaphors long before 'lean construction' was promoted as an essential component of best practice.

Several long-established national contractors went as far as re-branding (and re-listing) themselves as 'service companies'. But each incremental increase in the amount of work that was subcontracted also meant a corresponding loss of control over the production process. It is therefore understandable that the new hollowed-out contractors should try to reclaim as much control as possible while preserving the adopted model of numerical flexibility. Hence it is easy to understand how managers were easily seduced by the superficial attractiveness of 'modern' managerial improvement recipes such as partnering and supply chain management. Similar mechanisms were also at work within the construction sector's client base. This was certainly true of newly privatised clients such as the utility companies, Railtrack and BAA plc. It was also equally true for large influential retail clients such as Tesco, Sainsbury's and Asda. Even collectively, such clients admittedly account for a small proportion of the construction sector's workload. But, as we shall see, from the late 1980s onwards these were the very same clients who became increasingly active in the cause of 'industry improvement' (see Chapter 4).

3.3 Intensifying labour casualisation

3.3.1 *The incentivisation of self-employment*

It has already been argued that an increase in the level of self-employment is a logical consequence of the growth in subcontracting. This would not necessarily follow if it were only the main contractors which were resorting increasingly to subcontracting. But it does follow from an increased tendency to subcontract work throughout the supply chain. An increased reliance on labour-only subcontracting encourages a climate of self-employment. The major contractors which characterised the 1970s had previously offered a significant degree of stability in terms of employment regimes. This was often combined with a strong commitment to training, coupled with an active interest in the development of productivity improvement techniques. The major contractors had also previously employed large numbers of tradesmen (and occasionally, tradeswomen) who were supported by a full complement of apprentices. It is notable that the apprentice system was commented upon with broad satisfaction by both Emmerson (1962) and Banwell (1964), both of whom were firm believers in the benefits of direct employment. However, self-employment did not increase solely as a result of de-regulation and the 'hidden hand' of market forces; it was actively incentivised through government policy.

The primary mechanisms for the incentivisation of self-employment lay within the taxation and national insurance regimes, the foundations of which were first introduced during the early 1970s. Of particular importance was the 714 certificate launched in 1971. Possession of a 714 certificate was accepted by the Inland Revenue as proof of self-employed status, and thereby made workers responsible for submitting their own tax returns. An alternative route to self-employment was provided by the 'subcontract 60' (SC60) certificate, whereby tax was deducted at source by the main contractor. Access to a 714 certificate was initially relatively tightly controlled, but from the late 1970s onwards the criteria for qualification were progressively relaxed. Self-employed tax status in accordance with SC60 was consistently an easier option because in this case the authorities were less concerned about tax evasion.

Prior to engaging with the substance of the discussion, it must be conceded that the official statistics relating to the level of self-employment in the construction sector are notoriously unreliable (Briscoe, 2006; Cannon, 1994). Many of the difficulties relate to issues of definition; it is surprisingly difficult to derive any consistent test to determine whether an individual is self-employed or directly employed. Tax status alone cannot be taken as the determinant of whether an individual is genuinely self-employed or not. Even qualitative data are difficult to realise in that construction operatives are often understandably reluctant to provide information on their employment status. The situation is further complicated by the way in which self-employment is routinely conflated with labour-only subcontracting and the use of agency labour. Gang masters are also frequently active in the provision of migrant labour to major construction sites. The interactions between these various forms of 'contingent' labour comprise significant analytical difficulties. Despite the statistical difficulties, the broad consensus of opinion is that the percentage of self-employed operatives in the UK construction sector grew from around 30% in 1980 to over 60% in 1995 (Harvey, 2001; ILO, 2001).

3.3.2 Employment status: difficulties of definition

The legal employment status of construction workers has long been a matter of controversy, and despite a string of court cases it remains an issue of considerable complexity (Barker, 2007). Within the British legal system, the distinction between employment and self-employment rests on precedents established within case law. In essence, individual judgments rest on assessing whether there is a 'contract for services' (i.e. self-employment), or a 'contract of service' (i.e. employment). Yet the way in which the two parties label the relationship, or what they consider the relationship to be, is not in itself conclusive. It is the *reality* of the relationship that is important, although the intention of the parties can be decisive in circumstances where the relationship is ambiguous, or where the other factors are neutral.

Factors which are routinely taken into account in determining the relationship between two contracting parties include: (i) the degree of control exercised by the employer; (ii) the provision of tools and equipment; (iii) the arrangements made for tax, national insurance, value-added tax (VAT), statuary sick pay; and (iv) the presence or absence of 'mutuality of obligation'. These are derived from a list of 13 factors produced by Selwyn (2008) which it is recommended should be taken into account by employment tribunals. A further important consideration is the allocation of financial risk between the contracting parties; employees are not normally expected to risk their own capital. Part of the difficulty is that there is little consistency in either the contractual basis of, or the employment conditions which prevail within, catch-all categories such as labour-only subcontracting. The contractual reality of who is engaged by whom and on what basis becomes especially opaque when labour-only subcontracting becomes conflated with the use of labour agencies. What is clear is that many operatives who are routinely declared as self-employed within the construction sector could not be said to be in 'business on their own account'. In other words, many operatives who are nominally self-employed would fail any sensible 'economic reality test', and are hence more accurately classified as *falsely* self-employed. An independent report commissioned by UCATT (Harvey, 2001) suggested that there were between 300 000 and 400 000 thousand workers who were falsely classified as self-employed within the UK construction sector. At the time, this comprised between 30% and 41% of the directly-employed workforce as recorded by the official statistics.

There are of course many *bona fide* self-employed tradesmen in the construction sector. Many work in the domestic sector and undertake work directly for individual clients. As such, they act legitimately as entrepreneurs who are 'in business on their own account'. They negotiate a fixed price for the work, and they risk making a loss if unforeseen difficulties arise. Such circumstances account for a significant number of self-employed operatives in the sector. But there is also a significant grey economy at work whereby some projects 'go through the books' and others are performed on a 'cash-in-hand' basis. Homeowners often collude in tax evasion through their willingness to pay cash in return for avoiding VAT. The self-employed builders in turn also frequently use cash to pay others who help out with particular specialist trades. The informal economy in the domestic sector should in no way be condoned, but it is not the focus of this book. Of primary concern are the informal employment regimes which too often prevail on major construction sites, under the so-called management of major contractors. Of particular interest is the way contractors and clients alike consistently turn a 'blind eye' to false self-employment whilst at the same time advocating the benefits of 'integrated teams' and 'collaborative working'.

3.3.3   The benefits of self-employment for firms

The benefits of subcontracting have already been discussed. It has further been argued that in the context of the United Kingdom an increase in self-employment is an inevitable consequence of an increase in subcontracting. Given the increased reliance on subcontracting, the 'choice' between directly employed workers or self-employed workers is effectively pushed down the supply chain. Smaller main contractors also frequently engage individual self-employed tradesmen directly, but more frequently self-employed operatives are engaged by subcontractors. The point has already been made that individuals may often be unclear whether they are self-employed or not.

The financial benefits of self-employment to the 'employer' are on the surface reasonably clear. An operative who is declared self-employed immediately removes the employer's obligation to pay their share of the national insurance contribution with an immediate estimated saving of around 12% (Harvey, 2003). A shift to self-employment further removes the employer's obligation to provide statutory employment benefits such as holiday and sick pay. It also removes the obligation to contribute to the operative's pension scheme, the responsibility for which is transferred to the operative. It should be noted that such savings are achieved solely through changing the employees' tax status; it does not depend upon any productivity gain as a result of the operatives 'working for themselves'. A reliance on self-employed operatives therefore directly enhances the competitiveness of subcontractors. This raises the possibility of a symbiotic relationship between subcontracting and self-employment. Not only does an increased reliance on subcontracting encourage self-employment, but the incentivisation of self-employment also encourages an increased reliance on subcontracting.

But it must also be recognised that the cost benefits of self-employment are rapidly dispersed if contract labour has a higher unit price than the directly employed. And genuine self-employed workers are frequently able to earn enhanced rates – otherwise there would be no incentive for them to become self-employed. The price of contract labour is of course dependent primarily upon market mechanisms relating to supply and demand. In essence, labour-only subcontractors derive their profit from the difference between the price at which they supply labour and the cost at which they can secure it. Whilst reliable data is in scarce supply, the base rates for contract workers are undoubtedly routinely suppressed as a result of low barriers to entry. Regional imbalances and high unemployment throughout the 1980s meant that there were always significant numbers of workers from the North who were seeking work.

Even during periods of relatively high demand, a reliance on self-employed operatives can still serve a useful purpose in confining inflationary pressures to specific trades. The overriding issue is that the wages of directly employed operatives tend to be relatively stable, whereas the cost of contract labour is prone to significant fluctuations. The cost of labour supplied through labour-only subcontractors often declines dramatically during times of recession; this in itself is a significant benefit to those firms who base their competitive strategy on leanness and agility.

Consideration should also be given to the possibility that the widespread use of self-employed labour provided a useful strategy of off-setting the risk of industrial action (cf. Druker, 2008). Harvey (2003) is especially bold in attributing the origins of the growth in self-employment to the industrial unrest of the 1960s, which he characterises as:

'[a] period of major industrial strife, especially in the City of London where unionization and union power reached a high point. This period stirred employers and government to contemplate the encouragement of self-employment as a means of counter-balancing union power.' (Harvey, 2003; 201)

Conspiracy theorists of course always like to imagine groups of powerful plotters making secret deals in smoke-filled rooms. But the reality invariably tends to be much more prosaic, and the growth of self-employment is perhaps best read as an emergent characteristic of the enterprise culture. The major contractors had already insured themselves against industrial action through an increased reliance on subcontracting, and it is unlikely that the relatively transient labour-only subcontractors retained memories of industrial disputes from previous decades. The major contractors further insured themselves against industrial action by securing the services of blacklisting agencies who maintained records of individuals who were even suspected of militant track records. Such services of course were hardly necessary in the case of self-employed operatives, whose summary dismissal requires little justification.

3.3.4   Enterprising individuals

The preceding discussion sheds light on the reasons for the increased reliance on subcontracting, and for the way in which subcontractors made themselves more competitive through the use of self-employed operatives. It has been argued that the primary reason lies in the adoption of structural flexibility as the dominant model of competitive strategy. This was spurred by the increased volatility of the market, encouraged in part by extensive client diversification. An additional factor was the realisation amongst chief executives that the Thatcher administration of the 1980s was not interested in managing demand for the benefit of the construction sector. This led inexorably to the emergence of the hollowed-out firm as the dominant organisational model. Contracting firms placed a premium on leanness and agility in the market place. The shift to self-employment was often kick-started by the immediate cost savings which could be secured through the avoidance of employers' national insurance contributions and other statutory costs associated with direct employment. A contributory factor has also been the recurring lack of clarity in the legal definition of self-employment, although this is by no means unique to the construction sector. Rather more tentatively, it been suggested that government and employers sought to encourage self-employment as a means of emasculating trade union power.

What remains to be explained is the attractiveness of self-employment to the operatives themselves. The financial incentives for workers to become self-employed were certainly not as strong as the incentives on offer to employers. For those seeking self-employment on the basis of a 714 certificate there was a possibility of tax evasion. Ever since the scheme's initial introduction there has always been a thriving black market in 714 certificates, which may suggest a tendency towards tax evasion. However, the possibility of tax evasion only ever applied to a small minority, and was not even an option for those who were self-employed on the basis of SC60. Operatives shifting to self-employment do benefit from a small a reduction in the level of national insurance contribution, although this is hardly likely to be significant (Harvey and Behling, 2008). Those who are truly in business on their own account are also able to claim increased allowances against running costs, but this again would not account for the majority of those who opted for self-employment within the context of labour-only sub-contracting.

One of the reasons that labour-only subcontractors are able to attract high rates of remuneration relates to the absence of fixed overheads. However, the point has already been made that such factors are easily overridden by the mechanisms of supply and demand, much beloved of course by the advocates of the enterprise culture. It therefore follows that the additional earnings on offer during times of boom were sufficient to entice many into self employment. This would certainly seem to have been the case during the latter 1980s. Winch (1998) cites a calculation reported in the *Contract Journal* (6 August 1997) which suggests a mark-up of 50–100% for a self-employed operative in comparison to an operative employed on the rates set by the National Working Rule Agreement (NWRA). Such an uplift may indeed offer compensation for the loss of employment benefits such as holiday pay, sick pay and statutory pension contributions. This would be especially true for those who were willing to sacrifice long-term benefits for the sake of additional cash-in-hand.

The stereotypical image of building workers as a slightly irresponsible bunch who value autonomy and enjoy the 'crack' is one which continues to prevail. Given the pre-existence of such a culture, the argument is often promoted that self-employment is attractive because it offers a greater degree of autonomy. It might equally be argued that young men (and women) join the ranks of the British Army because they value the structure and discipline offered by military life. In both cases, a more convincing explanation hinges on the absence of attractive viable alternatives.

Many workers during the 1980s boom years were undoubtedly tempted to transfer to self-employment by the prospect of significantly higher cash-in-hand payments. And having opted for self-employed status, they then found it difficult to transfer back into direct employment. Indeed, they would only likely to be motivated to return to direct employment when the attractiveness of self-employment became eroded by a shift in the balance of supply and demand, i.e. by the onset of recession. And this of course would be the worst time possible to seek direct employment in the construction sector. In either case, work would be consistently easier to secure on a self-employed basis as a result of the sustained systemic bias towards self-employment among contractors. Word-of-mouth introductions to labour agencies would become a much more viable route to finding work on major building sites than submitting speculative applications to main contractors. The demise of the DLOs further served to reduce the opportunities which were available for direct employment. Such workers arguably benefited from the greater degree of flexibility which self-employment offered in comparison to direct employment. They could frequently work four-day weeks in London returning every Thursday evening for a long weekend at home. However, it would be wrong to suggest that self-employment was the preferred choice. Operatives may have been attracted initially to self-employment by the prospect of higher income during the 1980s boom. This does not necessarily mean that they *preferred* to remain self-employed. It was more often a case that self-employment was the only basis upon which work was available, and very often such workers were falsely self-employed. Many such workers made no conscious choice to become self-employed, and neither were they self-aware of their employment status. They just followed their mates. On the basis of any economic reality test they should have been directly employed.

3.3.5 Loadsamoney

The growth in self-employment undoubtedly exacerbated the pre-existing problems of industry fragmentation. It is therefore striking that the workers were positively encouraged to think of themselves as individual enterprises by the rhetoric of the enterprise culture (see Chapter 2). Of further note is the way in which the policy dimensions of the enterprise culture specifically targeted skilled workers, many of whom willingly traded in their trade union membership card for a mobile phone and a 714 certificate. The values being exhorted by government and industry leaders alike were the values of rugged entrepreneurial individualism. This was a significant shift from the worldview previously espoused by the likes of Emmerson (1962) and Banwell (1964), both of whom warned specifically against the dangers of labour casualisation. However, the advocated benefits of direct employment gave way to the ideas of the 'flexible economy'. Where once there had been employment stability there was now a euphemistic commitment to flexibility and innovation. The enterprise culture therefore undoubtedly legitimised the shift to self-employment, and served to bolster the self-identities of many construction operatives in the new, flexible economy. The individuals who were tempted into self-employment were those who fitted the 'Essex Man' stereotype; they were probably also exercising their 'right to buy' at the same time as they were opting for self-employment. The culture on offer was parodied memorably by comedian Harry Enfield in his 1988 hit song 'Loadsamoney'. Collectivist structures such as trade unions, council housing and direct employment were held to represent the failed values of the 1970s. Self-employment was also viewed by many skilled workers as a means of restoring eroded differentials, and many undoubtedly benefited in the short term. But the financial benefits on offer were severely reduced by the recession of the late 1980s, during which many self-employed workers experienced drastic reductions in their take-home pay (Harvey and Behling, 2008). Others became self-employed workers without any work, and in consequence many were lost to the construction sector forever.

The discussion above has focused on the reasons for the growth of self-employment from the mid-1970s through to 1997, when the partial re-imposition of barriers saw some success in shifting 185 000–210 000 workers back into direct employment (Harvey, 2003). However, since then the trend has again been inexorably upwards (see Figure 3.4). The discussion of self-employment will be revisited and brought up to date in Chapter 10, when particular attention will be given to the conflation of self-employment and migrant workers. In reading Figure 3.4, it should be borne in mind that the official statistics on the number of self-employed are widely held to under-estimate the true level by approximately 200 000 (Cannon, 1994). Hence the accepted assertion that at its peak in 1995 the majority of the construction workforce in the United Kingdom was nominally self-employed. This was a bizarre situation for a major industrial sector to find itself in; fragmentation of the construction sector is often cited as a major barrier to improvement. But at the same time, there remains little will to address the fragmented nature of the construction workforce.

3.3.6   Illegal blacklisting

It has been suggested that the primary reason for the systemic shift away from direct employment among contractors was the collective commitment to competitive strategy based on structural flexibility. More tentatively, it

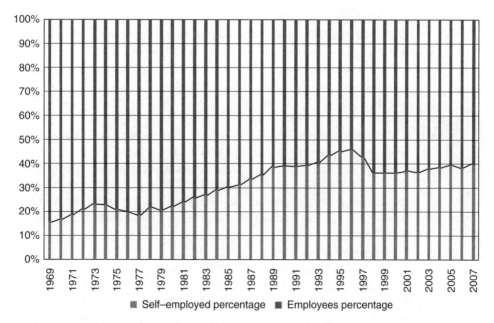

■ Self–employed percentage   ■ Employees percentage

Figure 3.4   Great Britain construction manpower: percentage of employees and self-employed (1969–2007) (*Sources: Housing and Construction Statistics 1980–1998; Construction Statistics Annuals 2000–2008*, Office for National Statistics, London)

has further been suggested that a contributory factor was the collective memory of industrial unrest from the 1960s and 70s. The national building strike of 1972 left a legacy of bitterness and many employers resented having been being 'held to ransom'. It is this background which goes some way towards explaining the widespread blacklisting of 'militant' trade unionists through a private enterprise blacklisting organisation known as the 'Economic League' (Hollingsworth and Tremayne, 1989).

The Economic League was originally established in 1919 to fight 'bolshevism' and remained shrouded in secrecy until subject to scrutiny by a House of Commons Select Committee in 1989 (Green, 2009). It habitually interfered in British industrial relations throughout the 1970s and 80s, before being finally abolished in 1994. In its latter years the League regularly attracted the attention of civil liberty campaigners who claimed that it held information on individuals illegally (Hencke, 2000). At its peak, the League was engaged by more than 2000 companies to weed out active trade unionists, or those who were suspected of being active trade unionists. Ironically, this was a system which would not have looked out of place in Erich Honecker's supposed socialist state of East Germany. The penalties of being exposed as a deviant in East Germany were of course much more severe, but illegally blacklisted workers in Britain undoubtedly suffered significant economic hardship. During the 1980s, the majority of the major contracting firms subscribed to the Economic League, and relied on its services for screening potential employees. The League had files on as many as 45 000 people it considered to be 'leftwing subversives' (Evans *et al.*, 2009). Individuals often found themselves blacklisted for entirely legal involvement in trade union affairs with no right to appeal or other form of redress. Some found themselves blacklisted simply for keeping company with trade unionists. Even complaining about safety standards or employment rights at work was enough to qualify for the blacklist.

To be clear, it is not being suggested that the Economic League was a contributory cause of the growth of self-employment. But the fact that the majority of chief executives of major contractors signed up to such an organisation is indicative of the mindset which prevailed. Direct employment was seen as a source of risk, and membership of the Economic League was a means of risk mitigation. If the League did not contribute directly to the growth in self-employment, then the growth in self-employment certainly contributed to the demise of the League. By the time of the Economic League's demise in 1994, self-employed workers were in the majority across the sector at large, and very often comprised the core workforce on major construction projects. Hence employers' concerns about employing 'trouble makers' were alleviated progressively by changing patterns of employment. At the risk of pointing out the obvious, self-employed operatives are not subject to the same degree of employment protection as directly employed workers. Hence they can simply be dismissed on the spot and escorted from the site. Self-employed operatives are of course rarely trade union members, and every incremental step towards self-employment serves to reduce the density of trade union membership.

3.4   Re-structuring in the client base

3.4.1   Demise of the Property Services Agency (PSA)

The preceding chapter described the wave of privatisations which characterised the 1980s and early 1990s. From the point of view of the construction sector, the privatisations of the public utilities, BAA and British Rail were especially significant. Privatisation was almost inevitably followed by extensive restructuring coupled with the reengineering of long-established processes. Procurement was no exception to these general trends, and privatisation frequently foresaw a radical shift in the way in which construction work was procured. In some cases, the newly privatised client organisations were liberated from the constraints of public sector lowest-cost tendering and moved towards procurement based on value and quality. Others chose a more cost-driven approach to reengineering their processes which resulted in a reliance on design-and-build architecture which prioritised efficiency of construction over design quality.

However, the privatisation which arguably did most damage to the accumulated procurement expertise was that of the Property Services Agency

(PSA). Prior to privatisation, the PSA played a mediating role between government departments and private sector suppliers (Burnes and Corum, 1999). Its demise therefore represented a significant change in the established mode of engagement between public sector clients and the construction sector. The intention to privatise was announced in 1989, although the process only started in earnest in 1992 under the stewardship of John Major. The privatisation of the PSA attracted little public attention at the time; the media was understandably fully occupied trying to make sense of the shambolic privatisation of the railways.

The origins of the PSA can be traced back to the Ministry of Works, formed in 1943 to take responsibility for the wartime procurement of property. After the war the Ministry of Works retained responsibility for government building projects. In 1962 it became the Ministry of Public Buildings and Works and was subsequently merged with the works directorates from within the Admiralty, War Office and Air Ministry (Burnes and Coram, 1999). The PSA formally came into being as an autonomous agency in 1972 following the creation of the Department of the Environment (DoE) in 1970. Its stated remit was to 'provide, manage, maintain, and furnish the property used by government'. In truth, creeping privatisation had been underway since 1981 through the enforced contracting out of construction services to the private sector. However, throughout the intervening period the PSA had retained an overall supervisory role, and it was this function which was to be effectively lost with full privatisation. The PSA had also been uniquely placed to adopt a through-life perspective on the management of the government's property stock, although attempts to strike an appropriate balance between capital and revenue were consistently hampered by Treasury rules.

Following the 1992 general election the PSA was split up into separate operating companies which were then sold as going concerns to private sector construction firms. The break-up and privatisation of the PSA potentially provided the opportunity for much closer, and more innovative, procurement relationships between government departments and the construction sector. But the downside was that the public sector lost much of its accumulated expertise in property procurement, including the PSA's much heralded database on construction and maintenance costs.

The acquisition of the privatised regions of the PSA by private sector firms offered the means of developing significant additional business streams based on the provision of facilities management (FM) services. It was in this context that the privatised operating companies began to emphasise the importance of 'partnerships' between themselves and public sector clients, with a particular emphasis on the benefits of long-term relationships. However, central government directives on competitive tendering and market testing continued to act against the possibility of long-term service contracts with the private sector (Erridge, 1996). While the rhetoric of 'partnership' prevailed on both sides, the reality was that any

shift towards a long-term service ethos was heavily mediated by a continued insistence on short-term contracts. The prospect of formulating cosy
partnership relationships with the privatised regions of the PSA was not
helped by memories of the corruption scandal of the early 1980s. Evidence
of corruption in PSA District Works Offices had prompted the appointment
of Sir Geoffrey Wardale to conduct a formal enquiry (Department of
Environment, 1983). The enquiry resulted in the removal of the PSA Chief
Executive, Montague Alfred, whose views were deemed to be 'contrary to
government policy' (Doig, 1997). The Wardale report undoubtedly revived
memories of the Poulson scandal of 1973. Corruption in public procurement may be relatively rare in Britain, but when it is exposed it casts a very
long shadow. To this day the shadow of corruption continues to undermine
the persuasiveness of those who argue in favour of 'collaborative working'.
As an aside, the notion of 'collaboration' also brings with it unpleasant connotations of the collaborationist Vichy regime in France during World War
Two. Others may recall Norwegian 'quislings' from the same era.

Returning to the central argument, it is further notable that since the
break-up of the PSA many government departments have chosen to take
responsibility for estate management back in-house, thereby establishing
their own property management departments. By far the largest of these is
Defence Estates, which continues to look after the military facilities operated by the Ministry of Defence (MoD). Defence Estates was destined to
play its own role in the construction sector improvement debate, especially
through the much heralded project known as *Building Down Barriers*
(see Cain, 2003).

3.4.2   Highways management and maintenance

The Highways Agency (HA) has played a pivotal role in the development
of the enterprise culture in the civil engineering industry. The 'trunk road'
system in Britain dates from the 1930s and from the outset was seemed to
be the responsibility of central government rather than local government.
The same applies to the motorway system which dates back to the opening
of the first section of the M1 in 1959. From the outset government was much
more willing to invest in the infrastructure of the nation's road system than
it was in the railways. This was especially the case in the 1980s when the
individualistic nature of the enterprise culture resonated much more
strongly with private car ownership than with public transport. The HA
was established as single executive agency of government in 1994 with a
remit of managing and maintaining the motorway and trunk road network
and delivering a programme of improvement works (Haynes and Roden,
1999). Its role has since evolved into that of network operator with the
declared mission of enabling 'safe roads, reliable journeys and informed
travellers'. Since its formation, the HA has progressively adopted a regional
procurement policy whereby they have contracted with fewer and fewer

contractors (Leiringer *et al.*, 1999). Initially, there were 91 so-called agency agreements, of which 85 were with local highway authorities. By 1996 the number of agency agreements had been reduced to 24, which in turn has since been reduced to 14 maintenance areas. The decreased reliance on local authority highway agencies equates in no small way to the progressive privatisation of the nation's highway maintenance capability. At the same time, central government pressures towards market testing and compulsory competitive tendering (CCT) encouraged local authorities to outsource their local highway maintenance work. Such trends often resulted in highway maintenance divisions within major contractors being comprised entirely of personnel who had been transferred on mass from local authorities. The expression which is commonly used is that of 'TUPE-transfer'.

The acronym 'TUPE' relates to the Transfer of Undertakings (Protection of Employment) Regulations 1981 (later superseded in 2006). The TUPE regulations ensured that the employees were employed by the new employer on the same terms and conditions, thereby offering an important degree of employment protection. The TUPE legislation thereby ensured that private operators did not make efficiency gains at the cost of employment conditions. Or at least this was the case in the short term. Critics such as Sachdev (2001) argue that in the longer term private contractors tend to gravitate towards a two-tier workforce. The first tier comprises those who have been TUPE-transferred from local authorities, and the second comprising those who have been subsequently recruited – invariably on reduced terms and conditions. Hence the 'efficiency gains' claimed by private sector operators are arguably achieved through the downgrading of employment terms and conditions, rather than by cunningly clever 'innovative practices'. The relative size of the first-tier workforce would of course gradually reduce over time through retirements and resignations, and their replacements would swell the ranks of the second-tier workforce. There would also be a tendency for the first-tier workforce to be much more heavily unionised. The different is that between a stable employment environment and an unstable employment environment. This is the enterprise culture in action; employment terms and conditions are set by the market. Hence, even in civil engineering, where the proportion of direct employment remains much higher than in building, mechanisms are continuously at work to undermine the possibility of a stable employment regime.

The situation within highway maintenance contractors is further complicated by the legacy of acquisitions. The tendency of the Highways Agency to want to work in 'partnership' with a smaller number of national contractors has encouraged growth through acquisition. The larger contractors are therefore not only juggling with the discrepancies in employment conditions presented by their TUPE-transferred workforce. They are at the time juggling with discrepancies between the employment conditions within different operating companies as a result of the ongoing legacy of acquisitions. Such discrepancies invariably require a disproportionate amount of attention from the contractors' 'human resource' (HR) departments such that they rarely progress much beyond short-term fire-fighting. As an aside, the re-branding of personnel departments with the ubiquitous HR badge can in itself be read as a manifestation of the enterprise culture (Legge, 1995).

3.4.3   Best value runs riot

At the risk of jumping ahead of ourselves, it is worth charting the more recent trends within the Highways Agency. In common with other executive agencies within government the Highways Agency has increasingly seen itself as the enabler and regulator for the private sector provision of services. The private sector has in turn increasingly aligned itself with the Highways Agency's stated mission of providing 'safe roads, reliable journeys and informed travellers' thereby creating opportunities for incentivised performance-based contracts (Leiringer *et al.*, 2009). The Highways Agency (2005) is further notable for adopting the Blairite notion of 'best value' whilst at the same time emphasising its commitment to 'collaborative partnerships'. However, in common with the experience of other government departments involved in the procurement of new buildings (post-PSA) the development of collaborative working remains curtailed by a lack of trust between public and private sectors. Once again, a key inhibitor is the public sector's inability to initiate long-term contracts as a result of the centrally imposed need for recurrent market testing. The Highways Agency did however win many plaudits for its Early Contractor Involvement (ECI) initiative. At least the Wood report's (Wood, 1975) recommendation in this respect eventually found a client willing to listen. Other procurement initiatives within the Highways Agency included the Managing Agent Contractor (MAC) contract which has sought to promote a more service-orientated engagement between the public and private sectors. MAC contracts were typically awarded for five-year periods, and these have more recently been supplemented by Extended Managing Agent Contractor (EMAC) contracts which can be renewed for a further two years (subject to performance) (Leiringer *et al.*, 2009).

The increasing emphasis on service provision within the Highways Agency also invoked a need to take a range of criteria into account over-and-above cost. This led firstly to the development of the Highway Agency's Capability Assessment Toolkit (CAT) followed soon thereafter by the Culture Assessment Framework (CAF) (Highways Agency, 2005). The emphasis throughout lay with the stated desire to promote a 'best value' culture based on partnerships. Consultants of course continue to have a field day advising contractors on how such assessments should be completed. Performance measurement and continuous improvement are key features of the Highways Agency approach. But given the history of prioritising agency agreements with private contractors rather than local highway authorities it would seem strange that the Highway Agency now finds it necessary to measure characteristics which were once seen as axiomatic. Had so-called 'soft-issues' such as employment terms and conditions, commitment to training and sustainability been prioritised in the mid-1990s, a greater proportion of motorway and truck road maintenance might still be undertaken by local highway authorities. But the net result of the Highways Agency's various initiatives has squeezed-out local authority DLOs while at the same time concentrating highways maintenance contracts with a small number of large contractors. Smaller independent regional contractors seemingly have no place in highways maintenance other than as subcontractors to the big players. They then become subject to the rigours of supply chain management which, as has already been demonstrated, is frequently a commercially uncomfortable place to be.

3.4.4 *The privatisation of research*

It is appropriate at this point to digress slightly from the theme of restructuring in the industry's client base to consider the implications of the enterprise culture for the industry's research base. In contrast to the privatisation of British Rail, the privatisation of the Building Research Establishment (BRE) in 1997 attracted little in the way of media attention. Nevertheless, it constituted an important shift in the landscape of the construction sector, especially for those with an interest in research and development (R&D). The changing status of BRE also provides an interesting microcosm of institutional change in the construction sector at large.

The BRE story commences with the Building Research Station (BRS), which was founded in 1921 in the wake of the First World War. Lansley (1997) describes how BRS was one of several governmental research organisations with 'a mission to bring industry into the modern world through the influence of applied science'. There was much evidence at the time to support the view that private industry lacked the capacity to innovate (as it might be phrased in today's terminology). Concerns about the 'damaging effects of competition' grew significantly as a result of the 1930s depression. It was this world view that subsequently made nationalisation synonymous with modernisation. In the early years of the BRS, research focused on the performance of building materials before soon expanding into geotechnical and structural engineering, building physics and various aspects of fire performance (Courtney, 1997). During the Second World War the BRS contributed to the Allied war effort in a range of areas, including the design of air-raid shelters and the effects of fire and explosions on structures. In addition to the main site at Garston, near Watford, a subsidiary location was opened near Edinburgh in 1949. The BRS originally fell under the wing of the Department of Scientific and Industrial Research, but in 1965 government reorganisation bought it under the wing of the Ministry of Public Buildings and Works. In 1970 the BRS found itself part of the newly created

Department of the Environment (DoE) where it was combined with the Fire Research Station and the Forest Products Research Laboratory, to form the Building Research Establishment (Courtney, 1997). Throughout this period, the BRE enjoyed a high reputation for technical excellence. It is difficult to avoid the imagery of a stable (and staid) governmental bureaucracy populated by research scientists with bowler hats and clipped moustaches. The BRS was very much rooted in the same era as the Banwell and Emmerson reports, and the BRE at its peak in 1975 was fully commensurate with the world view projected by the Wood report (Wood, 1975). But from 1976 the world started to change, and government departments came under sustained pressure to reduce costs, which invariably translated to reductions in employment. At its peak in 1975 the BRE had a staff of 1350, by 1989 this had reduced to 654 (Courtney, 1997). The 1980s of course had seen much change in the construction sector, not least that led by property developers such as Stuart Lipton of Stanhope. The *Building Britain 2001* reports also did much to articulate an industry research agenda which was consistent with the enterprise culture (see Chapter 4). The BRE remained largely isolated from these developments as it stuck to its traditional science base.

Lansley (1997) further describes the emergence of new research organisations which understood the changing needs of industry in a way which the BRE did not. The new organisations often grew from newly-formed links between leading industrialists and a new generation of industry-friendly academics. These collaborations were to lead to the establishment of bodies such as the European Construction Institute (ECI), the Centre for the Strategic Studies in Construction (CSSC) and Construct IT; born respectively at the universities of Loughborough, Reading and Salford. The CSSC later spawned the Reading Construction Forum which was to play a pivotal role in the industry improvement agenda through the 1990s. These bodies and their associated universities piloted a model of collaboration between academia and industry which still prevails today. However, none has ever evolved into a self-funding research organisation, and one of the reasons for this is that the dominant industry recipe of structural flexibility does not encourage investment in research and development.

Gradually a new model of research emerged whereby funding from the research councils and/or the Construction Sponsorship Directorate within the DoE was matched with 'resources-in-kind' contributions from industry. Such collaborations invariably invoke a tension between industry's need for knowledge which can be applied in the short term and the needs of academics to contribute to the international knowledge base as codified in refereed journals. In the context of construction management, such tensions have given rise to a model of research which is increasingly referred to as 'co-production' (Green and Harty, 2008). But such developments in research remained largely alien to the BRE during the 1980s.

However, in accordance with government policy elsewhere, the BRE was soon to receive its 'short, sharp shock'. In common with the PSA, in 1990 the BRE became an Executive Agency and was hence accorded a higher degree of managerial independence. In essence, it was required to operate much more along the lines of a business, although it was still subject to annual targets agreed with the DoE. In the years following 1990, the BRE focused increasingly on environmental issues. The launch of the first version of the BRE Environmental Assessment Method (BREEAM) took place in 1990. BREEAM subsequently cornered the market for environmental assessment and became a significant income generator. The increasing commercialisation of the BRE bought with it the need to do research for non-governmental clients. It was therefore required to re-orientate itself towards more instrumental research which was 'relevant to industry'. Courtney (1997) portrays the transition as moving from a centrally-funded organisation essentially concerned with scientific research to one which provided a range of professional services for a variety of customers. The cult of the customer seemingly knew no bounds.

BRE was subsequently privatised in 1997 although at the time it continued to derive in the region of 85% of its income from government sources. It continued to receive guaranteed income streams from the DoE for five years beyond privatisation. The BRE used much of this income to develop its expertise in near-market construction process research, presumably with a view to expanding its consultancy services.

3.4.5 Private finance initiative

John Major's Conservative government began to promote the use of private finance for public sector projects in the early 1990s. The private finance initiative (PFI) was formally announced by the then Chancellor, Norman Lamont, in his 1992 Autumn Statement. PFI was born in the Treasury and was advocated as a means of reforming the delivery of core state activities which did not lend themselves to outright privatisation. It took time for PFI to gain momentum and less than a billion in capital investment from the private sector had been secured by 1995 (NAO, 2009). Understandably, private sector firms were deterred by the high costs of bidding and were wary of the risks which they were being asked to take on. The early PFI contractors priced accordingly and were subsequently criticised for earning excessive profits, although these became much more normalised as the PFI market matured gradually.

PFI was undoubtedly an integral component of the re-energised 'enterprise culture' and was indicative of Major's willingness to push privatisation beyond the boundaries previously deemed acceptable by Margaret Thatcher. It is notable that PFI was born from an increasingly dominant Treasury, where John Major had previously spent the formative part of his political career. The PFI initiative under the Conservatives was underpinned by an ideological belief that private sector provision is more efficient than public sector provision. In essence, PFI combines the procurement of capital assets with the services associated with their operation. PFI involves the private sector in the designing, building, financing and operating (DBFO) of a particular facility in accordance with an 'output' specification devised by the public sector. Since its launch PFI has been most widely adopted in the procurement of schools and hospitals, although it has been applied across a wide spectrum of public projects including Second Severn Crossing and the Birmingham Northern Relief Road. There have also been numerous other non-construction examples, including several IT-based projects. Under PFI, the public sector does not own the asset, but commits to pay the PFI contractor a stream of revenue payments for the use of the facility over the life of the contract. Thereafter, ownership of the asset would remain with the private sector contractor, or revert to the public sector according to the terms of the contract. Contract durations typically vary from 25 to 35 years.

From its inception PFI has always been a controversial and hugely politicised issue. The advocates of PFI contend that it allows for more investment in public services than would otherwise take place. They also contend that the private sector is able to offer improved efficiency both in the delivery of new facilities and also in their operation. However, PFI has attracted extensive public criticism (Public Services Privatisation Research Unit 1997, Gaffrey *et al.*, 1999, Monbiot 2000, Pollock *et al.*, 2001). One of the main arguments against PFI is that it was motivated primarily by the need for off-balance sheet financing. The price for avoiding public sector capital borrowing in the short term is the commitment to guarantee revenue payments over the life time of the contract. This is seen to be a process akin to the mortgaging of public assets. Critics also point towards the consultancy gravy train associated with complex PFI deals. Fees for legal and financial advisors on a £25 million PFI can easily exceed £500 000. The argument is further made that the private sector cannot borrow money at the same preferential rates as the public sector, thereby increasing the overall cost of borrowing. A further concern relates to workforce issues, and the erosion of staff terms and conditions. It is notable that PFI was criticised repeatedly by Labour politicians Tony Blair and Gordon Brown while they were in opposition. Strangely, these criticisms were to be forgotten following the election of the New Labour government in 1997, which proceeded to implement PFI with even more enthusiasm than the Tories (see Chapter 9). There have also been many reservations about the architectural quality of PFI schemes, a cause which was adopted subsequently by the Commission for Architecture in the Built Environment (CABE).

Given the politicised nature of the debate regarding the merits of PFI, it is difficult to offer a neutral evaluation. Interested parties are best directed towards the plethora of reports produced by the National Audit Office.

The public sector trade union, Unison, has run a sustained campaign against PFI, supported by a succession of commissioned reports (e.g. Unison, 1997; Unison, 2001; Unison, 2002). For many critics, PFI provides an invidious extension of previous initiatives such as compulsory competitive tendering (CCT) and 'Best Value'. The common themes are privatisation, outsourcing and the introduction of market mechanisms into the provision of public services. Ultimately, such initiatives are seen to erode employment conditions, thereby leading to an increasingly casual-ised and vulnerable workforce. The counter argument of course is that too many public sector workers under-perform because they are sheltered from the invigorating influence of competition. The debate is therefore not really a debate about the technicalities of PFI, but more about the benefits and dis-benefits of the enterprise culture. As already noted, the momen-tum of PFI picked up significantly following the election of New Labour in 1997. The further development of PFI under New Labour will be described in Chapter 9.

3.5 Summary

This chapter has focused on the radical restructuring which characterised the UK construction sector throughout the 1980s and early 1990s. These changes can be best understood in terms of the material manifestation of the enterprise culture. However, it would be a mistake to attribute these changes entirely to Thatcherism; there were broader forces at work. Many of the trends were already apparent prior to the election of 1979. What can-not be denied is that the construction sector which was inherited by the Blair government in 1997 was very different from the sector which was bequeathed by the Callaghan government in 1979. Of particular note was the rapid and unprecedented growth in self-employment from the late 1970s through until the mid-1990s. During the late 1990s there was much talk of a 'return to direct employment', with firms such as Laing O'Rourke claiming to take a lead. In the region of 200 000–300 000 workers did indeed shift back into direct employment from 1996 to 1998, but this was largely in response to a half-hearted clampdown on tax evasion by the Inland Revenue. Such occasional clampdowns must be understood within the context of a more sustained incentivisation of self-employment through the tax and national insurance system. Furthermore, the acclaimed return to direct employment was as exaggerated as it was short-lived. There was never any danger of a return to the levels of direct employment which pre-vailed during the 1970s.

It is important to emphasise once again that self-employment has a legit-imate role in the construction sector. However, it is difficult to defend a system whereby such a sizeable proportion of the industry's workforce is nominally self-employed. The extent of fragmentation of the construction sector is cited consistently as a barrier to industry improvement. However, the extent of fragmentation increased dramatically over the period 1979–96 with significant implications for the construction worker. The reasons for this increase have been seen to be complex. In part, it relates directly to the diversification of client demand and an associated increase in subcontracting. A further contributory factor was the adoption of numerical flexibility as the dominant competitive strategy within the contracting sector. This in turn was stimulated by the retreat of government from any pretence of demand management for the benefit of long-term planning in the construction sector. A reliance on self-employment has further been argued to have been a risk management strategy in response to a perceived threat of industrial action. Another important part of the story is provided by the demise of public sector DLOs, which had previously provided an important degree of employment stability within the sector. In times of economic growth self-employed operatives also enjoy increased financial returns and a greater degree of flexibility to work as and when they prefer. It would be easy to suggest that the growth of self-employment was a direct consequence of the enterprise culture. However, it is perhaps more realistic to see the growth of self-employment as a constituent part of the discourse of enterprise whereby causes and effects are not so easily isolated.

References

Banwell, Sir Harold (1964) *The Placing and Management of Contracts for Building and Civil Engineering Work*. HMSO, London.

Ball, M. (1988) *Rebuilding Construction: Economic Change and the British Construction Industry*, Routledge, London.

Barker, J.C. (2007) Self-employment: legal distinctions and case-law precedents, in Dainty, A., Green, S, and Bagilhole, B. (eds) *People and Culture in Construction*, Spon, London, 56–69.

Briscoe, G. (2006) How useful and reliable are construction statistics? *Building Research & Information*, **34**(3), 220–229.

Burnes, B. and Corum, R. (1999) Barriers to partnerships in the public sector: the case of the UK construction industry, *Supply Chain Management*, **4**(1), 43–50.

Cain, C.T. (2003) *Building Down Barriers: A Guide to Construction Best Practice*, Spon, London.

Cannon, J. (1994) Lies and construction statistics, *Construction Management and Economics*, **12**, 307–13.

Courtney, R. (1997) Building Research Establishment: past, present and future. *Building Research and Information*, **25**(5), 285–91.

Department of Environment (1983) *Fraud in the Property Services Agency; System Controls in District Work Offices*. Twenty-sixth report from the Committee of Public Accounts. (The Wardale report), HMSO, London.

Doig, A. (1997) The Privatisation of the Property Services Agency: Risk and Vulnerability in Contract-Related Fraud and Corruption, *Public Policy and Administration*, **12**(3), 6–27.

Druker, J. (2008) Industrial relations and the management of risk in the construction industry, in Dainty, A., Green, S, and Bagilhole, B. (eds) *People and Culture in Construction*, Spon, London, pp. 70–84.

Emmerson, Sir Harold (1962) *Survey of Problems Before the Construction Industries*. HMSO, London.

Erridge, A. (1996) Innovations in Public Sector and Regulated Procurement, in A. Cox (ed.), *Innovations in Procurement Management*, Earlsgate Press, Boston.

Evans, S. and Lewis, R. (1989) Destructuring and deregulation in construction, in Tailby, S. and Whitston, C. (eds), *Manufacturing Change: Industrial Relations and Restructuring*, Blackwell, Oxford, pp. 60–93.

Evans, R., Carrell, S. and Carter, H. (2009) Man behind illegal blacklist snooped on workers for 30 years, *The Guardian*, Wednesday 27 May.

Gaffrey, D., Pollock, A.M., Price, D. and Shaoul, J. (1999) PFI in the NHS – is there an economic case? *British Medical Journal*, **319**, 116–9.

Green, S.D. and Harty, C.F. (2008) Towards a co-production research agenda for construction competitiveness, In: *Proc CIB Joint International Symposium*, 15–17 November, Dubai.

Green, S.D. (2009) The evolution of corporate social responsibility in construction: defining the parameters, in *Corporate Social Responsibility in Construction*, (eds M. Murray and A. Dainty), Taylor & Francis, Abingdon, pp. 24–53.

Harvey, M. (2001) *Undermining Construction*, Institute of Employment Rights, London.

Harvey, M. (2003) Privatization, fragmentation and inflexible flexibilization in the UK construction industry, in *Building Chaos: An International Comparison of Deregulation in the Construction Industry* (Eds. G. Bosch and P. Philips), Routledge, London, pp. 188–209.

Harvey, M. and Behling, F. (2008) *The Evasion Economy: False self-employment in the UK Construction Industry*, UCATT, London.

Haynes, L. and Roden, N. (1999) Commercialising the management and maintenance of trunk roads in the United Kingdom. *Transportation*, **26**(1), 31–54.

Hencke, D. (2000), Left blacklist man joins euro fight. *The Guardian*, September 9.

Highways Agency (2005) *Delivering Best Value Solutions and Services to Customers – HA Procurement Strategy Review 2005*, Highways Agency Publications Group, Wetherby, UK.

Hillebrandt, P.M. (1984) *Analysis of the British Construction Industry*, Macmillan, London.

Hollingsworth, M. and Tremayne, C. (1989) *The Economic League: The Silent McCarthyism*, National Council for Civil Liberties, London.

ILO (2001) *The Construction Industry in the Twenty-First Century: Its Image, Employment Prospects and Skill Requirements*, International Labour Office, Geneva.

Ive, G.L. and Gruneberg, S.L. (2000) *The Economics of the Modern Construction Sector*, Macmillan, Basingstoke.

Lansley, P.R. (1997) The impact of BRE's commercialization on the research community. *Building Research and Information*, **25**(5), 301–312.

Langford, D. and Agapiou, A. (2007) The impact of eastward enlargement on construction labour markets in European Union member states, in Dainty, A.,

Green, S, and Bagilhole, B. (eds) *People and Culture in Construction*, Spon, London, 205–221.

Latham, Sir Michael (1994) *Constructing the Team*. Final report of the Government/ industry review of procurement and contractual arrangements in the UK construction industry. HMSO, London.

Legge, K. (1995) *Human Resource Management: Rhetorics and Realities*, Macmillan, Basingstoke.

Leiringer, R., Green, S.D. and Raja, J. (2009) Living up to the value agenda: the empirical realities of through-life value creation in construction, *Construction Management and Economics*, **27**(3), 271–285

Monbiot, G. (2000) Captive State: The Corporate Takeover of Britain, Macmillan, London.

NAO (2009) *Private Finance Projects: A Paper for the Lords Economic Affairs Committee*, National Audit Office, London.

Pollock, A., Shaoul, J., Rowland, D. and Player, S. (2001) *Public Services and the Private Sector: A Response to the IPPR*. Working Paper, Catalyst, London.

Public Services Privatisation Research Unit (1997) *Private Finance Initiative: Dangers, Realities, Alternatives*, London: PSPRU.

Sachdev, S. (2001) *Contracting Culture: From CCT to PPPs*, Unison, London.

Selwyn, N. (2008) *Selwyn's Law of Employment*, 15th edn., OUP. Oxford.

Unison (1997) *PFI: Dangers, Realities, Alternatives*, Unison, London.

Unison (2001) *Public Service, Private Finance*, Unison, London.

Unison (2002) *Understanding the Private Finance Initiative*, Unison, London.

Winch, G. (1998) The growth of self-employment in British construction. *Construction Management and Economics*, **16**, 531–42.

Wood, Sir Kenneth (1975) *The Public Client and the Construction Industries* HMSO, London.

4   The Improvement Agenda Takes Shape

4.1   Introduction

The increased reliance on subcontracting coupled with the growth of self-employment was undoubtedly the big story for the construction sector during the four successive Conservative administrations from 1979–97. The growth in the number of self-employed over the period 1976–96 was unprecedented, and by the mid-1990s the situation had arisen whereby the majority of operatives in the construction sector were nominally self-employed. But strangely these were not the things which the advocates of industry improvement were talking about. The industry's leadership were at the time much more focused on the relative merits of different procurement methods. This was in itself indicative of a broader and prolonged trend whereby attention was increasingly directed away from construction workforce issues towards the industry's interface with its clients.

As the enterprise culture unfolded, the 'customer' was afforded an increasingly iconic status. The construction sector was seen to be characterised by outdated working practices which were insufficiently customer-focused. Much of the debate centred around the limitations of the traditional procurement method and the need to select the most appropriate procurement approach to suit the needs of individual clients. Design-and-build surged in popularity on the basis of single-point responsibility. It was also increasingly attractive to clients because it offered improved 'value for money' by subjugating design quality to the cause of functional efficiency. The rhetoric of value engineering was especially persuasive for clients who were uncomfortable with the way architects seemingly prioritised aesthetic design over customer satisfaction. Construction professionals at the time still defined themselves as having broader responsibilities to society at large. Such notions of professional élitism were of course anathema to the enterprise culture.

Of particular importance to the improvement debate during the early 1980s were the so-called management procurement methods. These were seen to offer the opportunity for much faster construction, while at the

---

Making Sense of Construction Improvement, First Edition. Stuart D. Green.
© 2011 Stuart D. Green. Published 2011 by Blackwell Publishing Ltd.

same time apportioning risk in a more appropriate way. Such approaches were also seen to be better at engendering teamwork than the traditional approach. At the same time, the major contractors were understandably keen to ensure that as much money as possible continued to pass through their books. Construction was – and still is – a cash flow business. Of particular interest is the heated debate which took place between the relative merits of management contracting and construction management, although neither ever accounted for more than a tiny proportion of the construction market.

This chapter commences with a discussion of the management procurement methods and the pivotal role played by the Broadgate development in London. Many of the firms – and individuals – who had played key roles on Broadgate subsequently became central figures in the improvement debate. Thereafter, attention is given to a succession of reports which were published during the years when the enterprise culture was radically reshaping the construction sector. *Building Britain 2001* (University of Reading, 1988) was especially notable for the way in which it pre-empted many of the recommendations which were to subsequently appear in *Rethinking Construction*. Attention is also given to *Faster Building for Commerce* (NEDO, 1988), *Partnering the Team* (Latham, 1994) and *Progress through Partnership* (OST 1995). Of particular note is the way these reports became increasingly reliant upon the managerialist discourse of the enterprise culture. Progressively, the construction improvement debate became less interested in the realities of what was happening on the ground, and much more interested in the restorative powers of modern management techniques. Career prospects were much more positive for those who focused on the latter. And those who insisted on talking about the former were cast progressively in the role of dinosaurs. The Latham report was especially notable for spawning new networks whereby the advocates of change could come together to swop ideas and fashionable buzzwords. Such networks can themselves be understood as essential components of the discourse of enterprise.

4.2 The rise of management procurement methods

The 1980s were characterised by a proliferation of different building procurement methods together with endless debates regarding which was 'best' (Masterman, 1992). Concerns regarding the increasing fragmentation of the construction sector were reflected in an increased interest in the integration mechanisms offered by project management (CIOB, 1982; Walker, 2002). Project management techniques such as work breakdown structure (WBS) further encouraged projects to be managed as a hierarchy of subprojects. Each subproject comprised a dedicated work package which would routinely be subcontracted. The hierarchical structure of the WBS further encouraged multi-tiered subcontracting such that the project manager had little interest in the employment status of those who were ultimately engaged to do the physical work of construction. This applied as much to the project manager employed by the contractor (previously known as a site agent), as it did to the project manager employed to act on behalf of the client organisation. Hence multiple tiers of management strove to keep themselves up-to-date with the latest management thinking, while at the same time becoming more and more divorced from the physical processes of construction. The tendency of construction contractors to increasingly see themselves as project managers was reflected in the rise in popularity of management procurement methods. However, the shift directly reflected the increasing focus on client satisfaction. The criteria which were used to decide which procurement method was 'best' all related to satisfying the needs of different clients (cf. Masterman, 1992). The long-term needs of the construction sector, or the aspirations of the construction workforce, were seemingly no longer relevant. The age of customer responsiveness had arrived.

4.2.1 *Management contracting*

Management contracting was probably at the peak of its popularity in the mid-1980s. In 1984 it accounted for 10% of new non-housing work, with over half being secured by the top ten firms (CCMI, 1985). Management contracting had originally been championed by Bovis Construction (now Bovis Lend Lease), especially in terms of their long-standing relationship with Marks and Spencer (Franks, 1996). By the early 1990s management contracting had become almost synonymous with fast-track construction. If clients wanted fast construction, they were encouraged to go for management contracting. At the time, almost all of the major construction companies were operating a management contracting division. The collective shift to management contracting evidenced the major contractors' progressive retreat from the physical activities of construction. It also enabled many contractors' staff to shed their muddy-booted image and to redefine themselves as sharp-suited management consultants. MacAlpine (1999) was one of the few voices to speak against the trend towards management contracting which he saw as turning contractors into nothing other than agencies for marshalling bids from a plethora of smaller firms. MacAlpine (1999) further lamented the way in which the new breed of manager sought to distance themselves from the workforce. Contractors were seen to be following the whims of influential clients to the detriment of any long-term responsibility for the workforce. The MacApline in question was a former chairman of the Conservative Party, and hence cannot easily be dismissed as a member of the 'loony left'. And the more management contractors focused their attention onto management and coordination the more they divorced themselves from any responsibility for training and what had by then become known as 'human resource development'.

However, the trade contractors within management contracting were invariably accorded the same lowly status as subcontractors. They were still in a subordinate position to the management contractor, and they were still treated like second-class citizens. Even more importantly, they were still in direct contract with the management contractor. Hence, they were still prone to being paid when the management contractor deemed it appropriate. Management contracting therefore acted in part to legitimise the increased reliance on subcontracting. Of particular note is the way in which employers refused to modify the working rule agreement to allow for site-wide trade union shop stewards on management contracts (Evans and Lewis, 1989). Clearly the employers considered it to be in their interests to fragment the industrial relations context. This was especially ironic, given the oft-repeated complaints about industry fragmentation from industry leaders. In amongst all the exhortations about integrated teams, it seems that the favoured strategy *vis-à-vis* industrial relations has long since been divide-and-rule. Management contracting therefore may well have succeeded in bringing managers from within the large contractors into line with increasing client expectations. But at the same time it increased the isolation between clients and the construction workforce. It also increased the distance between contractors' management and those who were ultimately engaged to perform the physical work of construction. This collective lack of interest in the construction workforce exacerbated established trends towards labour market casualisation.

4.2.2 Construction management

However, management contracting was not the only show in town. The late 1980s saw a rise in popularity of the procurement approach known as 'construction management' (CM). Its introduction into the United Kingdom was associated with a small number of sophisticated developer clients who had been inspired by US-style fast construction. The key characteristic of CM is that the trade contractors are in direct contract with the client, and are thus afforded a more privileged position. Perhaps most importantly, they are paid directly by the client, and in consequence are less likely to be subject to contractual abuses on the part of the main contractor, such as pay-when-paid. Most of the major main contractors had of course already retreated from the material activity of construction, preferring to limit themselves to the organisation and coordination of subcontractors. As has already been discussed, this accorded with the dominant strategy of numerical flexibility and enabled the main contractors to avoid the overheads associated with a directly employed workforce (see Chapter 3).

For the major contractors, the shift from management contractor to construction manager was less about the function performed, and more about cash flow and the allocation of risk. In common with management contracting, CM lent itself to 'fast-track' construction such that construction work could commence on earlier work packages without the need for all design details throughout the project to be entirely finalised. The approach also enables detailed design work to be undertaken by the trade contractors, thereby realising significant benefits in terms of buildability. However, a crucial distinction with management contracting is that the construction manager becomes part of the professional team in that management services are provided in return for a fee. CM also demands a much more hands-on approach from the client, who is required to accept a much greater degree of risk in terms of time and cost certainty. However, it must also be conceded that the time-and-cost certainty offered by traditional contracting was frequently a *false* certainty. This was especially true given the main contractors' prevailing propensity towards claims. In the context of CM, the supporting argument is that managerial personnel are released from the need to worry about cash flow management and thus are better able to concentrate on managing the works on behalf of the client. Construction managers are further released from the competitive need to document (or fabricate) claims in order to realise a profit. They are therefore much more able to concentrate on the organisation and coordination of the trade contractors in the interests of the client.

It can therefore be argued that construction management became even more client focused than management contracting. Gone is the role of the Architect as an independent 'professional' arbitrating between the interests of the client and the contractor. The construction manager became totally focused on meeting client requirements, constrained only by the client's commitment to treat the trade contractors fairly. Indeed, the prevailing approach to total quality management (TQM) served to define the responsibilities of all parties solely in terms of meeting client requirements. This is not to say that such a shift in emphasis was not badly needed, but it also served to undermine any sense of the need to maintain a balance between the conflicting needs of multiple stakeholders. The enterprise culture had arrived, and issues which fell outwith the remit of 'client requirements' became defined as externalities. Notions of sustainability and corporate social sustainability (SCR) found no place in the lexicon of construction management.

4.2.3   More lessons from Broadgate

It was the iconic Broadgate development which paved the way for the implementation of construction management in the United Kingdom. Broadgate is still widely acclaimed as setting new performance standards for UK construction. The client was committed to instigating change in the UK construction sector, and the experience of Broadgate provided clients with an exemplar of what could be achieved. As already noted, Peter Rogers, Stanhope's construction director, was later to play an important role as chairman of the Strategic Forum (see Chapter 9). Broadgate therefore exerted an important influence on the unfolding story of industry improvement. Some of the success of the Broadgate development was undoubtedly due to the strong leadership provided by the client. Of equal importance was the climate of collaborative working which was created by the construction management procurement team. However, Broadgate was not representative of the industry at large, and from the outset there were doubts regarding the extent to which the achievements of Broadgate could be more widely replicated. But there is also an argument that the Broadgate experience served to legitimise industry fragmentation and the decline of the main contractor. Certainly the Broadgate experience did little to prevent the growth in subcontracting and the rise of self-employment.

The CM appointment for the first phase of the Broadgate development was secured by a Bovis-Schal joint venture on the back of their extensive previous experience of large, fast-tract projects. As outlined above, the contractual arrangement enabled the CM team to concentrate on management and coordination without being distracted by the commercial risk of cost escalation. The construction managers were therefore able to concentrate on problem-solving, rather than having to allocate time and resources to demonstrate why problems were not their fault. Stanhope certainly placed great emphasis on the trade contractors being accepted as an integral part of the 'construction team', although they were still required predominantly to tender for each work package competitively. It should also be emphasised again that the CM procurement approach is heavily dependent upon a knowledgeable hands-on client who is willing to take on board a significant degree of price uncertainty. Stanhope undoubtedly lived up to this role with spectacular success. But clients of this nature are few and far between and are also crucially dependent upon their ability to raise the required capital. Lenders are likely to impose their own constraints on the degree of risk which such clients can accept; rarely are they faced with a free choice between procurement methods on the basis of their own assessment.

While the rhetoric of construction 'teamwork' has become a little jaundiced in recent years, there is little doubt that the adoption of the CM approach in Broadgate was highly successful in incubating a team approach. But as shall be explored at length, there is always a degree of ambiguity as regards who is included in the construction 'team' (see Chapter 9). Under CM, the trade contractors are recognised to be central to the construction team, but the construction workforce is no less marginalised from the overall management of the project than was the case under management contracting. In both cases, the fragmented employment regime cannot be talked away by warm words about teamwork together with the instigation of mandatory induction days. There is of course no reason to suggest that the choice of CM as a procurement method in Broadgate was influenced by memories of the Barbican dispute. But the adopted approach nevertheless served well as a divide-and-rule strategy for containing industrial relations problems. One of the rarely-acknowledged characteristics of CM is that neither the client nor the construction manager is required to take responsibility for industrial relations. Such issues, along with many others, were delegated conveniently to the trade contractors.

For the contracting sector at large, the transition from management contracting to construction management might on the face of it seem an attractive proposition. But in reality the attractiveness of CM is diluted significantly by the fact that the money no longer passes through the main contractor's books. At the time, this presented a direct challenge to the contracting sector's established business model which was based on efficient contract trading coupled with cash flow management. Main contractors' profitability may well be notoriously low if measured as a percentage of turnover, but it becomes much more attractive if measured as 'return on capital employed' (ROCE). Profitability in this respect can be improved dramatically by the imposition of onerous retention practices and delayed payments to subcontractors. Such an approach, coupled with the 'claims game' was central to the business model of many (arguably even most) main contractors throughout the 1980s. The subsequent Latham (1994) report was primarily about the eradication of such practices (see below). Many main contractors may well have been seduced by management contracting, but they were much less keen on switching to CM. This would have resulted in a huge reduction in turnover which was unlikely to have been well received by City analysts. It would also have meant a correspondingly dramatic reduction in the opportunity to 'work' the cash flow which has long been recognised as the industry's life-blood.

4.2.4 Schisms and distractions

The debate about the relative merits of CM and management contracting may well have caused a conundrum for the contracting sector at large, but it caused a particularly damaging schism within Bovis Construction. The directors of Bovis's 'B Division', largely as a result of their experience on Broadgate, were convinced that 'professional' CM was the way forward. Others within Bovis were less convinced, and were unwilling to give up on management contracting and the advantages it offered in terms of feeding turnover through the books. The outcome was a damaging schism whereby the directors of 'B' Division, under the leadership of Ian Macpherson, broke away from the parent grouping in 1990 to form a specialist construction management firm. The newly launched CM firm was named 'Mace', and it duly secured the construction management contract for Phase II of the Broadgate development.

Mace went on to be extremely successful and has since grown to become one of the UK construction sector's most influential firms. Closeness to the client, and closeness to the supply chain, is still central to their espoused philosophy. Bovis soon recovered from the schism and went on to espouse similar values, which rapidly become ubiquitous in accordance with the exhortations of the enterprise culture. However, the formation of Mace is of particular interest as the business was born from CM and was built explicitly on a commitment to collaborative working. It also aligned itself closely with the so-called 'can-do' attitude which had been central to Stanhope's philosophy of CM.

To this day, Mace still enjoys a close working relationship with Stanhope. But their business has diversified extensively and is now much more broadly based than CM. Mace currently market themselves as a 'global consultancy and construction group'. In time they progressed from offering CM for a fee to also offering generic project management on behalf of the client. Throughout all of this change the need to satisfy the requirements of the client remained paramount. Yet it is important to recognise that leading-edge firms such as Mace and Bovis did little to stem the structural shift within the industry towards the hollowed-out firm. Neither did they act to counter the increasing casualisation of the industry's workforce. Broadgate can rightly be credited with providing some credibility for the claims made in favour of 'collaborative working'. But this was collaborative working in a highly unusual context, presided over by a highly unusual client. Even at its peak of popularity in the early 1990s, CM accounted for less than 5% of the total market (see Bond and Morrison, 1994). This was a London-centric community which was largely divorced from the competitive realities experienced by the majority of clients. It should be further noted that Mace's origins did not prevent it from later offering fixed-price contracting for those clients who prefer to procure on this basis. Bovis likewise later reversed their animosity towards CM, and went on to be involved in numerous successful CM projects, including Legoland and EuroDisney. On the latter project one of Bovis's construction managers was interviewed for television, famously emphasising that it was categorically not a 'Mickey Mouse' project.

But all of the above is a long way from the issues faced by the industry at large. While the construction cocktail circuit in London was engaging in extravagant 'top-out' parties and debating the merits of CM, the rest of industry was facing a more prosaic reality. The industry as a whole was observing a huge increase in demand for design-and-build contracts, because this was what clients were demanding. They wanted price certainty and they wanted single-point responsibility. They also wanted designs which were functional; they were not interested in paying to enhance the quality of the public built environment. Neither were they interested in promising continuity of work so that contractors could invest in training and development. In response, contractors did what the increasingly influential diktats of TQM required them to do; they orientated their businesses towards satisfying their clients' requirements.

But the impact of CM was not limited to the extent of its use. Despite only ever accounting for a tiny percentage of industry turnover, the promotion of various 'management' procurement routes intensified the growth in subcontracting. The occasionally heated debate about the relative merits of construction management *vis-à-vis* management contracting disguised the fact that they both legitimised the main contractors' retreat from the physical activities of construction. Indeed, the progressive shedding of responsibility for the physical work of construction meant that main contractors frequently have little left to do other than engage in the rituals of 'supply chain management'. Rather more optimistically, Brady *et al.* (2005) have conceptualised the changing role of the main contractor as 'systems integrator'; but the model of systems integration which currently prevails depends primarily on contract trading. Packages of work are parcelled up and subcontracted to others, and the rationale is not primarily driven by technology. Even routine site cleaning is now routinely subcontracted in the form of a 'logistics' package. 'Integration' is frequently in rather short supply other than in terms of a much repeated rhetorical corrective. Simply put, the growth in subcontracting has been largely driven by competitive cost pressures rather than by any perceived need to provide 'high-value solutions'.

4.3   Bridging between eras

4.3.1   *Building Britain 2001*

Building Britain 2001 was published by the Centre for Strategic Studies in Construction (CSSC) at the University of Reading in 1988. The CSSC subsequently spawned the Reading Construction Forum, which went on to play a significant role in the quest for industry reform through a succession of influential reports (Gray, 1996; Bennett and Jayes, 1998; Flanagan *et al.*, 1998; Saad and Jones, 1998). In many ways, *Building Britain 2001* (University of Reading, 1988) was a direct forerunner of *Rethinking Construction* (Egan, 1998), with a similar remorseless emphasis on modernisation. But in contrast to *Rethinking Construction*, *Building Britain 2001* was essentially a manifesto for change written on behalf of the major contractors. The report had been commissioned by the National Contractors Group, which at the time comprised the 80 largest UK building contractors. The brief was to provide a stimulating and thought provoking report on where the industry then stood, and what it needed to do to remain competitive. *Building Britain 2001* was authored by an academic team led by John Bennett and Roger Flanagan. The report was directly informed by a two-day seminar held at Burnham Beeches which included 19 senior industry practitioners, together with representatives from the Building Employers Confederation, Department of the Environment (DOE), National Contractors Group and the Union of Construction, Allied Trades and Technicians (UCATT). Clients were represented by Capital & Counties and Nationwide-Anglia. It should be noted that *Building Britain 2001,* unlike other reports, was specific to the building sector, rather than about construction *per se.*

Building Britain 2001 was well-received by the industry and succeeded in stimulating debate. It seemed to tap into a mood for change, while at the same time offering a snapshot of the issues which were then considered important. The report emphasised the building industry's contribution to output and employment. It was further held to compare favourably with its international competitors; this was a view to which few clients subscribed to ten years later when *Rethinking Construction* was published. But *Building Britain 2001* captured the essence of the construction sector from the perspective of the main contractor. The dominant pattern of fragmented professional practice was perceived to be out of date and it was predicted that this would be replaced by large, multi-disciplinary practices. Attention was also drawn to the way in which the contracting sector was characterised by relatively few large contractors and a very large number of small, often one-person firms. This diagnosis is still broadly true of today's industry, and the basic structural characteristics were seen to be shared with many other building sectors internationally.

4.3.2   A backcloth of change

In contrast to many subsequent reports, *Building Britain 2001* made specific reference to the UK building sector having had to adapt to extensive change over the preceding two decades. One significant change was the shift in the source of demand from the public to the private sector. In 1970 over 51% of new work was publicly funded; by 1986 this it was claimed had declined to 29%. Particularly significant was the decline in public sector housing over the preceding decades, together with the significant increase in the proportion of repair and maintenance work. These trends were considered previously in Chapter 3, but for current purposes it is sufficient to note that they were soon to be commonly ignored by the industry improvement agenda.

Notwithstanding the shift towards private sector demand, *Building Britain 2001* also noted the dramatic change within the private sector itself. Reference was made to the increasing levels of direct foreign investment into the UK, thereby creating domestic clients with very different experiences, priorities and cultures. Reference was also made to the foothold gained in the UK market by foreign contractors, although subsequent trends in the globalisation of construction companies would soon render the presence of 'foreign' contractors entirely uncontroversial. *Building Britain 2001* further observed a shift in the types of buildings that were being demanded. Offices, leisure complexes and hotels were increasingly taking over from manufacturing facilities, again bringing new client organisations to the fore. Other observed trends included the increasing technological complexity of buildings, especially in terms of M&E services. *Building Britain 2001* also noted the increasing economic dominance of the South

East of England. Indeed, there was an element of truth in the joke in circulation at the time that Liverpool's rush hour took place at London's Euston station on a Friday afternoon. Such jibes should not deflect from the extensive social deprivation caused by the decline of the United Kingdom's manufacturing heartlands throughout the 1980s.

The issue of industry fragmentation was to be cited repeatedly as a problem which needed to be addressed through improved integration. The 'integration' storyline was especially prominent in the Strategic Forum (2002) report *Accelerating Change* (see Chapter 9). What was unusual about *Building Britain 2001* was the way in which it acknowledged the processes through which the industry was becoming increasingly more fragmented. Reference was made to the massive increase in the number of small firms at a time when the number of people working in the industry was falling. *Building Britain 2001* was also remarkably frank about the reasons for the growing shortage of skills in 1988, which was attributed in part to the industry's failure to invest in training. Other contributory factors included the supposedly new skills required as a result of the emergence of fast-track construction, although the report was unclear on the precise nature of the 'fast-track' skills which it considered to be in short supply.

4.3.3 *Employment and training*

It was noted that the level of construction employment fluctuated with the industry's workload. *Building Britain 2001* also highlighted the shift from direct labour to self-employment. This trend was seen to have been accelerated by the early 1980s recession when contractors turned to labour-only subcontracting as a means of reducing costs. The shift was seen to go some way toward explaining the dramatic increase in the number of small firms since 1975. In a sense, it is reassuring to learn that the industry was complaining about skills shortages in 1988. Even then, it was hardly a new complaint; the industry had also been complaining about skills shortages in 1908. Nevertheless in 1988 the economy was in danger of over-heating. Apparently in 1987 a survey by the Federation of Master Builders revealed that 70% of firms in the South were reporting difficulties. Warning was given that the skills shortage may constrain market growth, but this was not a warning that came to fruition. The authors had not factored in the increase in migrant workers from Eastern Europe that followed from the fall of the Berlin Wall in 1989. Nor had they factored in the onset of the subsequent recession in 1990. The authors of *Building Britain 2001* also expressed concern at the fall in the number of apprentices. The decline of training was linked to cuts made during the recession of 1981 as firms attempted to reduce their overheads. The shedding of directly-employed labour was also seen to be a contributory factor in that this reduced the number of skilled workers with whom apprentices could be placed, thereby indicating a degree of support for the diagnosis offered in Chapter 3.

In the expanding market that prevailed during 1988, the poaching of skilled labour was seen to provide further disincentive to investing in training. Why train your own workers when you could simply recruit those which had been trained by others. Such was the dominant prevailing mentality of free-loading.

While *Building Britain 2001* was clear on the negative impact on training caused by the growth of self-employment, the authors were also quick to celebrate the flexibility of successful contractors. Of particular note was the ability of contractors to adapt quickly to severe fluctuations in demand. Flexibility was seen to be enhanced by the widespread reliance on specialist subcontractors. And even for core operations, contractors were able to reduce overheads through the increased use of labour-only subcontractors. These trends were perceived to have resulted in contractors increasingly filling a higher-level managerial role, which the authors of *Building Britain 2001* saw to be a sensible response to changing circumstances. The National Contractors Group would not of course have sanctioned any support for CM, because this would have been detrimental to the interests of its members. But there is no denying the shift towards wholesale subcontracting.

4.3.4   Changing client requirements

Building Britain 2001 noted the increased sophistication of construction clients coupled with their increased focus on certainty of performance and value for money. Clients were seen to be especially keen on single-point responsibility, which was held to account for the growth in popularity of design and build. Clients were further seen to be in favour of a radical change in roles and responsibilities within the industry together with the introduction of a better capital base. The study notably foresaw the increase in client power that was to dominate much of the subsequent industry improvement agenda. Many expert clients were seen to despair of the industry putting its own house in order:

> '[t]oday, many such clients seek to dictate what is built. They determine the details of the technology used, the timing and sequence of its work, its costs, and the terms and conditions on which they will do business with the industry' (University of Reading, 1988; 40)

The seeds of the client movement were therefore already in place in the 1980s. Not only were major clients increasingly more sophisticated but they also displayed a greater willingness to flex their muscles in the marketplace. *Building Britain 2001* also referred to the emergence of new types of contracting firm which were more sensitive to client needs. They could have been referring to Mace. But the message of conformity to client requirements was destined to become widespread, at least in the rhetoric if not in the reality. Of further note was a reluctance to criticise clients for the way they engaged with the industry, and also for their retreat away from the previously accepted 'mutuality of responsibility'. In 1988 the doctrine of customer responsiveness had already rendered criticism of clients out-of-bounds, even if many contractors still relied on their well-tuned skills of 'claimsmanship' to return an honest profit.

Building Britain 2001 also looked East towards Japan and the way in which the supposedly unique Japanese approach to business had shaped their approach to the management of construction projects. The major long-term clients in Japan were seen especially to have had a significant influence in shaping the adopted approach:

> '[a]lthough the building industry has not directly copied manufacturing, its principles of planning in detail, of disciplined, controlled work, and of maintaining a pattern of work throughout each day are a direct reflection of manufacturing's requirements for certainty and reliability.'
> (University of Reading, 1988; 49)

Comparisons with supposed Japanese practices were therefore already part of the construction industry improvement debate some ten years before Sir John Egan advocated 'lean thinking' as a central component of *Rethinking Construction*. The antecedents were already in place within *Building Britain 2001*. But what was to be ignored was the recognition of structural trends which had already re-shaped dramatically the construction sector.

4.3.5 *Recommendations for action*

Building Britain 2001 set out an extensive list of actions which were held to be necessary if the UK building was to prosper. There were 18 detailed recommendations in total. It was seen to be essential that the industry continued to improve its productivity by the adoption of new technologies and by investing in plant and equipment. Better public relations and marketing were advocated to attract more school leavers into the industry; there was seen to be a particular need to target careers teachers and local authority careers officers, as well as students. It was also seen to be important to attract more women into the industry. The needs of clients received considerable emphasis, especially in terms of single-point responsibility, quality assurance and after-sales care. The industry was also criticised for valuing short-term profits rather than long-term relationships. Especially strong words were reserved for loss-and-expense contractual claims which were seen to have 'attacked British industry like a cancer'. The sponsorship of the report by the National Contractors Group did not prevent the authors of *Building Britain 2001* from offering some strong criticisms.

It was argued that clients needed much greater certainty, and the industry was exhorted to stop regarding the tender price as merely another stage of the negotiations. Further pleas were made in support of maximising the potential of information technology and for a greater investment in R&D. All of these recommendations were to be repeated by numerous other construction reports over the subsequent 20 years. The antecedents of both the Latham (1994) and Egan (1998) reports are readily observable within *Building Britain 2001*. But the final recommendation of *Building Britain 2001* was directed at government. The building industry was seen to require a stable economic environment in order to bolster business confidence. Reference was again made to the industry's remarkable ability to adapt in response to fluctuations in the size and nature of demand. However, this flexibility was also seen to have caused great weaknesses. Yet the authors also seemed to sense the pointlessness of demanding that the government should manage the economy in the interests of industry development. *Building Britain 2001* was published in an era when the doctrine of the free market was firmly in the ascendancy. The report therefore ended with a recognition that the industry's future was in its own hands and resisted placing the same with government. Pleas for government policies in support of stability and growth continued to be made at every opportunity. But the bottom line was the recognition that industry could not depend on any real assistance from government.

4.3.6 A disrupted legacy

A follow-up report to *Building Britain 2001* was published in 1989. Despite boasting a forecast by the Prime Minister of the time, Margaret Thatcher, *Investing in Britain in 2001* (University of Reading, 1989) never quite generated the same debate as its predecessor. The debate about industry improvement was in any case severely curtailed by the onset of recession in 1990. Certainly what *Building Britain 2001* lacked was any subsequent commitment to implementation. In this respect, it had much in common with many previous reports on the construction sector. Margaret Thatcher had of course made it quite clear beforehand that it was up to the construction sector to sort out its own problems. There was therefore never any prospect of a precursor to the bureaucratic Construction Industry Board (CIB) which followed the Latham Report (1994). There was even less chance of government support for a predecessor of the evangelical Movement for Innovation which followed *Rethinking Construction*. But it was not only the recession which curtailed enthusiasm for change within the construction sector. The prevailing political uncertainty distracted attention as the industry worried about the implications of Thatcher being ousted by John Major in 1990. The real concern of course was the prospect of Labour government under the leadership of Neil Kinnock rolling back the free-market policies which had prevailed since 1979. Others within the industry had very different concerns. The trade unions were concerned predominantly with declining membership and the damaging side-effects of labour casualisation. As such they would have warmly welcomed a Kinnock victory in 1992. But the electorate made its choice. Much to everyone's surprise John Major was elected and the enterprise culture moved forward as if on autopilot. It did not take long, however, for the Conservative Party to lose what remained of its tarnished reputation for economic management. The final blow was sterling's ignominious exit from the European Monetary System (EMS). Ironically, the unplanned 20% devaluation of sterling did much to benefit the economy in general, and the construction sector in particular. But by the time construction output returned to its previous level in 1993, *Building Britain 2001* was already five years out of date. In part its legacy was to live on in the form of the Reading Construction Forum which was formed as an independent industry think-tank in 1995.

In conclusion, *Building Britain 2001* was an important report for several reasons. It provided a quasi-neutral space for interested parties to come together to discuss issues of industry improvement. This was an approach which was to be championed subsequently by the Reading Construction Forum and the Design and Build Foundation, also spawned from within the University of Reading. However, *Building Britain 2001* was notable for the way in which it bridged between two eras. Its advocacy for new technologies and state-of-the art management practices – especially Japanese management practices – pre-empted *Rethinking Construction* by a decade. Yet the report also carried forward concerns from a previous era, especially those relating to labour-only subcontracting and the declining level of apprentices. The report was written at the height of the consumer-led boon presided over by Chancellor Nigel Lawson. The London skyline was rapidly being re-shaped as a direct consequence of the financial deregulation of the City of London. The Broadgate and Canary Wharf developments were to stand as epitaphs to the changing requirements of clients for fast-track construction. *Building Britain 2001* also acknowledged the backcloth of rapid industry change. In contrast to subsequent reports, it did not assume that the industry had essentially remained unchanged for the decades.

4.3.7 *Faster Building for Commerce*

However, *Building Britain 2001* was not a sole voice advocating change in the late 1980s. A further report worthy of mention is *Faster Building for Commerce*, published by the National Economic Development · Office (NEDO, 1988). *Faster Building for Commerce* was notable for the way in which it referred to customers rather than clients. The authors were especially clear that the 'customer' had a key influence on the outcome of building projects, thereby offering a contrast with the subsequent dominant view that buildings could be procured in the same way as any other manufactured commodity. The report also noted that regular and major customers received better service, and that such customers considered expertise in building procurement to be an important component of their in-house capability. This observation sat starkly ill-at-ease with prevalent trends towards outsourcing and the subsequent privatisation of the Property Service Agency (PSA) (see Chapter 3).

Faster Building for Commerce noted the substantial increase in specialist participants, but did not see this to be a particular problem provided there was a collective spirit of confidence and partnership. Briefing was seen to comprise a dialogue between the customer and the construction team, although it was also noted that several experienced customers were making use of standard briefs to cover aspects of procurement policy as well as design issues. It was further noted that many customers required last minute changes to meet market demands, and that very few projects have sufficient flexibility in design and procurement method to enable change without time and cost implications. The so-called management procurement methods such as CM and management contracting probably perform best on flexibility, whereas design-and-build makes it very difficult for clients to implement late changes.

Faster Building for Commerce was perhaps most striking in its observations relating to subcontractors:

> 'Subcontractors carried out all or most of the work on site and were in one way or another responsible for supplying to the site most labour and materials. On site, they were expected to be self-contained in their operations.' (NEDO, 1988; 8)

Of particular note is the recognition given to the growing tendency of subcontractors to sublet work to labour-only subcontractors (LOSCs):

> 'Many subcontractors let most of their fixing to labour-only operatives because of the difficulties of coping with their variable workload and the need to reduce overheads. Supervision and quality control were often inadequate.' (NEDO, 1988; 8)

The passage above describes with almost perfect succinctness the arguments promoted in Chapter 3. It also links a reliance on LOSCs with inadequate supervision and poor quality control; a connection that was to be ignored systematically in subsequent reports. *Faster Building for Commerce* further observed that many LOSCs claimed to have initially been driven into this form of employment because of the difficulty of 'finding work on the cards'. But LOSCs saw themselves as workers capable of high output, and in consequence they expected to be attended by the main contractor's directly-employed labourers. In light of the ever-increasingly reliance on subcontractors, it is notable the report specifically linked good performance with a smaller number of subcontractors. Perhaps most telling of all was the observation that:

'Labour-only subcontractors form a significant part of the workforce, exacerbating the problem of operatives working together on a site having different employers and variable terms of employment. Over half the workforce is now self-employed. Shortages of skilled labour are not infrequent. Control over the quantity and quality of labour supply is diffuse'. (NEDO, 1988; 8)

The high proportion of labour-only contractors was further observed to be creating tensions within the workforce. Directly employed operatives were held to feel increasingly marginalised and often felt they were only there to support the self-employed operatives. These were the very same concerns which caused Emmerson (1962) and Banwell (1964) and to argue in support of direct employment, but they of course were writing in an age which preceded the onset of the enterprise culture. What is so notable about *Faster Building for Commence* is that it was published in the wake of the 1987 general election, just at the time when the enterprise culture was starting to be defined as a retrospective justification of government policy (see Chapter 2). NEDO was of course very much a hangover institution from the age of state planning, and was consistently ignored by Margaret Thatcher before being finally abolished by John Major in 1992. The truth is that the industry's *cognoscenti* cared about the recommendations of NEDO only if they might be acted upon by an incoming Labour government. By the time Tony Blair was elected Prime Minister in 1997, New Labour had embraced the enterprise culture enthusiastically such that *Faster Building for Commerce* could safely be forgotten. In the wake of the financial crisis of 2007–10 arguments in favour of unrestrained enterprise have begun to lose some of their gloss; the recommendations of NEDO might yet become fashionable once again. This is especially true given the increasing importance attached to sustainability (see Chapter 10).

Given the ongoing fixation with lean thinking (see Chapter 8), it is pertinent to point out that the interpretation of 'lean' in *Faster Building for Commerce* was entirely consistent with the diagnosis offered in Chapter 3:

'[c]ustomers should be wary of "lean" resourcing in calibre and members of site staff and economies in overhead costs.' (NEDO, 1988; 7)

Unfortunately, neither customers nor industry leaders were wary enough of the dangers associated with lean resourcing, which was soon to be embraced as an essential component of 'best practice'. This provides a good example of the way in which best practice is entirely dependent upon the prevailing dominant discourse, thereby discounting the possibility of neutral, quasi-scientific objectivity.

Faster Building for Commerce is further dated by the recommendation that trade unions and employers should cooperate to rationalise the grading and pay structure of operatives. This is a proposal which might had been met with some acceptance in the previous decade. But in 1988 it was a recommendation which had already been bypassed by events. The only surprise is that Thatcher had not seen fit to abolish NEDO earlier.

The closing comment is perhaps best left to Male (2003), who observes that the rise of the self-employed contractor can be linked to the entrepreneurial spirit espoused by Margaret Thatcher:

> 'what is clear, however, is that it shaped the industry then and continues to shape the industry today, As such, the *Faster Building for Commerce* report highlighted a fundamental shift in the management of site operations and defined or at least re-defined new roles in the industry....The new role of the main contractor set out in the report is the forerunner of the "prime contractors" and principal supply chain managers of today.' (Male, 2003; 142)

In short, the industry was already changing and 'best practice' was being formulated on the ground. This is an observation which is in direct contradiction to the commonly held assumption that it was the Egan report which galvanised an outdated industry into change. The truth of course is much more complex; one possible counter argument is that *Rethinking Construction* served only to legitimise changes which were already happening. Some of the changes were not necessarily for the better.

4.4   Constructing the Team

4.4.1   *Latham in context*

Latham was charged with 'making recommendations to Government, the construction industry and its clients regarding reform to reduce conflict and litigation and to encourage the industry's productivity and competitiveness'. In reviewing *Constructing the Team* (Latham, 1994) we should remind ourselves that the early 1990s comprised a severe and prolonged recession. Following the boom in development activity during the late 1980s, construction sector output peaked at £55 billion in 1990. By 1993 it had fallen to £46 billion before recovering to almost £53 billion in 1994 (ONS, 2010). The recession had undoubtedly forced many contractors to batten down the hatches in order to survive. It is unlikely that contractors which had previously competed on the basis of 'bid low, claim high' had many opportunities to break the mould during the early 1990s. Many contractors continued to bid for work below cost with the specific intention of subsequently realising a profit through claims (Cahill and Puybaraud, 2003). Latham (1994) acknowledged specifically the consequences of the flow of work being less than available capacity. These were seen to include uneconomically low tender prices, a lack of investment in training and education and a lack of funds for research and development.

Latham (1994) further observed that there were at the time around 200 000 contracting firms of which 95 000 were private individuals or one person firms. The declining influence of the public sector was also noted, together with the dramatic reduction in local authority housing. Of particular interest is the following observation:

> '[p]rivatisation has also involved the transfer of many professional services to the private sector which were previously carried out "in house" by Government departments or local authorities.' (Latham, 1994, 7)

Latham stated unequivocally that the industry remained dependent upon wider economic stability. But he was equally clear that it was not within his remit to advise the government on economic policy. This would have been anathema to the mood of the times. But it must also be stated that the government's reputation for economic management had been destroyed by the events of 1992, and the industry saw no escape from the seemingly endless cycle of boom-bust. Few at the time would have predicted that from 1994 the construction market would enjoy 14 years of continuous growth. And a continuously expanding market provided a rich context for Latham's recommendations relating to team working and partnering.

4.4.2 Key recommendations

Latham highlighted from the outset that many clients do not always get what they ask for. But the report also conceded that clients need better advice on how to approach the construction process, especially in respect of briefing. There was no acknowledgement that the privatisation of the PSA and the widespread outsourcing of professional services had contributed to the perceived lack of expertise. Even if the clients' criticisms were unfair, it was still made clear that it was ultimately the construction industry's problem.

Few could argue that the construction industry in 1994 needed to focus more on satisfying its customers. But Latham also made clear that clients have responsibilities as well, and that government should commit itself to being a best practice client. Public sector clients were seen to have a responsibility to commission projects which contributed to the quality of the built environment. Good design was seen to be crucial to the provision of value for money, both in terms of total cost and cost-in-use. Energy and maintenance issues were further stressed as being important. Latham was therefore not quite as focused on short-term customer responsiveness as the subsequent Egan report (1998). Government departments were even exhorted to use their spending power to assist the productivity and competitiveness of the construction industry, although the report was much vaguer on the responsibilities of private sector clients. Of further note was the recommendation that private sector clients should come together to form a Construction Clients' Forum (see below).

Constructing the Team was prescient of future trends in the way it advocated the integration of designers and specialists. Also of note was the endorsement of Knowledge-based Engineering (KBE). Rather more prosaic was the recognition of the need for common standards for the exchange of electronic data. Latham also provided a strong endorsement of construction management (CM), although this was seen to be dependent upon a level of close engagement by the client. It was further conceded that not all clients would wish to accept the associated financial risks.

The main thrust of *Constructing the Team* related to the choice of contract and selection/tendering procedures. Latham noted that many clients and contractors either heavily amended or did not use standard forms of contract such as JCT. The endless refining of existing forms of contract was not seen to be the way to address adversarial practices. A call was made for a set of basic principles upon which an interlocking family of documents could be based. The New Engineering Contract (NEC) was seen to be a good starting point and public and private sector clients were exhorted to use it. It was further recommended that the Department of the Environment (DoE) should develop a quality register of approved main contractors and sub-contractors interested in bidding for public work. On the same theme of quality, it was recommended that tenders should routinely be evaluated on quality as well as on price.

Particular emphasis was given to the selection of subcontractors, with a recommendation that subcontractors should be engaged in accordance with the same principles as main contractors. The same recommendation was extended to the engagement of sub-subcontractors. The use of 'Dutch auctioning' was ruled out explicitly. It was also advocated that subcontractors should coordinate their activities effectively with each other to assist in the achievement of the main contractor's programme. This can be read as a further legitimisation of the main contractors' retreat from the physical activities of construction, with an onus on the subcontractor to expand their remit to include coordination and planning. Latham strangely had nothing to say on the corrosive effects of labour-only subcontracting, or on the growth of self-employment. These issues were presumably beyond his scope of interest.

Constructing the Team should also be remembered for its enthusiastic endorsement of partnering, which was defined as:

> '…a contractual arrangement between two parties for either a specific length of time or for an indefinite period. The parties agree to work together, in a relationship of trust, to achieve specific primary objectives by maximising the effectiveness of each participant's resources and expertise. It is not limited to a particular project.' (Latham, 1994, 62)

Whether contracting parties are entities which are capable of forging a 'relationship of trust' was not addressed (see Chapter 7). For some reason, process and power station construction were seen to be especially suitable for partnering, although there was also seen to be potential in building. However, it was notably emphasised that the principal aim of partnering should be improving performance and reducing costs *for clients*. From the very outset it seems that the benefits of partnering where orientated mainly towards clients; this was an emphasis which was to become increasingly apparent as time went by.

In the age of customer responsiveness, partnering was destined to become a central component of construction best practice, and was subsequently even more strongly endorsed by the Egan report (1998). But Latham was responsible for flagging several themes which were later more commonly accredited to Egan. It was Latham who pre-empted Egan in promoting unflattering comparisons between construction and the automotive sector. It was also Latham who advocated the use of benchmarking, together with off-site prefabrication and modularisation (later to be rebranded as 'modern methods of construction'). The examples of Stanhope and McDonald's Restaurants were both cited in respect of prefabrication. In the case of McDonald's they had apparently reduced cost and time of construction by 60% over a five-year period. But calls for a greater degree of pre-fabrication had a much longer track record and for many still carried the stigma of Ronan Point and timber-framed housing (see Chapter 1). And McDonald's of course were building repetitive buildings and as such were hardly a typical client.

But there was much within *Constructing the Team* which is to be admired. Latham deserves considerable credit for consulting widely across the sector and especially for taking care to understand the divergent views of multiple interest groups. In no small way, the review was driven by a vision of a dispute-free industry. Latham listened sensitively to the frustrations of the industry's clients, but was acutely conscious that the subjugation of specialist sub-contractors to the traditional main contractor was no longer appropriate in a modern industry. This was especially true given the increasing techno-logical complexity of M&E services which were already accounting routinely for in excess of 50% of building cost. He listened to endless complaints from specialist sub-contractors regarding the main contractors' abuse of power through the use of contractual malpractices such as 'pay-when-paid'.

Further important recommendations from Latham included his plea for a Construction Contracts Bill to give statutory backing to the revised Standard Forms, including NEC. He also called for specific unfair contract clauses to be banned, and that adjudication should become the normal means of dispute resolution.

4.4.3 *Main contractors take the hump*

When *Constructing the Team* was published in July 1994 it was met with widespread critical acclaim. It contained 30 recommendations which spanned a wide range of issues. The main messages to impact upon the industry were that the client should be at the core of the construction process, and that the best route to achieving client satisfaction was through teamwork, trust and cooperation. In truth, there was little that was new in either of these themes; the key role of the client had been emphasised repeatedly in several previous reports (e.g. Wood, 1975; NEDO, 1983; NEDO, 1988). The case for teamwork, trust and cooperation has an even longer heritage, dating back at least to Simon (1944), who argued that 'efficiency and success are dependent on an honest desire for co-operation'. Phillips (1949), Emmerson (1962) and Banwell (1964) were all equally focused on the need for better teamwork. Buchanan (2000) has referred elsewhere to the ongoing rediscovery of teamworking as a management idea, and nowhere does this apply to a greater extent than construction. And if the label 'teamwork' becomes a little jaundiced, a sense of novelty can always be maintained by referring to 'collaborative working'.

Where the Latham report differed from previous reports was that the industry (or at least parts of it) listened, and sought to do something about it. However, *Constructing the Team* did not by any means receive universal endorsement. Many main contractors felt especially threatened by what they saw as a challenge to their position as the principal agent in traditional forms of contract. They also feared that their ability to control the supply side was in danger of being eroded. In the words of Adamson and Pollington (2006):

> '...the federation of Civil Engineering Contractors (FCEC) and the Major Contractors Group (MCG), within the Construction Industry Employers Council (CIEC), would have been happy to have seen Constructing the Team join Simon and Banwell on the shelf.'

Despite the reservations of main contractors, the Latham report remains one of the best considered reports of the modern era; it identified real issues and advocated practical solutions. *Constructing the Team* is perhaps best remembered for its advocacy of the elimination of unfair contract conditions and the establishment of fair methods of payment together with dispute resolution procedures. The Construction Act of 1998 stands as a legacy to Latham's endeavour. This of course is not to say that unfair methods of payments disappeared overnight. Many main contractors continued to flout the Construction Act's stipulations for the timely payment of subcontractors (Contract Journal, 2000). Just because a particular practice is made illegal does not necessarily mean that it ceases to occur.

In summary, the central message of *Constructing the Team* was that the client should be at the core of the construction process. The general route recommended to achieve the desired client satisfaction was through teamwork and cooperation. Tired clichés perhaps, but this was an age when main contractors' staff where programmed to search out claims at every opportunity. Latham was also notable for the way in which he emphasised that clients were an integral part of the team, and that this brought responsibilities on their part if they wanted to get the best from the construction industry. Unfortunately, clients were less interested in being reminded about their responsibilities, and rather more interested in procuring construction at minimum cost. Of particular significance was Latham's clarion call for a 30% real cost reduction by the year 2000, coupled with an agreed benchmarking system. This last recommendation was especially prescient of *Rethinking Construction*, which was to launch seven targets together with an even more extensive benchmarking system. Of course, by 2000 everybody had forgotten about Latham's target of 30% cost reduction. There were new and much more complicated targets to learn about by then.

4.4.4 Construction Industry Board

The Construction Industry Board (CIB) was established as a direct result of the Latham Report (1994) with the mission of providing 'strategic leadership and guidance for the development and active promotion of the UK construction industry' (Adamson and Pollington, 2007). The CIB sought to improve efficiency and effectiveness throughout the construction and procurement process. Of particular note was its ambitious aim to secure 'a culture of co-operation, teamwork and continuous improvement in the industry's performance'. Its principal objectives were to implement, maintain, monitor and review the recommendations of *Constructing the Team*. The CIB was born of Latham's consultative style, and claimed for a while to represent the interests of the sector at large. In essence, it comprised six member bodies:

- the *Construction Industry Council* representing professionals and consultants;
- the *Construction Industry Employers' Council* representing main contractors;
- the *Constructors' Liaison Group* representing subcontractors and specialist trade contractors;
- the *Construction Clients' Forum* representing regular and occasional/one-off clients;
- the *Alliance of Construction Product Suppliers* representing the materials and product suppliers
- the *government* with the Department of the Environment as the lead department and including the DTI, the Scottish Office and the Health and Safety Executive.[1]

Adamson and Pollington (2007) describe in detail the political machinations that led to the eventual demise of the CIB. Its main problem was that it was

---

[1] Descriptions taken from original CIB publicity material.

never really supported by the major clients. Perhaps the most important legacy of the CIB is the set of reports produced by its 12 working groups. These stand as a lasting record of the way in which 'best practice' is negotiated by disparate interest groups and thereafter legitimised and constituted within inter-organisational networks. This will be a recurring theme throughout the remainder of this book. And if the Latham report itself was largely free of the influence of popularised management fashions, this was not true of the reports produced by the 12 working groups. Simply put, the working groups were looking for ideas to demonstrate their worth; but they needed ideas which did not threaten the delicate alliance that constituted the CIB. They essentially needed ideas that preserved the status quo, while at the same time positioning themselves as being harbingers of change.

The working groups of the CIB therefore provided the networks within which a range of management ideas became constituted as an essential part of best practice. They were not of course the only networks which prevailed at the time; another concurrent example was provided by the networks which supported the 1995 Technology Foresight Report *Progress through Partnership* (see below) The concurrent existence of more than one network raises a further challenge for the participants: not only must they appear progressive without disturbing the *status quo*, they must also align their output with that of other similarly-constituted networks. Clearly there is no benefit in disagreement; if best practice is to be persuasive, it must be unequivocal.

4.4.5   *Construction Clients' Forum*

In common with the Construction Industry Board, (CIB), the Construction Clients' Forum (CCF) was established at the end of 1994 as a direct result of Latham's recommendations. The CCF sought to bring together both public and private sector clients so that they could achieve a better collective impact on construction sector policy. But the Forum was also interested in encouraging change in the construction sector through the influence of the buying power of its members. The benefits which could be achieved through proactive client leadership had already been demonstrated by Stanhope plc on the trailblazing Broadgate Development. In essence, the Forum operated as a loosely organised grouping of large clients who wished to exchange information and experience of how to get the best from the construction sector. It strove to offer an advisory service to its members together with examples of how value for money could be obtained through 'new approaches'. In the longer term, it proposed a *Which*-type approach to the review of the construction industry's performance. This was an idea which was later picked up by the authors of *Rethinking Construction* and is indicative of the trend among clients to think of construction projects as if they were consumer goods which are traded in the marketplace. This is a further telling indication of the way the enterprise culture had incubated the construction improvement debate.

Table 4.1   Membership of the Construction Clients' Forum

British Property Federation
Capital Projects Clients' Group
Central Unit on Procurement (later to become the Office of Government Commerce)
Chartered Institute of Purchasing and Supply
Committee of Vice Chancellors and Principals
Confederation of British Industry
Construction Round Table
Highways Agency
Ministry of Defence (MOD)
National Housing Federation
NHS Estates
Department of the Environment (observer)

(*Source*: Adamson and Pollington, 2006)

Although the formation of the CCF was initiated as a direct result of the Latham report, the antecedents of the Forum lay in years of client dissatisfaction with the services offered by the construction sector. The publication of the Latham report, coupled with the easing of the early-1990s recession, renewed the appetite for change among the industry's major clients. The CCF was funded through client subscription with office space and support services provided by the British Property Federation (BPF) at preferential rates.

The original membership of CCF is listed in Table 4.1. In is notable that most members comprised pre-existing representative bodies. Private-sector clients with a direct involvement in the procurement of construction were limited to BAA, National Power, Northumbrian Water and British Telecom. These were all relatively recently privatised companies striving to shrug off the vestiges of public sector bureaucracy. In other words, they were striving to become much 'leaner' organisations in the way they procured construction work. The pressure to improve efficiency resulted in part from the need to compete in the marketplace, but in practice performance targets were often dictated by the respective regulators. The existence of such clients owed much to the government's ideological preference for private sector enterprise over public sector bureaucracy. And yet the environment within which such firms initially operated remained highly regulated, and certainly fell a long way short of perfect competition. The subsequent omnipotence of 'key performance indicators' on the policy level owed much to those clients being used to having such indicators imposed upon them by their respective regulators.

The CCF subsequently became fond of claiming to represent 80% of identified client expenditure on UK construction. The reality was that the CCF failed to provide a single representative body for the industry's major clients. In reality it never really progressed beyond a loose grouping of pre-existing representative bodies. According to Adamson and Pollington (2006), the CCF struggled from the outset to engage client organisations which were directly involved in the procurement of construction. Influential clients such as McDonald's Restaurants, Marks & Spencer, Tesco and Sainsbury's remained aloof and only engaged indirectly through the pre-existing Construction Round Table (CRT). The major clients seemed to prefer to consult with each other in the informal environment provided by the CRT and were much less willing to commit themselves to a more formal structure. Attendance at CCF meetings was apparently in serious decline by 1996. Many clients remained stubbornly resistant to paying the membership subscription. One function that the CCF did provide was the formal structure through which the clients provided an input into the guidance documentation produced by the Construction Industry Board (CIB), although whether this is anything to be especially proud of is a moot point.

4.5   Progress through Partnership

4.5.1   Realising our Potential

The Technology Foresight report *Progress through Partnership* was published in 1995 as a direct consequence of the UK government White Paper *Realising our Potential* issued in 1993. The aspiration was to replicate the perceived success of 'Japan plc' by developing a much stronger sense of mutual dependency amongst industry stakeholders (Green, 2003). *Realising our Potential* can be read as an important component of the enterprise culture, with significant implications for research and innovation policy in the United Kingdom. One of its key messages was the need for a greater alignment between government, industry and academia.

The Construction Panel of the Technology Foresight Programme was one of 15 sector panels that reported to an overall steering group. In contrast to the CIB working groups, it was not concerned primarily with offering advice to industry on how to improve its performance. Technology Foresight was a higher-class game altogether, and was first and foremost concerned with guiding government and industry decisions on research funding and development. Nevertheless, *Progress through Partnership* serves as a good example of how research agendas are shaped by popularised management fashions. It also illustrates the recurring tendency among construction sector's policy makers to ignore the radical structural changes which were already underway.

4.5.2   Outdated attitudes and the need for culture change

Despite the criticisms above, *Progress through Partnership* provides a useful summary of the UK construction industry's strengths and weaknesses as they were perceived at the time. The United Kingdom was said to benefit from some of the lowest input costs in Europe in terms of labour, while at the same time suffering from some of the highest output prices. It therefore becomes immediately apparent that the authors were concerned mainly with reducing the cost of construction rather than improving working conditions for the industry's workforce. Reference was also made to the decline of the United Kingdom's national competitiveness relative to its competitors - a decline which the much vaunted enterprise culture notably failed to reverse. A recurring assumption throughout the report seemed to be that British managers had failed to keep up-to-date with the latest management techniques. Certainly there was little sympathy with broader arguments that trace the United Kingdom's decline in productivity to structural factors rooted in the nineteenth century. Advocates of the enterprise culture are rarely interested in historical analysis, preferring instead to focus on the need to eradicate outdated attitudes through 'culture change'. *Progress through Partnership* is indicative of many subsequent sources in that it seemingly assumes that the UK construction sector is a static entity. The need for change is advocated without any recognition of the pre-existing dynamics of change which characterise the sector. In light of the extensive re-structuring which was in full swing at the time, this was a significant failing.

The Technology Foresight Report followed a further well-established grove in criticising the low level of investment in R&D in the UK construction sector, which was contrasted unfavourably with the offshore oil & gas industry. Again, the implication is that the UK construction sector is in some way characterised by outdated attitudes which can be overcome by a culture change. However, the comparison between the construction sector and the off-shore oil & gas industry made no reference to the different structural characteristics of the two sectors, or to their respective degrees of capitalisation. Neither was any consideration given to the differing relationships that exist between different sectors and central government. Such institutional issues are consistently ignored in favour of culture change. This vague reliance on 'culture change' was to become a recurring feature of the industry's improvement agenda. Each of a seemingly endless succession of improvement recipes was held to be dependent upon a culture change. This was as true for BPR as it was for partnering and lean thinking. The underlying theme was that culture is some sort of independent variable which is subject to managerial manipulation. If only life were so simple.

4.5.3 Identified key challenges

Progress through Partnership was notable for setting out six key challenges with which the construction sector was seen to be faced. The report could be construed to be ahead of its time in advocating a 'holistic' approach whereby the industry needs to be seen to be acting responsibly in terms of its effect on the environment, society and the workforce. But others might argue that the emphasis on social consequences was harking back to the policy agendas of the previous era. As with many such documents, there is a sense of a struggle taking place between different agendas. The emphasis on social implications certainly sat ill-at-ease with the uncritical acceptance of the prevailing 'process-orientated' approach. The latter was undoubtedly influenced by the fashionable status of business process re-engineering (BPR), which certainly paid scant attention to the social externalities of efficiency improvement (see Chapter 7).

The following summary of the six key identified challenges in *Progress through Partnership* is adapted from Green (2003):

1. **To reduce costs, add value and sharpen international practice**
    The mantra of cost reduction can be seen to be an ever-present element of construction sector policy. Exhortations for 'major improvements in productivity, cost and quality' seem almost obligatory. Calls for greater efficiency are always uncontroversial, but the difficulty comes in identifying precisely how the advocated efficiencies are to be achieved. What is apparent throughout *Progress through Partnership* is the repeated rhetorical device of linking 'reduced costs' to 'added value'. Such a linkage served to deny use of the word 'value' to those who might have felt uncomfortable with the relentless focus on cost reduction. The discourse of enterprise had appropriated the word 'quality' in a similar way.

2. **To pay greater heed to environmental and social consequences**
    The authors of *Progress through Partnership* were indeed ahead of their time in advocating the need for the industry to pay greater attention to environmental and social consequences. The representation of these issues is indicative of the success of Technology Foresight in escaping from the immediate constraints of the dominant enterprise culture. Nevertheless it was to be several years before such issues found their way into the mainstream debate on construction improvement.

3. **To strengthen technological capability**
    The recommendation to strengthen technological capability was entirely predictable, given the remit of Technology Foresight. The report duly noted the increased technological intensity of the industry in terms of the automation of construction processes and the extended use of IT. The authors also called for a greater convergence between construction and manufacturing in the cause of 'modular and prefabricated components and subassemblies'. The call for increased modularisation was of course not new, but such debates were destined to continue unabated through to the present day. Interestingly enough, no mention was made of the potentially adverse social consequences of pre-assembly in terms of de-skilling local communities.

4. **To improve education and training**
    The need to improve education and training is always an easy call, and the publication of *Progress through Partnership* coincided with a period

of relentless educational reform. For the champions of the enterprise culture, the professions in general were held in low esteem. This was especially true of university academics, who were seen to be 'out of line' with the economic needs of UK plc. The precise details of the required reforms were vague, beyond a plea for more versatile people who are able to draw on a wide range of technical specialisms. Elsewhere in the report, reference was made to the need to change education and training programmes to ensure improved effectiveness from those entering the industry. This again was prescient of an increasingly instrumental view of education which would become progressively more dominant.

5. **To upgrade existing buildings and infrastructure**
    The report referred to the decay and obsolescence of the United Kingdom's infrastructure, and the corresponding need to upgrade buildings, structures and infrastructure in an affordable manner. Certainly in 1995 the United Kingdom possessed an infrastructure which had suffered through a systemic long-term failure in public investment. The failure to invest in the nation's infrastructure was one which had been repeated by a succession of governments since World War Two, and it was a failure which was to be repeated by future governments. Upgrading existing buildings of course depends more upon traditional skills and is much less prone to efficiency improvement through modularisation and pre-assembly.

6. **To re-engineer basic business processes**
    The authors of *Progress through Partnership* were confident throughout of the need 'to re-engineer basic business processes to provide lean, rapid and effective performance improvement'. Such practices were seen to be commonplace throughout other industries, although it was contended that they rarely occur in construction. The introduction of such 'advanced business techniques' was exhorted as a matter of urgency. There was an obvious influence of the mantra of BPR, which at the time was at the height of its popularity (see Chapter 7). The focus on 'lean, rapid and effective performance' was also strikingly prescient of the rhetoric of the subsequent Egan report.

The challenges above were seen to be facing a number of obstacles to progress. In common with many previous reports, the industry's fragmented structure was lamented whereby a large number of small firms are organised routinely into project coalitions. However, industry fragmentation was cast predominantly as a static condition; there was no explicit recognition of the extensive re-structuring which had taken place over the previous two decades. Brief reference was made to the industry's casual workforce, but there was no acknowledgement that this was in part a direct result of government policy since 1980 in providing tax incentives to the self-employed (see Chapter 2). The espoused purpose of *Progress through*

Partnership was to look to the future, but there was little recognition of the extent to which the future is shaped and constrained by the past.

4.5.4 Engines of change

In addition to the identified 'key challenges' and 'obstacles to progress', *Progress through Partnership* also identified four 'engines of change' which were also discerning of future directions:

- Promoting learning and learning networks
- Reaping the benefits of the information revolution
- Establishing a favourable fiscal regime
- Creating a culture of innovation

Maximum enthusiasm was reserved for the last of the topics above. The need to create a 'culture of innovation' was destined to become central to the mantra of construction sector improvement. Especially notable was the perceived relationship between innovation and research together with the importance of a continuous stream of fresh ideas:

> '[a] flourishing resource base needs to be built up in the UK, which is constantly refreshed and enriched by research and cross-fertilised by ideas from different disciplines and cultures.'

This indeed was a very different style to the measured language which dominated in previous reports, including Latham's *Constructing the Team*. Innovation was increasingly being accorded an iconic status, undoubtedly deriving its legitimacy from the prevailing rhetoric of the enterprise culture. Industry bodies were exhorted to provide a culture that was more conducive to innovation, and companies were encouraged to appoint innovation directors. Clients were further tasked with evaluating companies' 'innovation profiles' as part of their supplier assessment profiles. Clients of course would have preferred an off-the-shelf *Which* report rather than trouble themselves with innovation profiles. Nevertheless, innovation had arrived as a preferred alternative to the slow-to-change bureaucratic procedures of the past. Particular emphasis was given to the importance of ideas derived from different disciplines and cultures. Unfortunately, such arguments in favour of diversity were soon to be replaced by an overriding emphasis on commitment and compliance.

4.5.5 Opportunities for wealth creation and the quality of life

Progress through Partnering was concluded by a list of key opportunities for increasing wealth creation and the quality of life. The focus on wealth creation was straight from the lexicon of the enterprise culture. Given the emphasis of the discussion, the reference to the quality of life would appear to have been added on as an afterthought. Among the opportunities identified were the 'huge potential savings' to be achieved through the adoption of customised solutions from standard components. Innovation was also seemingly favoured on the basis of cost reduction, with little emphasis on the less tangible aspiration of 'added value'. But the most striking opportunity highlighted within *Progress through Partnership* relates to the application of 'business processes'. The authors paraphrased directly the mantra of BPR in calling for a 'step change' in industry performance. Certainly there was little sympathy with the notions of incremental improvement advocated by Total Quality Management (TQM). The reliance of *Progress through Partnership* on the rhetoric of BPR is indicative of the way in which policy makers were tapping increasingly into the rhetoric of the latest management fashions for up-to-date ideas.

The extent to which the expressed faith in 'business processes' was justified will be examined at length in Chapter 7. Other suggested opportunities for increasing 'wealth creation' included adopting a 'constructing for life' approach, which was seen to hinge on the importance of operating costs as opposed to capital costs. Further emphasis was given to the need for a more holistic approach to assessing the impact of construction on the environment and society. Such calls were indeed ahead of their time and it would be several years before they became mainstream. At the time, they were strangely out-of-line with the discourse of the enterprise culture with its monotonous emphasis on business processes and the need to eliminate activities which do not add value to the narrowly-construed customer. The fact that these two discourses coexist within *Progress through Partnership* illustrates the pragmatic accommodations that often need to be made between seemingly contradicting positions. Indeed, it could be argued that the ability of the two discourses to coexist is essential for the longevity of both. But the report undoubtedly owes much more to the discourse of the enterprise culture than it does to the discourse of sustainability. This is especially notable on the topic of competitive infrastructure, where the emphasis lies strangely on developing countries and the associated export opportunities for British companies.

Although not without its merits, the real points of interest within *Progress through Partnership* lay more in what was omitted, rather than what was included. Certainly the authors had little interest in reviewing the dynamics of change which had already radically reshaped the construction sector over the previous two decades. They noticeably failed to mention the privatisation of the utility companies in the 1980s, or the even more disastrous privatisation of the railways in 1993. The possibility that government policy had had a direct and negative impact on the nation's 'competitive infrastructure' did not even seemingly enter into consideration. The overriding tendency was to ignore the implications of important policy decisions in favour of the supposed healing powers of 'modern management techniques'.

4.6 Summary

Rethinking Construction (Egan, 1998) is often credited with initiating the quest for radical change in the construction sector. But in truth, the calls for radical reform which appeared within the Egan report were predated by those which appeared in *Progress through Partnership* (OST, 1995). And the calls within *Progress through Partnership* were at least in part predated by those which appeared within *Building Britain 2001* (University of Reading, 1988). There is undoubtedly a very real sense of continuity in the rhetoric which characterises these three reports. For example, the following sentence is taken from *Progress through Partnership*, but would have been equally at home within *Rethinking Construction*:

'[a] vibrant and up-to-date modern construction industry providing a comprehensive "best-in-class" (with low cost but high value) service to clients in the UK and throughout the world.' (OST, 1995)

Faster Building for Commerce (NEDO, 1988) was perhaps less in tune with the spirit of enterprise and in many ways was more reflective of the social contract which prevailed during the 1970s. But even here there was a strong emphasis on the need to satisfy customer requirements. However, *Faster Building for Commerce* was more notable for its observations on the negative implications of the growth in subcontracting. It was especially critical of the role of labour-only subcontractors in exacerbating the fragmented nature of the construction workforce. Even more notable was the way in which *Faster Building for Commerce* warned against the trend towards 'lean' resourcing. Such arguments were not what the government wanted to hear at the time, nor were they of any interest to industry leaders who had already committed themselves to a competitive strategy based on leanness and agility in the market place. Hence *Faster Building for Commence* sank quickly into obscurity prior to the abolition of NEDO in 1992. The only surprise was that it had not been abolished earlier.

In contrast, the Latham report of 1994 had a significant impact on the industry improvement debate – at least until it was outflanked by the Egan report in 1998. Latham echoed the prevailing emphasis on customer satisfaction, but he also emphasised that customers have responsibilities as well. This was decidedly off-message from the diktats of the enterprise culture, and perhaps reflected Latham's position on the left wing of the Conservative Party. However, *Constructing the Team* (Latham, 1994) is best known for its enthusiastic endorsement of partnering, although Latham was probably surprised by how influential the rhetoric of partnering would subsequently become. He was ostensibly unworried by the rapid expansion in labour-only subcontracting and self-employment. It seemed that by 1994 these had become part of the landscape, and their corrosive side effects could seemingly be talked away by warm words about teamwork and partnering.

The Latham report also led to the formation of the Construction Industry Board (CIB) and the Construction Clients' Forum (CCF), neither of which was to be long-lived. Given the long-running criticisms of clients about the fragmented nature of the construction sector, it was especially ironic that the clients themselves were unable to maintain a stable representative structure.

Progress through Partnership (OST, 1995) also noticeably ignored the way in which the industry had been radically reshaped by the strictures of the enterprise culture. But the report readily adopted the rhetoric of enterprise; the endorsement of the 're-engineering of business processes' distinguished it from the more gentle language of *Constructing the Team*. Modular and prefabricated components were also seemingly in fashion once again, and their recommendation was subsumed into a broader discourse of enterprise and innovation. But the most notable achievement of *Progress through Partnership* was the way it combined an emphasis on re-engineering with a plea to pay greater heed to environmental and social consequences. The authors even managed to bring together a focus on wealth creation with a consideration of the quality of life. This combination was prescient of the embryonic discourse of New Labour whose leader at the time, Tony Blair, was working hard to distance the Labour Party from the discredited socialism of the 1970s.

References

Adamson, D.M. and Pollington, T. (2006) *Change in the Construction Industry: An Account of the UK Construction Industry Reform Movement 1993–2003*, Routledge, Abingdon.

Banwell, Sir Harold (1964) The Placing and Management of Contracts for Building and Civil Engineering Work. HMSO, London.

Bennett, J. and Jayes, S. (1998) *The Seven Pillars of Partnering*, Thomas Telford, London.

Bond, C. and Morrison, N. (1994), 'Contracts in use: a report for RICS by Langdon Davis & Everest', *Chartered Quantity Surveyor*, December/January, pp. 16–17.

Brady, T., Davies, A. and Gann, D.M. (2005) Creating value by delivering integrated solutions. *International Journal of Project Management*, **23**(5), 360–365.

Buchanan D. (2000) An eager and enduring embrace: the ongoing rediscovery of teamworking as a management idea. In *Teamworking* (Procter S. & Mueller F., eds), Macmillan, London, pp. 25–42.

Cahill, D. and Puybarand, M.-C. (2003) *Constructing the Team: The Latham Report (1994)*, in *Construction Reports 1944–98*, (eds. M. Murray and D. Langford) Blackwell, Oxford, pp. 161–177.

CCMI (1985) *Survey of Management Contracting*, Centre for Construction Market Information, London.

CIOB (1982) *Project Management in Building*, Chartered Institute of Building, Ascot.

Contracts in Use (RICS)

Contract Journal 2000 Contractors still defy law. 19 July 3.

Egan, Sir John. (1998) *Rethinking Construction*. Report of the Construction Task Force to the Deputy Prime Minister, John Prescott, on the scope for improving the quality and efficiency of UK construction. Department of the Environment, Transport and the Regions, London.

Emmerson, Sir Harold (1962) *Survey of Problems Before the Construction Industries*. HMSO, London.

Evans, S. and Lewis, R. (1989) Destructuring and deregulation in construction, in Tailby, S. and Whitston, C. (eds), *Manufacturing Change: Industrial Relations and Restructuring*, Blackwell, Oxford, pp. 60–93.

Flanagan, R., Marsh, L. and Ingram, I. (1998) *Bridge to the Future: Profitable Construction for Tomorrow's Industry and its Customers*, Thomas Telford, London.

Franks, J. (1996) *Building Procurement Systems*, 2nd edn., The Chartered Institute of Building, Ascot.

Gray, C. (1996) *Value for Money*, Reading Construction Forum, Reading.

Green, S. (2003) Technology Foresight Report: Progress through Partnership, in *Construction Reports 1944–98*, (eds M. Murray and D. Langford) Blackwell, Oxford, pp. 161–177.

Masterman (1992) *An Introduction to Building Procurement Systems*, Spon, London.

Male, S. (2003) Faster Building for Commerce: NEDO (1988), in *Construction Reports 1944–98*, (eds. M. Murray and D. Langford) Blackwell, Oxford, pp. 130–144.

MacAlpine, A. (1999) Fashion victim, *Building*, 23 April, p. 33.

NEDO (1983) *Faster Building for Industry*. National Economic Development Office, HMSO, London.

NEDO (1988) *Faster Building for Commerce*. National Economic Development Office, HMSO, London.

ONS (2010) *Construction Statistics Annual 2010*, Office for National Statistics, London.

OST (1995) *Technology Foresight Report: Progress through Partnership*, Office of Science and Technology, London.

Phillips (1949) Working Report to the Minister of Works. HMSO, London.

Latham, Sir Michael (1994) *Constructing the Team*. Final report of the Government/industry review of procurement and contractual arrangements in the UK construction industry. HMSO, London.

Saad, M. and Jones, M. (1998) *Unlocking Specialist Potential*, Reading Construction Forum, Reading.

Simon, Sir Ernest (1944) *The Placing and Management of Building Contracts*. HMSO, London.

Strategic Forum (2002) *Accelerating Change*, Rethinking Construction, London.

University of Reading (1988) *Building Britain 2001*, Centre for Strategic Studies in Construction, Reading.

University of Reading (1989) *Investing in Britain in 2001*, Centre for Strategic Studies in Construction, Reading.

Walker, A. (2002) *Project Management in Construction*, 4th edn, Blackwell, Oxford. [first published in 1984]

Wood, Sir Kenneth. (1975) *The Public Client and the Construction Industries* HMSO, London.

5 Rethinking Construction

5.1 Introduction

The general election of 1997 had brought New Labour to power after many years in the political wilderness. John Major's government had ultimately run out of ideas. New Labour politicians had been working hard to create links with business leaders, and it was therefore no surprise that they should select somebody like Sir John Egan to chair a review of quality and efficiency in the construction sector. First and foremost it was important to keep business leaders, and especially the City of London, firmly on board with the New Labour project. Quality and efficiency were therefore safe topics for debate, and it was certainly important to signal that there was to be no return to the interventionist policies of previous Labour governments. The construction sector was held responsible for implementing its own reforms, and the role of government was limited to encouraging progress through an endless succession of annual improvement targets. This was of course the New Labour way, and the construction sector was happy to play along provided it was left alone to enjoy the opportunities of steady economic growth.

The Egan report – otherwise known as *Rethinking Construction* – was published in July 1998. This was a very different kind of report from those which had gone before. *Rethinking Construction* made no attempt to balance the needs of different interest groups – this was a client-led agenda for change. And the clients involved were already fully signed up to the enterprise culture. Certainly, there was little hint of the rhetoric of the 'Third Way' which characterised New Labour's first term in office. *Rethinking Construction* was the enterprise culture applied to construction. Its focus was on efficiency improvement, with little recognition of the way the industry worked. Great faith was placed in the quest for wide-scale culture change implemented through fashionable management techniques. The overall tone was strident; this was not a call for incremental improvement – this was a call for radical

Making Sense of Construction Improvement, First Edition. Stuart D. Green.
© 2011 Stuart D. Green. Published 2011 by Blackwell Publishing Ltd.

change. *Rethinking Construction* was nothing short of an agenda for re-engineering the UK construction sector on a big scale.

The messages within *Rethinking Construction* went way beyond the managerialism advocated by previous reports. It took the construction improvement debate into entirely new territory. Perhaps even more interesting than the messages within *Rethinking Construction* were the actions which were taken to implement the report's findings. These again were entirely unlike anything which had gone before. The institutions set up in the wake of the Latham report were rapidly abandoned in favour of a new set of improvement bodies. Many of the new agencies were tasked with promoting 'best practice' in the absence of any supporting evidence. Private sector enterprise had on the face of it not been quite enough to encourage the construction sector to become more customer-focused; it was seemingly also necessary for government to promote the benefits of new ways of working. It is of course a strange kind of enterprise which requires government support.

5.2   Background

5.2.1   Tories exit stage right

The Conservatives would almost certainly have suffered electoral defeat in 1992 had Margaret Thatcher not been ousted from power by the proverbial 'men in grey suits'. The electorate seemed to like John Major because he wasn't Margaret Thatcher. But equally important was that he wasn't Neil Kinnock, who had the misfortunate to succeed Michael Foot as leader of the Labour Party. As described in Chapter 3, John Major had initially succeeded in injecting new life into Thatcherism. Ultimately, however, his government ran out of ideas. Thatcher had resided over a period which saw extensive centralisation of power within the corridors of Westminster. Many within local government felt themselves to have been under a state of siege as they were starved consistently of funds and responsibilities (see Chapter 2). As local government became weaker, central government become stronger. The Treasury became especially powerful, and held sway over the spending plans of all government departments. Jenkins (2006) describes how the Treasury was positioned at the centre of a new governmental 'cult of audit'. Enterprise, it seems, was fine when applied to industry and the workings of business. But for government departments, and public services which could not be tested through market competition, the fall-back position was audit – often relentless audit. John Major had previously served as Chancellor of the Exchequer and his appointment as leader of the Conservative Party was in part due to Thatcher's personal patronage, but he was also dependent upon the patronage of the Treasury.

Major was a weak Prime Minister, with a small parliamentary majority. Some would give him the credit for reinvigorating Thatcherism and pushing privatisation policies into areas where Thatcher feared to tread; others would say he followed the course previously set by Thatcher because he lacked his own policy directions. In either case, Major's election victory in 1992 was rapidly followed by sterling's ignominious exit from the European Monetary System (EMS). Some £5 billion had been wasted in a vain effort to defend an unsustainable exchange rate. Ironically, the enforced 20% devaluation of sterling undoubtedly benefited the economy through-out the mid-1990s. Nevertheless, the Conservative government's reputa-tion for economic competence had been destroyed. Norman Lamont was unfortunate to be Chancellor of the Exchequer at the time. He is perhaps best remembered for facing the cameras in Downing Street, flanked by a youthful David Cameron, in the wake of the United Kingdom's humiliat-ing exit from the ERM. Cameron was destined to become leader of the Conservative Party more than a decade later.

John Major's weakness was undoubtedly due in part to his small parlia-mentary majority. His reputation had also been severely damaged by the EMS debacle. But his biggest problem was his colleagues in government. The Conservative Party at the time was riven by divisions over Europe, and Major had to cope with constant carping about his leadership style, forcing him to resign the leadership in June 1995 and to stand again for re-election under the slogan of 'put-up, or shut-up'. John Redwood stood as the only challenger and was soundly beaten in the leadership ballot. But Major's victory did little to stem the constant moaning amongst his parlia-mentary colleagues. The Major government was also rocked by a series of scandals involving sex and money changing hands in brown envelopes. Tory ministers became synonymous with sleaze as scandal followed upon scandal. The Conservative Party seemed to have grown tired of power as Major's government limped tamely to electoral defeat in 1997. The only surprise was the extent of his defeat. The Labour Party won a landslide vic-tory with 418 seats, the most seats the party has ever held. The Conservatives in turn ended up with 165 seats, the fewest seats they have held since the general election of 1906, and with no seats at all in Scotland or Wales. John Major withdrew immediately to watch cricket and drink warm beer. His relief was almost palpable.

5.2.2   New Labour takes charge

Tony Blair was elected Prime Minister on the 2 May 1997. The election ended the Conservative Party's run of four consecutive electoral victories spanning 18 years. People struggled to remember what life was like under a Labour government. For the first time ever, a serving British Prime Minister wore jeans and played the guitar. Britain it seemed was fashiona-ble once again, over the 'Cool Britannia' summer of 1997, as a succession of rock stars visited 10 Downing Street. Even the invitation extended to Margaret Thatcher failed to dampen the prevailing climate of optimism.

But this was a very different Labour government from those which were in power during the 1970s. This was a *New* Labour government.

Tony Blair undoubtedly deserves credit for making the Labour Party elect able once again. Neil Kinnock also warrants recognition for confronting the quasi-Trotskyite Militant Tendency within the party which previous leaders had chosen to ignore. John Smith was elevated to party leader in the aftermath of Neil Kinnock's disastrous electoral defeat in 1992. Smith also won important battles in the quest to return the Labour Party to the mainstream of British policies. It should also be said that the mainstream of British policies had moved significantly to the right following the election of Margaret Thatcher in 1979. The Labour Party therefore had a long way to move. Of particular note was Smith's success in breaking the bloc vote of the trade unions, achieved with the crucial support of John Prescott. Left-wingers, of course, such as Tony Benn maintain a different version of events. In their view the Labour Party was hijacked by so-called 'entryists' such as Tony Blair, Gordon Brown and Peter Mandelson, credited collectively as the prime creators of New Labour.

Blair was elected party leader in 1994 following the premature death of John Smith. His vision was to create a modern European-style social demo-cratic party. After four electoral defeats the view was taken that the Labour Party's lingering commitment to 'old style' state socialism had rendered it unelectable. The aim was to make the party less threatening to the voters of middle England by ditching left-wing policies such as nationalisation, high taxation and unilateral disarmament. The totemic singing of the socialist anthem the 'Red Flag' was also dropped from the annual party conference. At the heart of the New Labour project was the campaign to repeal Clause IV of the constitution which committed the party to the 'common owner-ship of the means of production'. Blair, Brown and Mandelson courted big business assiduously while at the same time weakening the party's links with the trade union movement. The party's links to the trade union move-ment had of course previously placed Neil Kinnock in an impossible posi-tion during the 1984 miners' strike. The discourse of New Labour embraced progressively the rhetoric of enterprise. The previous emphasis on wealth distribution was replaced by a new emphasis on wealth creation, and a new era of political history began. A revised version of Clause IV, without the offending commitment to common ownership, was accepted at a special Easter party conference in 1995. The way was clear for the Labour Party to be re-elected, and the electorate at large was relieved at last to have a viable alternative to the sleaze on offer from the Conservative Party. There was, however, no denying that power within the Labour Party had become sig-nificantly more centralised. Party conferences were increasingly stage-man-aged to avoid any embarrassing policy reversals. Many of Labour's core voters felt alienated, but the electoral calculation was made that they had nowhere else to go. Old-style trade unionists become 'dinosaurs' who were resisting progress, albeit a different species of dinosaur from those who were perpetuating adversarial practices within the construction sector.

Notwithstanding the above, Tony Blair cut a reassuringly presidential figure as he grinned for the cameras on the steps of 10 Downing Street on 3 May 1997. Gordon Brown was appointed as Chancellor of the Exchequer, and Peter Mandelson accepted the role of Minister Without Portfolio in the Cabinet Office, where his job was to coordinate policy messages within government. A few months later, Mandelson also acquired responsibility for the Millennium Dome. John Prescott was appointed Deputy Prime Minister, acquiring responsibility for a large portfolio which included the newly created Department for Environment, Transport and the Regions. Prescott at least looked and spoke like a trade unionist, and was never quite programmed only to deliver Mandelson-approved sound bites. But the former full-time official of the National Union of Seaman had cultivated his own contacts amongst senior industrialists, one of whom was Sir John Egan.

5.2.3   Construction Task Force appointed

Although there was never any prospect of a return to the 'planned economy' of the 1960s, New Labour were nevertheless well aware that the construction sector lay at the heart of their various policy initiatives relating to health, education and transport. Housing also remained at the top of the political agenda, and short of re-creating the Direct Labour Organisations (DLOs) of the past, the provision of new housing depended upon the private sector. Prescott was further aware of an underlying concern regarding the ability of the construction sector to deliver value for money, and wasted no time in setting up a 'Construction Task Force' (CTF) to review the state of the industry. Sir John Egan was approached to chair the Task Force. Egan at the time was chairman of BAA plc, previously British Airports Authority which had been privatised in 1987. In 1997 BAA was one of the construction sector's largest and most influential clients. Egan's background lay in the automotive sector with an impressive reputation as the saviour of Jaguar Cars. He was a declared New Labour supporter and his appointment as chair of the CTF reflected Prescott's enthusiasm to demonstrate their business-friendly credentials.

The CTF was tasked with reporting on the scope for improving quality and efficiency in UK construction. The members of the Task Force (see Box 5.1), were chosen primarily for their expertise as construction clients, or for their experience of other industries which were seen to have improved their performance. It is notable that there was no attempt to represent the different interest groups within the construction industry; neither did any of the participants have any previous involvement with the working groups of the CIB, or with any other such pre-existing groupings (Adamson and Pollington, 2006). Contractors were especially noticeable by their absence. The inclusion of a trade union representative would appear to have been a sop to the Labour Party's traditional support base, although it is nevertheless difficult to see any trade union influence in the text of the final report.

> **Box 5.1   Members of the Construction Task Force**
> (*Source*: Egan, 1998)
>
> Sir John Egan (Chairman), Chief Executive, BAA plc.
> Mike Raycraft, Property Services Director, Tesco Stores Ltd.
> Ian Gibson, Managing Director, Nissan UK Ltd.
> Sir Brian Moffatt, Chief Executive, British Steel plc.
> Alan Parker, Managing Director, Whitbread Hotels.
> Anthony Mayer, Chief Executive, Housing Corporation.
> Sir Nigel Mobbs, Chairman, Slough Estates and Chief Executive, Bovis Homes.
> Professor Daniel Jones, Director of the Lean Enterprise Centre, Cardiff Business School.
> David Gye, Director, Morgan Stanley & Co Ltd.
> David Warburton, GMB Union.

The Egan report claimed to be building upon the foundations laid by Latham, but in truth it provided a significant break from what had gone before. This was a very different kind of document to its predecessors, produced in almost total isolation from the existing machinery of industry consultation. In essence, this was the view of influential private-sector clients on how they wished the construction sector to respond. The message was customer responsiveness writ large; the cult of the customer was given full vent.

5.3 The Egan report

Since its publication in 1998, the Egan report has dominated the improvement debate in the construction sector. At the time of writing, *Rethinking Construction* continues to be cited heavily and the 'Egan Agenda' remains synonymous with industry improvement. The report begins on an up-beat note by stating:

> '[t]he UK construction industry at its best is excellent. It capacity to deliver the most difficult and innovative projects matches that of any other construction industry in the world.'

But this opening compliment was not the sentiment which drove the rest of the report, and was quickly followed by a heavy caveat:

> '[n]onetheless, there is deep concern that the industry as a whole is under-achieving. It has low profitability and invests too little in capital, research and development and training. Too many of the industry's clients are dissatisfied with its overall performance.'

Given these opening statements it might have been expected that the CTF was focused on spreading the acclaimed excellence of the construction industry's leading-edge throughout the sector at large. But it was clear from the outset that the members of the CTF were more interested in learning from other industries, especially from what they saw to be striking improvements in the automotive sector.

5.3.1 The need to improve

The need for the construction industry to improve was justified on the basis of client dissatisfaction and the contention that it rarely provides best value for clients and taxpayers. Reference was made to the sector's low level of profitability, and alleged 80% decline in in-house R&D investment since 1981. Further reference was made to the prevailing crisis in training, with a 50% decline in the number of trainees since the 1970s. Discussions with City analysts had indicated that effective barriers to entry coupled with 'structural changes' could result in higher share prices. Fragmentation was seen to be a barrier to performance improvement and the extensive use of subcontracting was held to have brought contractual relations to the fore. There was no diagnosis of the reasons for the growth in subcontracting, or of its broader implications beyond 'contractual relations'.

Notwithstanding the above, the report also cited a number of promising developments, including the *Pact with Industry* published by the Construction Clients' Forum and the Construction Best Practice Programme (which following the 1997 election now fell within the DETR). Praise was also reserved for the trends towards standardisation and pre-assembly and new technology such as 2-D object orientated modelling. Certainly there was little here to challenge the conclusions of *Progress through Partnership*. Partnering and framework arrangements were also praised, and were seen to be used by the 'best firms' in place of traditional contract-based procurement. Explicit reference was made to the Reading Construction Forum's best practice guides to partnering *Trusting the Team* (Bennett and Jayes, 1995) and the *Seven Pillars of Partnering* (Bennett and Jayes, 1998).

The CTF was especially impressed that partnering when used over a series of projects could account for 30% savings. Tesco were cited as having reduced the capital cost of their stores by 40% since 1991 through partnering. The report further recorded its approval of:

> 'increasing interest in tools and techniques for improving efficiency and quality learned from other industries, including benchmarking, value management, teamworking, Just-in-Time, concurrent engineering and Total Quality Management.'

Sir John Egan's enthusiasm for instrumental improvement techniques seemingly knew no bounds. As can be imagined, the passage above was seized upon by a plethora of management consultants seeking to sell the advocated techniques to an allegedly backward construction sector. Management consultants were of course awarded an especially privileged status within the enterprise culture. If management consultants were advocating a particular improvement this in itself was seen to be evidence of its potency.

5.3.2 Drivers for change

Perhaps the greatest legacy of the Egan report lies in its advocacy of performance targets for the construction sector. The mania for performance targets lead subsequently to an ongoing obsession with key performance indicators (KPIs) within construction sector quangos, thereby reflecting wider trends amongst government policy makers. The increased use of centralised performance targets had long since become an established component of the 'cult of audit' as cultivated by the Treasury. The trend towards performance targets was evident across the full range of government policy areas, including health, education and crime. Such trends continued to characterise the government's approach to privatised sectors such as the utilities, telecoms and the railways. Public sector bureaucracy had given way to private sector enterprise held subject to monitoring through the use of ubiquitous performance indicators. The centralising tendencies of New Labour were perhaps even stronger than those of the preceding Tory administrations. The approach advocated within *Rethinking Construction* was the epitome of this broader trend. The argument emphasised the need to drive dramatic performance improvement through the use of clear measurable objectives. The focus on the need for 'dramatic' improvement reflects the strident tone throughout the Egan report; this was not a call for incremental continuous improvement; this was a call for a step-change in industry performance. The plea for dramatic improvement seemed to have been inspired by business process re-engineering (BPR). Yet the accompanying emphasis on performance measurement against pre-determined targets was a direct manifestation of the 'cult of audit' as championed by the Treasury. The enterprise culture contained many such paradoxes which enabled different versions of the discourse to be played out differently in different contexts. BAA plc was of course still subject to performance indicators at the hands of the regulator.

However, the authors of *Rethinking Construction* were not overtly interested in performance measurement as advocated by government-controlled bureaucracies. They set out their ambitions for the construction sector by noting the increases in efficiency and transformations of private-sector companies within the manufacturing and service industries. They believed that the observed radical changes within these sectors had been driven by a series of five fundamentals, the first of which comprised the need for committed leadership. It was emphasised that management needed to be totally committed to driving the improvement agenda forward if the required 'cultural and operational changes' were to be made. The second fundamental was held to be a focus on the customer:

> '[i]n the best companies, the customer drives everything. These companies provide precisely what the end customer needs, when the customer needs it and at a price that reflects the products value to the customer. Activities which do not add value from the customer's viewpoint are classified as waste and eliminated.'

This definition of waste prevailed throughout the Egan report, and serves to undermine those who were to claim subsequently that *Rethinking Construction* was compatible with the principles of sustainability, to which the report contained only two fleeting references.

The third fundamental concerned the need 'to integrate the process and the team around the product'. This was a plea against the tendency within the construction sector towards fragmented processes. The point was made that the most successful companies work back from the customer's needs and focus on the product and the value it delivers to the customer. There is a strong echo here of the same allegiance to the diktats of BPR expressed previously by the authors of *Progress through Partnering*. Certainly the lack of integrated teams had been a longstanding criticism of the construction sector. One problem was the main contractors' arms-length management of subcontractors alluded to previously by Latham (1994). Within the professional team, difficulties were further exacerbated by the longstanding existence of different professions. Each profession has its own sense of self-identity, reinforced continuously by the respective professional institutions together with their separate educational frameworks.

The fourth fundamental set out by *Rethinking Construction* related to the need for a quality driven agenda, which was taken to imply not only zero defects but also delivery on time and to budget. Quality was further seen to include stripping out waste, after-sales care and reduced costs in use. Prevailing industry concerns about the difficulty of providing quality when designers and contractors are routinely appointed on the basis of lowest cost were recognised explicitly. But the industry was exhorted to understand what clients mean by 'real quality'. It is of course the cult of the customer which dictates a client monopoly when it comes to defining quality; the role of the construction sector is to doff its cap and deliver customer requirements. Societal requirements *vis-à-vis* sustainability were not (quite) yet on the agenda.

The fifth and final stated fundamental of success concerned the need for the sector to demonstrate a commitment to people. This was taken to imply not only decent site conditions, but also fair wages and attention to the health and safety of the work force. Training and the development of capable, committed managers and supervisors were further flagged as important. In addition, reference was made to the need to 'respect all participants in the process and to involve everyone in sustained improvement and learning'. The overriding espoused aspiration was for a no-blame culture based on mutual interdependence and trust.

The fundamentals above were seen to provide the basis for the required dramatic improvement in the construction industry's performance. If this was to be achieved, *Rethinking Construction* advocated that the construction industry should set itself qualified targets, milestones and performance indicators. It was further advocated that the industry should put in place a means of measuring progress towards its objectives and targets. Such an industry-wide performance measurement system would, it was contended, provide clients with the means of differentiating between the best firms and the rest. Emphasis was placed on the industry producing its own structure of objective performance measures. Nevertheless, it is notable that the proposed structure ultimately had to be agreed with the industry's clients; clearly, the industry was not to be entirely trusted to set its own performance targets. It was further advocated that construction companies should share their performance data with clients and with each other to aid an ongoing process of performance improvement through benchmarking. The fashionable status of benchmarking was already assured (Pickrell *et al.*, 1997), and was to become even more popular within the context of partnering (see Chapter 8).

5.3.3   *Improving the project process*

The Construction Task Force (CTF) claimed to have learnt much from its visit to Nissan UK in Sunderland. The visitors also included a number of construction industry representatives, one of whom enthused:

> 'we see that construction has two choices: ignore all of this in the belief that construction is so unique that there are no lessons to be learned; or seek improvement through re-engineering construction, learning as much as possible from those who have done it elsewhere.'

The construction industry representative in question had undoubtedly picked up on the language of BPR, given vent previously in *Progress Through Partnership* (OST, 1995). The term 're-engineering construction' was to become much repeated in subsequent years. If the required end was to be achieved, the Task Force believed that the industry had to re-think the process through which it delivered its projects. It was maintained that up to 80% of its input into buildings is repeated. The parallel was drawn not with the building of cars on the production line, but with the design and planning of a new production model. The plea was made for the definition of an integrated project process in construction, thereby giving rise to much ensuing enthusiasm from the construction research community about the benefits of process modelling (cf. Cooper *et al.*, 2005).

The observed tendency towards largely separate processes for planning, designing and constructing projects was seen to reflect the fragmentation of the sector and to sustain the contractual culture targeted previously by Latham (1994). The authors of *Rethinking Construction* were especially critical of the propensity of construction clients towards choosing a new team of designers, contractors and suppliers competitively for every project. In making this recommendation, they were undoubtedly mindful that the leading clients at the time were leaning towards the use of framework agreements, which was certainly true of Sir John Egan's very own BAA plc. The key premise of the CTF's integrated project process was that teams of designers, constructors and suppliers work together throughout a series of projects, developing the product and the supply chain continuously, eliminating waste in the delivery process, innovating and learning. The benefits of serial contracting had of course been espoused previously by Emmerson (1962) and Wood (1975) during the age of the planned economy. BAA had relatively recently escaped from public ownership and at the time enjoyed the certainty of stable investment plans.

Especially telling was Egan's stark observation that most of the clients of construction were interested only in the finished product, its cost, whether it had been delivered on time, its quality and functionality. There was little sense of the assumed joint responsibility for industry improvement between the construction sector and its clients that had characterised the Emmerson (1962), Banwell (1964) and Wood (1975) reports (see Chapter 2). Such assumptions had been ruthlessly cast aside by the enterprise culture and the ideas of corporate social responsibility (CSR) had yet to become fashionable. CSR was certainly ignored by Sir John Egan and his acolytes. The overriding message was to concentrate on the needs of the customer; other stakeholders in the construction process did not figure in the analysis. Given the subsequent developments in respect of the Commission for Architecture in the Built Environment (CABE), it is notable the Construction Task Force accorded no emphasis to the need to maintain the quality of the built environment. Any notion of design quality beyond satisfying the business needs of the client would again have been classified as waste, and eliminated.

The Construction Task Force further divided the overall project process into four complementary and interlocked elements:

(1) product development;
(2) project implementation;
(3) partnering the supply chain; and
(4) production of components.

The first of the above was seen to relate to the means of developing a generic construction product – such as houses, roads, offices or a repair and maintenance service – to meet and inform the needs of clients and customers. This was perceived to parallel the research into customer needs that was seen to be characteristic of most other industries. Particular emphasis was given to listening to the voice of the customer and understanding their needs and aspirations. Much of what is described would seem to align with the established practise of client briefing, although the implication is that construction firms should develop dedicated business streams around the needs of targeted clients. Reference was made to the existing tendency to do this in response to opportunities provided by the Private Finance Initiative (PFI), and the plea was that such arrangements should be available to all clients.

The second of the listed elements – project implementation – concerned the task of translating the generic product into specific projects in order to meet the needs of specific clients. The use of computer modelling was advocated to test the performance of the end-product and to minimise the problems of construction on site. This is an area where parts of the industry have since made significant progress through the use of building information modelling (BIM), although the impact of such technologies will always be limited in the absence of associated changes in embedded working practices (see Eastman *et al.*, 2008,; Harty, 2008). Appeals were also made for greater use of standardised components and pre-assembly; the Construction Task Force had not yet learnt the newspeak trick of labelling such ideas as 'modern methods of construction' (MMC) to avoid previous connotations of poor quality.

The third element of the proposed process – partnering the supply chain – was seen to be critical to driving innovation and sustained improvement in performance. It was emphasised that partnering is not an easy option and that it can be more demanding than conventional tendering. Tesco would certainly not have approved of approaches that approximated towards an easy option. Reference was made to open relationships and an ongoing commitment to improvement. The effective measurement of suppliers' performance was given further emphasis. Performance measurement was held to be central to the advocated notion of partnering, there were therefore obvious limitations to relying on a no-blame culture based on mutual interdependence and trust. Mention was, however, made of the importance of sharing the rewards of improved performance. Partnering might not be an easy option, but at least construction firms could take solace from the willingness to share the benefits of partnering, subject of course to the effective measurement of performance.

The final element to be addressed in the quest to improve the project process related to the production of components. The construction sector was exhorted to follow the example of leading manufacturers of consumer products. The slogan here called for detailed planning, management and sustained improvement of the production process. The overall stated aim was to eliminate waste (repeated yet again) and to ensure that the right components are produced and delivered at the right time, in the right order and without any defects. These are all classic messages derived from the Toyota Manufacturing System and its just-in-time (JIT) principles. Retail and distribution, together with vehicle manufacturing, were further cited as exemplar industries with extensive experience of effective logistics management. The implied sub-text was that if only the construction sector had kept up to date with the latest management thinking, quality would be much better and clients would be much happier.

5.3.4   Targets for improvements

Despite the expressed preference that the industry should set its own targets, the authors of *Rethinking Construction* were not shy in setting out seven targets as an illustration of what they would like to see. The targets were offered with the caveat that they were based on an assessment which was necessarily 'impressionistic and partial'. Strangely, this caveat was quickly forgotten as the suggested seven targets became the accepted drivers for industry change. The targets comprised annual performance benchmarks for seven indicators. The expectation of 'dramatic' increases in efficiency and quality was articulated once again, although the authors went on to suggest that in their experience the greatest value was to be obtained through significant sustained improvement rather than one-off advances. There would seem to have been something of a battle going on here between the advocates of two different philosophies; one derived from TQM and one derived from BPR. But the advocates of BPR won the day more often than not, with a corresponding repeated emphasis on the need for 'dramatic' improvements.

The targets specified in Table 5.1 raise a number of important questions. They contain repeated references to improvements achieved by 'leading companies', but none of the claims is supported by any evidence. The New Labour government went on to be much criticised for commissioning research in support of pre-determined policy, and *Rethinking Construction* was indicative of this broader trend. Claims made in respect of capital cost and construction time seem especially dubious, given that they are so dependent upon the initial baseline adopted for comparison. Many of the achievements quoted appear to be based on hearsay rather than any objective assessment. For example, it is impossible to verify the claim that 'leading UK clients and design and build firms in the USA are currently achieving reductions in construction time for offices, roads, stores and houses of 10%-15% per year'. But even more starkly, it is appropriate to ask for how many years the suggested targets could be achieved. For example, if capital cost were to be reduced by 10% for five years, the overall reduction in cost would be 33.26%. This would seem to be an implausibly high target to be achieved through the removal of inefficiencies. Improved efficiency had previously been assumed to be a direct derivative of private sector competition. This had at least been the espoused motivation for

Table 5.1   Targets for Improvement

The Scope for Sustained Improvement		
Indicator	Improvement per year	Current performance of leading clients and construction companies
Capital cost All costs excluding land and finance.	Reduce by **10%**	Leading clients and their supply chains have achieved cost reductions of between 6 and 14% per year in the last five years. Many are now achieving an average of 10% or greater per year.
Construction time Time from client approval to practical completion.	Reduce by **10%**	Leading UK clients and design and build firms in the USA are currently achieving reductions in construction time for offices, roads, stores and houses of 10–15% per year.
Predictability Number of projects completed on time and within budget.	Increase by **20%**	Many leading clients have increased predictability by more than 20% annually in recent years, and now regularly achieve predictability rates of 95% or greater.
Defects Reduction in number of defects on handover	Reduced by **20%**	There is much evidence to suggest that the goal of zero defects is achievable across construction within five years. Some UK clients and US construction firms already regularly achieve zero defects on handover.
Accidents Reduction in the number of reportable accidents	Reduce by **20%**	Some leading clients and construction companies have recently achieved reductions in reportable accidents of 50–60% in two years or less, with consequent substantial reductions in project costs.
Productivity Increase in value added per head	Increase by **10%**	UK construction appears to be already achieving productivity gains of 5% a year. Some of the best UK and US projects demonstrate increases equivalent to 10–15% a year.
Turnover and profits Turnover and profits of construction firms.	Increase by **10%**	The best construction firms are increasing turnover and profits by 10–20% a year, and are raising their profit margins as a proportion of turnover well above the industry average.

(*Source*: Egan, 1996)

opening-up the DLOs to private-sector competition. But if *Rethinking Construction* is to be believed, this had not been successful enough in improving industry efficiency. Returning to the suggested targets, it is perhaps coincidence that the compounded reduction over five years accords roughly with the target of 30% improvement in productivity previously set by Latham (1994), which was of course equally unsubstantiated. In respect of defects, the case is again made that 'there is much evidence to suggest that the goal of zero defects is achievable within five years'. Given that so much evidence was available, it is notable that an annual target of 20% reduction per year for five years would still leave nearly 33% of defects unaddressed. Either the authors' skills of compounding were underdeveloped, or the evidence in favour of zero defects within five years was not as convincing as had been suggested.

The target of increasing turnover and profitability was undoubtedly welcomed by many in the contracting sector, and yet it seems strange to combine them into a single measure. In an expanding economy, it might feasibly be possible for all firms to expand turnover by 10% per year, but even this would depend upon an economic growth rate of at least 10% coupled with the imposition of significant barriers to entry. However, the target of 10% growth in turnover for all construction firms would clearly be unachievable in the context of a stagnant or declining economy. Such issues are clearly beyond the control of the contracting sector. As a further comment, it is worth pointing out that the Construction Task Force measured profitability as a percentage of turnover. On this basis, contractors are not especially profitable; many indeed operate in the region of 2–4% profitability. But there is another measure of profitability which is more significant for contracting firms: return on capital employed (ROCE) (Gruneberg and Ive, 2000). If ROCE is used than the contracting sector becomes much more attractive. On this measure, many contractors record profitability levels of 25–30%. The need to maximise ROCE also says much about contracting as a cash-flow business. Hence it offers a better explanation of the competitive behaviours of construction firms. For example, viewing contractors through the lens of ROCE explains the long-standing allegiance to the late payment practices targeted previously by Latham (1994). It also explains the industry's reluctance to give up management contracting in favour of construction management (see Chapter 4).

Comment must also be made in respect of the *Rethinking Construction* target for accidents. Firstly, it should be noted that the specified target relates to the number of reportable accidents, thereby ignoring longstanding problems of systemic under-reporting (Gyi *et al.*, 1999; 197–204). The reporting rate for reportable (non-fatal) injuries has been estimated to be as low as 40% (HSC, 2003). Reporting rates are especially low among the self-employed who have little incentive to comply with statutory procedures. Furthermore, there is a general consensus that prevalent subcontracting arrangements have negative implications for health and safety as a result of blurred responsibility demarcations (Gyi *et al.*, 1999; Haslam *et al.*, 2005). Clarke (2003) goes further and agues that the possibility of a positive safety culture is seriously undermined by workplaces that comprise a reduced number of permanent employees supplemented by contract and contingent workers. Needless to say, none of these issues was addressed by *Rethinking Construction*. This reflects a recurring theme whereby issues of substance are ignored in favour of a managerialist process of target setting. Reference was made to a number of unnamed clients and construction companies who had apparently achieved reductions in reportable accidents of 50–60% in two years or less. Such achievements of course do not deserve to be denigrated, but the authors of *Rethinking Construction* notably found it necessary to emphasise that the reductions were achieved *with consequent substantial reductions in project cost*. The inference here is that the industry would not be motivated by a reduction in accidents alone; it seemingly has to be convinced that any such reduction would also result in reduced costs. Advice to this end was provided by the Heath and Safety Executive, whose morale had undoubtedly suffered as a result of £15 million worth of budget cuts instigated between 1994 and 1996 (Beck and Woolfson, 2000). The discourse of the enterprise culture railed consistently against government regulation of business practices, and health and safety was no exception. The preferred alternative to regulation was to emphasise the importance of self-regulation, whilst at the same time continuing to encourage a reliance on self-employment.

In reviewing the seven targets as a whole, it is useful to remind ourselves that one of the suggested fundamentals was a 'commitment to people'. Unfortunately – other than the target for reducing the number of reportable accidents – this fundamental commitment does not filter through to the targets for improvement. For example, there was no target for improving site conditions despite the observation that the facilities made available to workers on site were 'typically appalling'. Tesco Stores was cited as an exemplar of how enlightened clients could act to improve conditions on site. Tesco was further described to have introduced visitor centres, on-site canteens, changing rooms and showers on all its construction sites. Workers were apparently required to wear branded overalls with both Tesco and their employer's name on them. These innovations had seemingly engendered team spirit and commitment, which in turn contributed to Tesco's achievement of a 40% reduction in construction costs. As was the case with health and safety, it would seem that improving site conditions alone is not enough. It has to be justified in terms of the extent to which such improvements contribute towards reducing costs. Such is the omnipotent nature of arguments in favour of the 'business case' for change. Western commentators in the 1970s used to (quite rightly) poke fun at the need within the Soviet Union for policy decisions to be compatible with Leninist/Marxist doctrine. And yet the enterprise culture incubated a similar effect within the UK such that new initiatives must always be justified with reference to the 'business case'. Keeping people alive on construction sites and the eradication of appalling site conditions should not be dependent upon arguments relating to cost reduction.

It is further notable that *Rethinking Construction* lacked any specific target for ensuring fair wages. Given the emphasis on improving profitability, it could reasonably have been expected that a share of the gains would have been apportioned to the workforce through increased bonus payments. This is especially pertinent given the statement that the Task Force wished to see:

> '[a]ll the players in the team sharing in success in line with the value that they add for the client. Clients should not take all the benefits: we want to see proper incentive arrangements to enable cost savings to be shared and all members of the team making fair and reasonable returns.' (Egan, 1998; 33)

There is of course a slight ambiguity about who is included within the 'team'; but given the expressed emphasis elsewhere on the need for commitment to people it is difficult to believe that the Task Force were purposely excluding building operatives from the accrued benefits. Unfortunately, the subsequent 2002 Strategic Forum report *Accelerating Change* (see Chapter 9) specifically limited membership of the 'integrated project team' to those firms judged to be 'pivotal' in meeting the client's requirements. Hence all those firms not deemed to be pivotal would presumably be denied a fair share of success. By extension, employees not judged to be pivotal would similarly fall outside the 'proper incentive arrangements'. Hence the bogusly self-employed stood no chance; they would remain entirely subject to market mechanisms relating to the cost of labour, in isolation of any assessment of 'added-value' to the client.

Of further note is the absence of any targets relating to environmental impact, or any other aspect of sustainability. Training and development were similarly not judged sufficiently important to warrant an improvement target. When training was addressed within *Rethinking Construction*, emphasis was given to the need for top management to display the right balance between technical and leadership skills. A plea was also made in favour of multi-skilling; modern building methods were judged to require fewer specialist craftsmen. The view was also expressed that training would only be given the recognition it deserved if major clients gave preference to firms who could demonstrate the use of trained workers. In this respect, *Rethinking Construction* endorsed the Construction Skills Certification Scheme, although it nevertheless stopped short of setting a target for the number of trained workers.

It would of course be unfair to be too critical of the suggested targets; they were after all only based on an 'impressionistic and partial assessment'. Unfortunately, this was quickly forgotten as the suggested targets soon acquired an iconic status in the form of the '5-4-7' diagram shown in Figure 5.1. This provided a pictorial representation of the supposed linkages between Egan's suggested five drivers for change, the four identified elements of the project process and the seven annual targets for improvements. Although this diagram did not appear in *Rethinking Construction*, it was widely used following publication to illustrate the key principles. It was later published in the 2002 Strategic Forum report *Accelerating Change* (see Chapter 9).

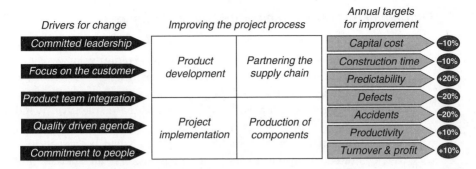

Figure 5.1   Rethinking construction: annual targets for improvement (*Source: Accelerating Change*, Strategic Forum, 2002)

Irrespective of the feasibility of the targets suggested, the use of industry-wide targets seems strangely at odds with the New Labour government's stated committment to free-market economics. In light of this it seems especially bizarre to be advocating an industry-wide bureaucracy of performance targets that was in many respects reminiscent of Stalin's five-year plans in the former Soviet Union. But as has already been observed, this constituted a broader paradox which went to very heart of the enterprise culture and its manifestation within government policy. Devotees of free-market economics would have us believe that inefficient firms go out of business due to market forces. But, contrary to the suggestion of *Rethinking Construction*, contractors survive not only through satisfying their clients, but also through effective risk management. Well-managed construction firms avoid becoming over-reliant on a limited number of clients, and thus limit the amount of turnover which is dependent upon partnering arrangements with single clients. The limitation of risk exposure is one of the key skills of contracting, and is of central importance to being successful in the marketplace. Contractors who bought into Egan's suggested regime of performance management too enthusiastically for the purposes of securing work from a limited number of clients would ironically therefore put themselves directly at risk.

As a final comment, it is worth noting that it was an ideological belief in the benefits of the 'cold wind' of market competition had driven the extensive privatisation programme throughout the 1980s. One of acclaimed successes of the privatisation programme was British Airports Authority, which became BAA plc in 1987. It must therefore be remembered that in 1998 Sir John Egan was the Chief Executive of a privatised quasi-monopoly which was subject to a regime of performance targets set by a government-appointed regulator. This was hardly the most credible platform from which to preach to the construction sector on the need for dramatic change. But construction sector representatives were of course too polite to raise such concerns at the time.

5.3.5 *Endorsement of modern management techniques*

The authors of *Rethinking Construction* demonstrate a consistent faith in the efficacy of modern management techniques. The report was heavily influenced by notions of business process re-engineering, as evidenced by the repeated advocacy of the need for 'dramatic' improvements. But a broad range of other techniques was also endorsed explicitly, including benchmarking, value management, partnering, TQM, just-in-time (JIT) and effective logistics management. These techniques had been seen to have demonstrated their effectiveness in other sectors; all that seemingly remained to be done was to apply them in construction. The commitment to the specified techniques was combined with an overriding commitment to a regime of performance management based on a range of generic targets. Even if the specified targets were to be accepted uncritically, there was still a broad assumption that a single set of targets could be meaningful across the heterogeneous terrain of the construction sector. As previously observed, differences between particular niches within the construction sector are often no less pronounced than differences with firms from within other sectors. The combination of these ideas can in part be understood to be a constituent part of the enterprise culture, but it would again be unrepresentative to portray such storylines as a static and homogeneous discourse. Paradoxes and ambiguities abound.

The authors of *Rethinking Construction* undoubtedly reserved their maximum enthusiasm for the ideas of 'lean thinking'. Clearly they were in the market for new ideas, and the alleged success of lean thinking in the automotive sector made it an obvious candidate. The endorsement of lean thinking was in no small way due to the web of relationships which surrounded the Construction Task Force. Ian Gibson, at the time Managing Director of Nissan UK, had previously worked with Dan Jones, co-author of the hugely influential *Machine that Changed the World* (Womack *et al.*, 1990). Jones was a highly convincing advocate of the benefits that could be accrued through lean thinking. The Construction Task Force also sent a delegation to the United States to attend a conference organised by the Lean Construction Institute (LCI), and to visit some of the US companies that had been championing the application of lean techniques. However, presumably there was a significant amount of additional research which justified the following claim:

> '[w]e have investigated the emerging business philosophy of "lean thinking" which has been developed first in the car industry and is now spreading through the best manufacturers and into retailing and other industries. Lean thinking presents a powerful and coherent synthesis of the most effective techniques for eliminating waste and delivering significant sustained improvements in efficiency and quality.' (Egan, 1998; 25)

In common with the mantra which recurs throughout the Egan report, primary emphasis was given to lean thinking as a means of eliminating waste which is, seemingly, separate from its contribution to delivering sustained improvements in efficiency and quality. The elimination of waste could of course be construed as a contribution to efficiency, so the sentence is in part tautological. Presumably the much repeated emphasis on waste elimination was deliberate just to make sure that the reader had indeed got the message.

The authors followed sources such as Womack *et al.* (1990) in describing lean production as the generic version of the Toyota Production System. Lean thinking is contended to describe the core underlying principles that can be applied to every business activity. The advocacy of supposedly universal principles of management which can be applied irrespective of context dates back to the principles of scientific management as advocated by Taylor (1911). According to *Rethinking Construction*, the key to success lies in defining value from the end customer's perspective. Once this has been done, all the non-value activities can be targeted for removal. Such a view rests on the assumption of a unitary client organisation which has a fixed and uncontested interpretation of value. Such a view was in harsh contrast with much of the prevailing literature on the briefing process, which increasingly conceptualised construction clients as pluralistic entities (cf. Cherns and Bryant, 1984; Blyth and Worthington, 2001; Green, 1996). Where *Rethinking Construction* differed from Taylor was in the emphasis placed on the need for new relationships to eliminate inter-firm waste and to manage the value stream as a whole. Such storylines are reminiscent of supply chain management (SCM) which consistently emphasises the need to manage efficiency across organisational boundaries. SCM was of course an important part of the Toyota Manufacturing System and is often cited to be an integral component of lean production. Nevertheless, the extent to which practices developed in the highly-consolidated automotive sector could be transferred to construction remained hugely contested (see the extended discussion in Chapter 8).

The Egan report cites two examples of firms who have applied lean thinking to construction. Both were based in the United States. The Neenan Company in Colorado had apparently used lean construction to reduce the time it took to produce a schematic design by 80% and project times and costs by 30%. Pacific Contracting had used lean construction to increase productivity and turnover by 20% in 18 months. Both companies were at the time active within the Lean Construction Institute (LCI) which had been founded in 1997. Presumably, the authors of *Rethinking Construction* had been unable to find any exemplary companies of lean construction within the UK, although it is clearly much safer to cite exemplars from elsewhere in that they are much more difficult to discredit. And as with previous exemplar companies cited within *Rethinking Construction*, there was no attempt to verify the performance figures quoted. Indeed, there is every reason to believe that the figures were sourced from the individuals who were leading the lean construction initiatives within the two companies. Clearly the individuals concerned would have a vested interest in 'talking up' their achievements. The claims made would of course have been warmly endorsed by others active within LCI, who would have a similarly vested interest in advocating the benefits associated with lean construction. Simply put, the Construction Task Force were unlikely to receive an independent assessment of lean construction from the LCI. But of course, an independent assessment was not what they were looking for.

Rethinking Construction was undeniably a very different style of report from those published previously. In contrast to the Latham report, the Egan report was not produced on the basis of extensive consultation with a multitude of interest groups from within the construction sector. It was produced remarkably quickly by a small tightly-knit cabal of New Labourfriendly industrialists with a remarkable faith in the restorative powers of modern management techniques. The composition of the Construction Task Force was heavily biased towards large client organisations and completely by-passed pre-existing industry consultation structures. The report was very much a child of its time; it reflected the institutionalisation of the enterprise culture in construction, but crucially it also reinforced it. The rhetoric of business process re-engineering (BPR) pervades throughout the entire report, with a repeated emphasis on the need for 'dramatic' improvements. But there were plenty of other recommendations to adopt other management techniques sprinkled right through *Rethinking Construction*, including benchmarking, value management, partnering, TQM, just-in-time (JIT), supply chain management (SCM). The first three of these had previously been endorsed by the by now dormant working groups of the Construction Industry Board (CIB). Lean thinking was the big new idea which caught the headlines, and the need for the construction sector to model itself on the Japanese automotive industry. Lean thinking was itself somewhat vague and ill-defined, but can certainly be seen to subsume previously advocated techniques such as TQM, JIT and SCM. There was therefore little in the recommendations which can be understood to be especially new; the antecedents of the recommendations can be found in previous reports including: *Building Britain 2001, Faster Building for Commerce* and *Progress through Partnership*. There was therefore a direct continuity with what had gone before, even though the strident tone of *Rethinking Construction* suggested a radical break with previous reports. But what were new were the resources which were mobilised to support the implementation of *Rethinking Construction*. What was also new was the associated emphasis on being Egan-compliant. A climate very quickly developed whereby it became commercially impossible to disagree with Egan's message. If you were not in tune with the Egan agenda you risked being exposed as an outdated dinosaur who was perpetuating adversarial practices. The climate was not dissimilar to that portrayed in the 'blasphemer' sketch in Monty Python's *Life of Brian*. Client pressure was brought to bear such that contractors and consultants had little choice other than to pledge themselves as 'Egan compliant'. Such a commitment very quickly became an essential prerequisite to pre-qualification. It would, however, be overstating the case to describe this as 'radical culture change'; in most cases resigned behaviour compliance would have been a more realistic diagnosis.

5.4   Eganites on the march

5.4.1   *Commitment to the cause*

The final chapter of *Rethinking Construction* laid out recommendations for moving forward. Despite the title of the report, there was little emphasis given to rethinking, or even thinking *per se*. The main focus of the recommendations was on harnessing commitment to the way forward, as if the proposed direction had already been universally agreed. The rallying call focused primarily on the need for commitment from all parties. The overriding message was that the construction sector should fall into line and concentrate on implementing the Egan principles - in all their glorious ambiguity. The implementation project was seen to require commitment from major clients, the construction industry and government. The respective requirements were as follows:

- *major clients, to fulfil their responsibility to lead the implementation of our agenda for dramatically improving the efficiency and quality of construction;*
- *the construction industry, to work with major clients to deliver the significant performance improvements that are possible, and offer these to the occasional and inexperienced clients; and*
- *Government, to create and sustain the environment that is needed to enable dramatic improvements in construction performance, and encourage the public sector to become best practice clients.* (Egan, 1998; 35)

The major clients represented on the Construction Task Force pledged their own commitment to improving performance by undertaking 'demonstration projects'. These were intended to develop and illustrate the ideas set out in the report. It is notable that demonstration projects were not advocated for the purposes of evaluating objectively whether the ideas were worth pursuing; this had seemingly already been decided. Other clients were invited, together with their contractors, designers and suppliers, to offer similar projects on which innovation could be tested and developed. This at least left a little room for innovation which had not been envisaged within *Rethinking Construction*. The stated ambition was to commence with £500 million worth of projects.

The report went on to recommend a 'movement for change and innovation in construction'. The stated aim was to pool experience, develop ideas and drive improvements in quality and efficiency. This was seen to be the way the construction industry could benefit from the lead being given by the major clients, despite the fact that the industry had not been consulted as to whether or not the major clients were heading in the right direction. It was emphasised that the envisaged 'movement for change' should be open to all who are able to demonstrate commitment to a range of aspirations, including the advancement of knowledge and practice of construction best practice. As suggested previously, the notion of 'best practice' implies a single best way of performing any particular task, thereby shifting management theory back to the days of Taylor's (1911) scientific management. The more recent ideas of contingency theory, which advocate the most appropriate way contingent upon the circumstances at the time, had seemingly passed the Construction Task Force by (cf. Lawrence and Lorsch, 1967). *Rethinking Construction* also neglected the possibility that what is 'best' for the client may not necessarily be optimal for all other parties. Given the long history of factional interests in the construction sector, this assumption is highly dubious at best. But the rhetoric of the Egan report did at least resonate with the 'win-win culture' promoted by the discourse of enterprise. The dominant unitary perspective was further emphasised through the plea that participants in the movement for change should develop a 'culture of trust and respect' within their own organisations and throughout their supply chains. It will be argued in Chapter 6 that such pleas are best understood in terms of their metaphorical connotations, rather than as expressions of literal language.

Other qualifying requirements for membership of the 'movement for change' included a commitment to 'training all staff fully and providing them with conditions of employment and facilities that enable them to give of their best'. The expression 'staff' seemingly limited the commitment to white-collar employees who were on the direct payroll. There was no corresponding commitment in respect of employees generally, and certainly there was no commitment to self-employed operatives. But we can be clear that the Construction Task Force was in favour of benchmarking; participating firms were required to be committed to measuring performance against other member's projects and project processes, and sharing the results with the wider industry. The clients on the Construction Task Force strove for the equivalent of a *Which* report which they could use to understand the industry's products and enable them to choose between them. They went so far as to state:

> '[w]e see no other practical strategy that the industry can adopt to escape from the debilitating cycle of competitive tendering, conflict, low margins and dissatisfied clients.' (Egan, 1998; 37)

The debilitating cycle in question was of course not new; but it had undoubtedly been exacerbated by the enterprise culture. It retrospect, it seems highly naïve to suppose that the competitive dynamics described in

Chapter 3 could be unpicked through the use of a *Which*-style report aimed at clients. At the time of writing, the proposed report which clients could use to inform their choices has yet to materialise.

Other recommendations outlined in the final chapter of *Rethinking Construction* included the establishment of a 'knowledge centre' through which knowledge gained from the demonstration projects could be shared with all parties. The DETR's embryonic Construction Best Practice Programme was recommended as the basis for the proposed knowledge centre. With this sole exception the Latham legacy in the form of the Construction Industry Board (CIB) was notably ignored. The idea of a 'knowledge centre' was seemingly based on an assumption that knowledge can be captured and codified in a form which makes it easily transferable. This is an assumption which has been repeatedly challenged (cf. Fernie *et al.*, 2003), and needless to say the knowledge centre never quite materialised in the form envisaged. The authors of *Rethinking Construction* clearly did not rate the universities as suitable hubs for the proposed knowledge centres, presumably because the Construction Task Force could not easily control the knowledge that was on offer. Universities at the time were lagging behind the enterprise culture and had not yet completely subjugated the generation of knowledge to the cause of enterprise. But the New Labour government continued to pursue this course of action with perhaps even more vigour than the preceding Major administration. At the time of writing 'enterprise' has become established as a central component of academic life, and relatively few academics remain resistant to its imposition.

The final call to action within *Rethinking Construction* worth noting is that which was directed at public sector clients. In 1998 the public sector remained the largest client of the construction industry, despite the preceding two decades of privatisation. The Construction Task Force exhorted government to commit itself to leading public sector bodies becoming best practice clients. Particular attention was given to the need to improve public sector procurement, with Public-Private Partnerships and PFI being cited as examples of government's ability 'to make radical and successful changes in its procurement policies'. Emphasis was given to the need to define precisely what is required and then to allow the construction industry to respond in innovative ways. After having directed so much criticism at the construction sector it comes as a surprise so late in the report to learn that the sector's 'rich seam of ingenuity' had been previously stifled by traditional processes of prescriptive design and tendering.

5.4.2 Movement for Innovation

Following the launch of *Rethinking Construction* in July 1998, attention shifted immediately towards implementation. Latham-style consultation across the industry's representative structures was not on the agenda.

It was quickly agreed that existing industry representative bodies, such the Construction Industry Board, could not be tasked with moving forward the change agenda. This was largely due to lobbying by the industry's large repeat clients. The overriding framework was provided by the prevailing cult of customer responsiveness; clients were no longer interested in the 'mutuality of responsibility'. When they shouted 'jump', they expected the construction industry to jump. They no longer felt able to support an improvement agenda which, it their view, had more to do the construction industry's interests than their own. The task of initiating the £500 million programme of demonstration projects fell to the Movement for Innovation (M4I) which was specifically created for the purpose. At the behest of the Construction Minister, Nick Raynsford, Alan Crane agreed to help establish a set of key performance indicators (KPIs) related to the *Rethinking Construction* targets (see Table 5.1). From this point on, performance measurement through the use of KPIs was to be a central component of the agenda for change. The slogan became "*if you can't measure it, you can't manage it*". The adopted approach aligned directly with the Treasury-instigated 'cult of audit', but at the same time sat ill-at-ease with the notion of private sector enterprise. Government was now investing significant sums of money in the infrastructure of the construction industry and the head of the DETR Construction Directorate, John Hobson, would have crucially been aware of the need to demonstrate its impact to Treasury. However, from the outset, there seemed to be a fundamental confusion of purpose. Were the KPIs intended primarily to help individual companies improve their performance through benchmarking? Or were they intended primarily to measure the extent of improvement across the sector at large? Ironically, neither of these objectives seemed to relate to the clients' expressed desire for a *Which*-style report to help them discriminate between the industry's products. If the construction sector really did possess a 'rich seam of ingenuity', the imposition of a generic set of KPIs would seem to have been a good way of stifling it. Adamson and Pollington (2006) describe how the importance of clarity in presentation became fully recognised, leading to the development of the spider's web chart (or radar chart) of the form illustrated in Figure 5.2. At least there was a general agreement on the importance of clarity in presentation, if not on clarity of purpose.

The industry's clients, however, were not satisfied with the original seven targets specified in *Rethinking Construction*. They demanded the addition of an eighth target relating to client satisfaction. Apparently improvements in capital cost, construction time, predictability and defects were not enough; they also wanted a measure relating to client satisfaction with the product and with the level of service. The possibility of conflicting or transient views within the client organisation was not acknowledged; this was characteristic of the impoverished conceptualisation of organisation which characterised *Rethinking Construction* as a whole.

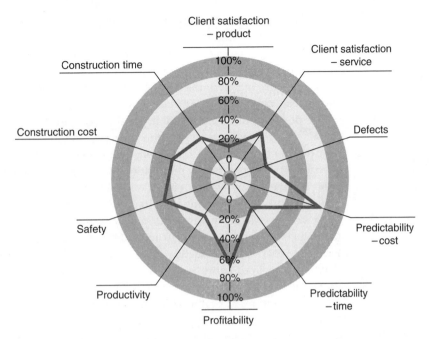

Figure 5.2 KPIs for benchmark performance (*Source*: *Constructing Excellence*, 2006)

The KPIs were launched with much fanfare in November 1998, after which Alan Crane took over the leadership of M4I. The government pledged £500 000 in financial support for the first year, the M4I team was formed and a programme of work was outlined. Many firms were keen to be involved, recognising the commercial importance of being seen to be an active supporter of the Egan agenda. The M4I also provided important networking opportunities; contacts were made and reputations were forged. Representatives of different organisations came together to share ideas with a common commitment to innovation. Within many organisations there was undoubtedly a genuine enthusiasm to support the M4I mission. Particularly notable is the way in which the industry was now signed up to the *Rethinking Construction* agenda despite having been marginalised by the Construction Task Force. The industry's major clients had flexed their muscles to push forward their own agenda for change. They made it quite clear that they only wanted to work with construction firms who were 'Egan compliant'. Cultural change also remained very much on the agenda, and the 5-4-7 model was seen to be the instrument through which the espoused culture change would be achieved. But few were clear on how this was going to work in practice. Notwithstanding the buying power of the major clients, Alan Crane also deserves significant credit for engaging the construction sector in the M4I. Alan Crane was a convincing advocate for change and one who was not disinclined to air his own opinions. He was also seen by the contracting sector to be 'one of their own'. The task of engaging support was especially tough in the regions, which were naturally suspicious of yet another initiative emanating from London. For the more critically inclined, the M4I started to accumulate the trappings of evangelistic cult (Murray, 2003).

The portfolio of demonstration projects continued to grow, although there were consistent doubts about the rigor with which some of the improvements were being measured. Claims were made that the demonstration projects consistently outperformed the rest of the industry, supposedly on the basis of applying the 'Egan principles'. An alternative explanation is that the demonstration projects were a self-selecting group and were hence likely to out-perform the rest of the industry irrespective of the principles being applied. Arguably, it is the very process of making localised sense of 'best practice' on specific projects which creates a lived reality of improved performance. Demonstration projects which were seen to be under-performing could always be quietly dropped, thereby maintaining the supposed performance gap. There was also a prevailing suspicion that neither the DETR nor M4I were looking for an objective evaluation of the Egan principles; what they were looking for was evidence in *support* of the Egan principles. The New Labour government was subsequently to apply the same approach of policy-biased research to promote the dangers of weapons of mass destruction (WMD) in Iraq. The Blair administration became increasingly orientated towards spin, and the implementation of the change agenda in construction was no exception. Spin consistently overshadowed substance. Meanwhile, the Construction Industry Board (CIB) looked increasingly marginalised and moribund.

5.4.3 Shifting alliances

The marginalisation of the CIB from the task of implementing the Egan Agenda undoubtedly undermined its credibility. Its demise was subsequently ensured by the withdrawal of support by the Construction Clients' Confederation (CCC). The CCC was launched in December 2000 and was the successor organisation to the Construction Clients' Forum (CCF) and the Construction Round Table (CRT). The unfolding politics of the client movement are described in detail by Adamson and Pollington (2006). But beyond all the politics is the stark reality that the clients struggled collectively to hold any alliance intact for any significant period of time. The prime difficulty lay in the fact that construction was not their prime business activity. In-house property and construction departments were at the time being consistently downsized in accordance with the principles of business process re-engineering (BPR). Ironically, the clients were themselves the victims of the same doctrines they were seeking to impose on the construction sector. In consequence, many clients struggled to justify the time and resources which were necessary to keep representative bodies intact. Certainly the major clients had no interest in supporting a successor body to the Construction Industry Board (CIB). Such support was too easily judged as peripheral to core business. Exceptions did occur during periods when particular clients were focused on major infrastructure investments, as was the case with BAA plc and Terminal 5.

The public sector offered slightly more coherence in the form of the Government Construction Clients Panel (GCCP), established in March 1997 with a membership drawn from some 50 government departments, Executive Agencies and Non-Departmental Public Bodies. Similarly the Local Government Task Force was launched in March 2000 under the auspices of the wider Rethinking Construction initiative to promote innovation and best practice in construction. But the public sector client movement was hugely fragmented with a deeply embedded commitment to the procedures recommended by Banwell (1964) and Emmerson (1962). Many within the public sector remained deeply suspicious of partnering in respect of a perceived potential for corruption. Considerable procurement expertise had also been lost through the privatisation of the Property Services Agency (PSA) (see Chapter 3). Public sector clients were much leaner than they had been in previous decades, with a consequently reduced capacity for absorbing new ideas. And adversarial attitudes were by no means limited to the private sector. The public sector had its own share of dinosaurs who saw their prime purpose in terms of preventing private contracting firms from making a profit. The challenges of 'culture change' were spread evenly across supply and demand. Nor should it be assumed that the self-appointed 'enlightened' clients of the private sector were paragons of virtue (see Chapter 7). This was not christened the age of spin without good reason.

5.4.4 Constructing Improvement

The tortuous existence of the Construction Clients' Forum (CCF) has already been described in brief in the preceding chapter. In common with the Construction Industry Board (CIB), the CCF was undoubtedly a child of the Latham era. As such it was rapidly marginalised by *Rethinking Construction* and the long march of the Eganites. Tangible outputs from the CCF were few and far between, but it did manage to produce a short document entitled *'Constructing Improvement – The Clients' Pact with the Industry'* (CCF, 1999). Despite the CCF being a 'Latham organisation', *Constructing Improvement* echoed many of the themes of the Egan report which had been published the previous year. In truth, the reports were largely concurrent and the evidence points towards a degree of collusion. *Constructing Improvement* commenced with the oft repeated claim that clients are too often let down by cost and time overruns, poorly performing technical solutions and contractual disputes. It was conceded that clients had not previously been well enough organised collectively to call for better service from the construction industry, but the CCF was seen to now provide them with the necessary platform:

> '[r]ecognition of the clients' strong influence, resulting from their purchasing power and the increased focus in business generally on customer requirements, allows them to press for rapid change in the industry to achieve the world class solutions that they seek.'

The quote above does of course somewhat overstate the ability of the CCF to exert a strong collective influence. The Construction Round Table (CRT) had continued to pursue its own agenda and the retail sector clients who were increasingly acknowledged as lying at the forefront of the best practice agenda consistently kept their distance. *Constructing Improvement* is further notable for reducing the complexity of construction clients to unitary notions of 'customers' with pre-existing requirements (see the further discussion in Chapter 6). Nevertheless, the clients of the CCF committed themselves nobly to working with the supply chain to adopt 'best business, design and construction practice'. They also endorsed the codes of practice produced by the Construction Industry Board (to which they had contributed formally) and welcomed the industry's commitment to the Construction Best Practice Programme.

Other commitments laid out in the Forum's *Pact with the Industry* included the following:

(1) Set clearly defined and quantified objectives for each project and realistic targets for achieving it.
(2) Pool information about construction and benchmark performance across the industry.
(3) Promote relationships based on teamwork and trust, and work jointly with all our partners to reduce costs.
(4) Where unanticipated savings in project costs result from innovative thinking, in appropriate circumstances share them with the relevant parties.

Constructing Improvement is awash with such warm words, but the evidence that leading clients progressed to put them into practice is less than overwhelmingly strong. The clients represented on the Forum also committed themselves to work together with the supply side of the industry to achieve change in a number of specific areas, including:

(1) Eliminate waste, streamline processes and work towards continuous improvement.
(2) Working towards standardisation in components where this provides efficiency gains.
(3) Using a properly trained and certified workforce and keeping skills up to date.
(4) Improving management of supply chains.

Neutral calls to eliminate waste, streamline processes and work towards continuous improvement always go down well with an audience, and in essence they differ little from the core messages of scientific management (Taylor, 1911). What is starkly noticeable is the way in which these challenges were set out in absence of any recognition of the challenges created by a lack of stability in workload. The commitments were sufficiently bland not have encountered any resistance amongst the Forum's membership, and yet there was a pointed absence of detail on *how* the specified goals were to be achieved. The endorsement of increased standardisation in the cause of greater efficiency was especially well-travelled ground, and was to be much repeated over the following decade (see Chapter 9). While the pledge to favour contractors with a properly trained and certified workforce was to be welcomed, there was little willingness to engage with – or even acknowledge – the underlying root causes for the lack of training in the sector. In practice, there was little evidence that clients would favour a properly trained workforce if it resulted in extra cost.

Adamson and Pollington (2006) describe *Constructing Improvement* as the high point of the client movement. But it should be remembered that the CCF fell a long way short of being representative of the industry's major clients. Many client organisations could not commit themselves to turn up to the meetings, or even to pay the modest subscription fee. Nevertheless, the sentiments expressed in *Constructing Improvement* characterised the improvement agenda as advocated at the time. Yet the underlying conceptualisation of the industry's engagement with client organisations was essentially mechanistic. The overriding message was that objectives should be set at the outset and that all parties should then work together as a team to achieve them. The assumption seemed to be that if all parties were committed to fulfilling their allocated roles then the mutually desirable ends of eliminating waste and reducing costs would be achieved. The dominant mindset of instrumental rationality remained firmly intact, and the more often the mantra was repeated the more it seemingly became unchallengeable. However, the text of *Constructing Improvement* is also notable for what it did not include. There was no mention of the growth of self-employment in the construction sector, nor was there any mention of the proliferation of procurement methods. Previously well-aired arguments about the benefits of flexibility and the disbenefits of fragmentation had also slipped off the agenda. Perhaps most tellingly, *Constructing Improvement* failed to articulate any strategy for sustainability in the construction sector. In retrospect, Latham's recommendation for a separate Clients' Construction Forum served to further fragment the debate into 'demand side issues' and 'supply side issues'. The very terminology is demonstrative of the iconic status of the marketplace, where the customer is king. Rather than promoting the client as being of central importance to the task of industry improvement, *Constructing Improvement* represented a further step towards the legitimisation of arms-length management.

There is much commonality between *Constructing Improvement* and *Rethinking Construction,* although the latter relies more heavily on the supposed benefits of 'modern management techniques'. But in truth, the antecedents of both reports can be found in *Building Britain 2001* and *Faster Building for Commerce.* Unfortunately, the industry's clients could no longer endorse the former because it had been sponsored by the National Contractors' Group. The latter was equally problematic in that it had been published by the corporatist National Economic Development Office (NEDO). It also made the mistake of pointing out the dangers of labour-only subcontracting and thus was decidedly 'off-message'. As noted in Chapter 4, NEDO had been ignored consistently by Margaret Thatcher before finally being abolished by John Major in 1992.

Throughout the years of its existence there was a further tension within the CCF in terms of whether clients should follow the Latham line of engaging wholeheartedly with the full construction team, or whether they should limit themselves to a one-on-one engagement with the main contractor. *Constructing Improvement* strove hard to remain true to the Latham philosophy, although there were clearly constraining voices which saw little value in clients engaging with the complexities of the construction supply chain. The latter voices certainly seemed to dominate within the CRT. Ultimately, the dilemma was resolved when the CCF failed to mobilise sufficient support to remain in existence. It was later replaced by the Confederation of Construction Clients (CCC) which, in contrast to the CCF, enjoyed the full support of CRT. The CCC was undoubtedly more aligned with the Egan agenda which saw the needs of the client as paramount. The 'mutuality of responsibility' was indeed gone for good, and the advent of New Labour had done little to dislodge the dominant discourse of the enterprise culture.

5.4.5 *Construction Best Practice Programme*

Although often cited as being one of the outcomes of the Egan report, the Construction Best Practice Programme (CBPP) was already under development prior to the publication of *Rethinking Construction.* It had been originally established as a joint initiative between the DETR and the CIB (*Construction News,* 1999). Nevertheless, in contrast to the CIB, it was endorsed specifically by both *Rethinking Construction* and *Constructing Improvement.* The CBPP was hence adopted as one of the main conduits through which the Egan Agenda was to be disseminated. The Construction Task Force had called for the CBPP to be used as the basis for a national knowledge centre for construction. It was seen to be important that the knowledge centre was 'objective, impartial and efficient'. It could be argued that such aspirations were unlikely to be achieved given that the recommendations of *Rethinking Construction* fell significantly short of being either objective or impartial. These words might conceivably be used to describe Latham's *Constructing the Team*, but not *Rethinking Construction* which offered a highly partisan view of how the industry should seek to improve.

The Construction Best Practice Programme was funded by the DETR to the tune of £2 million per year and aimed to raise awareness of the supposed benefits of best practice. It further sought to provide guidance and advice to the UK construction sector and client organisations so that they had the 'knowledge and skills required to implement change'. The programme was nothing if not ambitious. Its main focus was stated to be 'the transformation of outmoded management practices and business cultures'. Here was a slogan around which the Eganites could unite.

The key objectives of the programme, as stated on its website in October 1999, were to:

- create a desire for improvement by identifying, publicising and supporting the use and benefits of adopting improved business practices;
- offer an initial point of contact for organisations wishing to improve;
- facilitate links between such organisations and those with the knowledge of how to improve;
- provide tools, techniques, advice and knowledge to those wishing to improve.

The programme was supported by the Department of the Environment, Transport (DETR) and is commonly attributed to John Hobson, who was at the time head of the DETR's construction directorate. The CBPP provided information on the 15 business improvement themes listed in Box 5.2:

Box 5.2   Business improvement themes of the Construction Best Practice Programme

- Benchmarking
- Briefing the Team
- Choice of Procurement Route
- Culture and People
- Health and Safety
- Information Technology
- Integrating Design and Construction
- Lean Construction
- Partnering and Team Development
- Risk Management
- Standardisation and Pre-assembly
- Supply Chain Management
- Sustainable Construction
- Value Management
- Whole Life Costing

Each of the business themes specified in Box 5.2 was supported by a fact sheets together with a plethora of dissemination events. The selected topics were consistent with the Egan Agenda, although the influence of Latham is also apparent. The need for targeted advice on briefing certainly owed more to the recommendations of Latham than it did to *Rethinking Construction*. The choice of the most appropriate procurement route also had a long legacy, and was a topic upon which Egan had made little comment other than endorsing PPP/PFI. Given the reliance of *Progress through Partnership* and *Rethinking Construction* on the metaphors and imagery of BPR, it might have been expected that re-engineering would have been included amongst the advocated business improvement themes. But in many respects, the ethos of BPR permeated across all of those themes; the declared focus on the 'transformation of outmoded management practices and business cultures' is immediately reminiscent of the doctrines of BPR. The only theme for which this is categorically not true is that of sustainable construction, which displays a significant divergence from the mainstream improvement agenda of the time.

The CBPP fact sheet on sustainable construction defined it as the application of sustainable development to the construction industry. Emphasis was given to the 'triple bottom line' in terms of achieving a balanced performance as measured against economic, environmental and social criteria. *Progress through Partnership* had previously emphasised the need to pay greater heed to environmental and social consequences. Such issues were given scant attention by *Rethinking Construction*. Indeed, the dominant message about eliminating activities which did not 'add value' from the customer's viewpoint can be seen to be in direct contrast to the principles of sustainability. Notions of corporate social responsibility (CSR) had been marginalised specifically by the discourse of the enterprise culture (see Green, 2009). Margaret Thatcher had famously questioned the very existence of society, and the factoring in of so-called 'externalities' was anathema to the 'cult of the customer'.

Certainly there was little willingness among private sector clients at the time to leverage construction expenditure to achieve broader social goals. Such arguments were seen to belong to the discredited age of state planning. Aspirations of demand management for the sake of maintaining stable levels of employment were similarly perceived to have perished together with the 'Social Contract' of the 1970s. Throughout the 1970s and 80s the need for a stable workforce consistency took second place to the perceived advantages of labour market flexibility (see Chapter 3). Post-Egan arguments in favour of sustainability were perhaps the first indication that the high point of unadulterated enterprise had been passed. The discourses of enterprise and sustainability have been unfolding in mutual interaction ever since.

The subsequent increasing emphasis on sustainability undoubtedly reflects changing values in society, together with undisputable scientific evidence of climate change. But in the late 1990s such debates were in their infancy and sustainability was at best a marginal interest. At least at the time the Eganites could point towards the CBPP improvement theme of sustainable construction as evidence that 'something was being done', even if it was lip-service. With this sole and notable example, the rest of the advocated improvement themes were entirely 'on message' with the main thrust of *Rethinking Construction*. Certainly for the more critical observer, the CBPP's description of sustainable construction appeared starkly out of place with the emphasis on narrowly-construed performance improvement which prevailed elsewhere. It should be noted, however, that the outputs of CBPP never came close to the Construction Task Force's recommendation that it should turn itself into a knowledge centre that provided access to information and learning from the demonstration projects. In truth, the demonstration projects were never that rigorous.

The eleventh-hour inclusion of the theme of sustainable construction was perhaps due to a belated recognition by John Prescott and his advisors that *Rethinking Construction* had failed to portray the government's credentials *vis-à-vis* sustainability. Adamson and Pollington (2006) credit the then Construction Minister, Nick Raynsford, with calling for sustainability to have a much higher profile following the publication of the report in July 1999. If true, Nick Raynsford deserves considerable credit for his vision in extending the debate beyond the consensus which prevailed at the time. There is notably little evidence that the inclusion of sustainable construction emerged from within the client movement, nor is there any evidence that at the time it was taken especially seriously within the contracting sector. The members of the Construction Task Force certainly did little to promote the cause of sustainability, and arguably set back the prospect of any sensible debate by at least a decade. They were especially blind to prevailing criticisms of lean thinking in respect to its alleged negative impact on sustainability (see Chapter 8). *Rethinking Construction* similarly did little to promote CSR, and arguably deflected attention away from the dangers of labour casualisation through the power of omission. The CBPP was equally silent on ethical employment practices, despite its supposed theme on 'Culture and People'. The M4I was subsequently to launch its *Respect for People* initiative, together with a set of proposed Environmental Performance Indicators. These initiatives will be dealt with separately in Chapter 9.

5.5 Summary

This chapter has summarised the main arguments promoted in *Rethinking Construction*. It has also described the subsequent emphasis on implementing the 'Egan agenda' in practice. However, despite the strident nature of the report, it is ultimately difficult to be clear on what precisely what was being advocated. Ultimately, the message on offer was rich in metaphor, but rather vague in terms of its substantive content. The tone of the report was quite unlike anything that had gone before. Egan and his disciples were not interested in incremental improvement; they demanded dramatic improvement. Words such as 'radical' and 'dramatic' are peppered throughout the report. *Progress through Partnering* had undoubtedly paved the way in terms of its managerialist rhetoric and in its advocacy of 'leading-edge' management techniques. However, *Rethinking Construction* carried both of these tendencies to an extreme rarely observed previously or since. Certainly the tenor of Egan's message was very different from that of Latham, as was the methodology by means of which the report was produced. It is notable that the 1988 NEDO report *Faster Building for Commerce* took four years to be researched and published (Male, 2003). In contrast, the Construction Task Force was established in October 1997 and *Rethinking Construction* was published in July 1998. In consequence, it would be unfair to criticise the report for its underpinning lack of research. *Rethinking Construction* was not a research-based report; neither was it based on extensive consultation across the industry as had been the case with *Constructing the Team*. The Egan report is perhaps best understood as a clarion call for change issued by the industry's major clients.

Rethinking Construction was especially notable for its emphasis on meeting client needs, thereby reflecting the dominant discourse of customer responsiveness. Activities which did not 'add value' for the client was classified as waste; and waste was to be eliminated. In this respect, *Rethinking Construction* was undoubtedly a product of the enterprise culture. Yet this was a particular interpretation of enterprise which emphasised managerialism much more strongly than enterprise on the part of the individual. Neither was this a plea for the restorative powers of the free market; it was much more of an argument in favour of a ruthless focus on performance measurement. In many respects, *Rethinking Construction* represents the epitome of managerialism. The authors displayed considerable faith in instrumental improvement recipes such as lean thinking, partnering, value management and benchmarking. Remarkably little attention was given to the need for supporting evidence; these were seen to be the latest management techniques, and thus it was taken as axiomatic that their adoption would be beneficial. It is notable that all of these techniques were advocated primarily in terms of their contribution to efficiency; the machine metaphor reigned supreme. The Egan Report is further notable for promoting the cause of standardisation and pre-assembly. Once again, such technology-based solutions were evaluated in terms of their contribution to narrowly defined efficiency. There was no attempt to explore the implications of a shift towards pre-assembly on the traditional craft skills within local communities.

In summary, the Egan report did much to promote the managerialist discourse of performance improvement within the construction industry. The report also popularised the concept of key performance indicators (KPIs), thereby reinforcing the industry's predilection for instrumental management techniques. It also created an unhealthy climate whereby individuals were afraid to speak out against the advocated approach. If you were not identified as being 'Egan compliant' you ran the risk of being classified as a 'barrier to progress'. The impact of *Rethinking Construction* is summed up neatly by Murray (2003) as the 'Eganisation of construction'. Murray further describes how the disciples of *Rethinking Construction* became known as the 'Eganites', who displayed collectively many of the attributes of an evangelical cult. Whatever it was that was being offered, it clearly owed more to rhetoric than it did to science.

References

Adamson, D.M. and Pollington, T. (2006) *Change in the Construction Industry: An Account of the UK Construction Industry Reform Movement 1993–2003*, Routledge, Abingdon.

Banwell, Sir Harold (1964) The Placing and Management of Contracts for Building and Civil Engineering Work. HMSO, London.

Beck, M. and Woolfson, C. (2000) The regulation of health and safety in Britain: from old Labour to new Labour, *Industrial Relations Journal*, **31**(1), 35–49.

Bennett, J. and Jayes, S. (1995) *Trusting the Team: the Best Practice Guide to Partnering in Construction*, Centre for Strategic Studies in Construction/ Reading Construction Forum, Reading.

Bennett, J. and Jayes, S. (1998) *The Seven Pillars of Partnering*, Thomas Telford, London.

Blyth and Worthington (2001) *Managing the Brief for Better Design*, Spon Press, London.

Cherns, A.B. and Bryant, D.T. (1984) Studying the client's role in construction, *Construction Management and Economics*, **2**, 177–184.

Clarke S. (2003) The contemporary workforce: Implications for organisational safety culture, *Personnel Review*, **32**(1), 40–57.

Cooper, R., Aouad, G., Lee, A., Wu, S., Fleming, A., and Kagioglou. M. (2005) *Process Management in Design and Construction*, Blackwell, Oxford.

Constructing Excellence (2006) *UK Construction Industry Key Performance Indicators Handbook*, Constructing Excellence, London.

Eastman, C., Teicholz, P., Sacks, R. and Liston, K. (2008) *BIM Handbook: A Guide to Building Information Modeling for Owners, Managers, Designers, Engineers, and Contractors*, John Wiley & Sons, Hoboken, New Jersey.

Egan, Sir John (1998) *Rethinking Construction*. Report of the Construction Task Force to the Deputy Prime Minister, John Prescott, on the scope for improving the quality and efficiency of UK construction. Department of the Environment, Transport and the Regions, London.

Emmerson, Sir Harold (1962) *Survey of Problems Before the Construction Industries*. HMSO, London.

Fernie, S., Green, S.D., Weller, S. and Newcombe, R. (2003) Knowledge sharing: context, confusion and controversy, *International Journal of Project Management*, **21**(3), 177–187.

Green, S.D. (1996) A metaphorical analysis of client organisations and the briefing process, *Construction Management and Economics*, **14**(2), 155–164.

Green, S.D. (2009) The evolution of corporate social responsibility in construction: defining the parameters, in *Corporate Social Responsibility in Construction*, (eds. M. Murray and A. Dainty), Taylor & Francis, Abingdon, pp. 24–53.

Gruneberg, S. and Ive, G. (2000) *The Economics of the Modern Construction Firm*, Macmillan, London.

Gyi, D.E., Gibb, A.G.F., Haslam, R.A. (1999) The quality of accident and health data in the construction industry: interviews with senior managers, *Construction Management and Economics*, **17**, 197–204.

Harty, C. (2008) Implementing innovation in construction: contexts, relative boundedness and actor-network theory, *Construction Management and Economics*, **26**(10), 1029–1041.

Haslam, R.A., Hide, S.A., Gibb, A.G.F., Gyi, D.E., Pavitt, T., Atkinson, S., Duff, A.R. (2005) Contributing factors in construction accidents, *Applied Ergonomics*, **36**, 401–415.

Jenkins, S. (2006) *Thatcher and Sons*, Allen Lane, London.

Latham, Sir Michael (1994) *Constructing the Team*. Final report of the Government/industry review of procurement and contractual arrangements in the UK construction industry. HMSO, London.

Lawrence, P.R. and Lorsch, J.W. (1967) *Organization and Environment*, Harvard Press, Cambridge, Mass.

Male, S. 2003 in *Construction Reports 1944–98*, (eds. M. Murray and D. Langford) Blackwell, Oxford, pp. 161–177.

Murray, M. (2003) Rethinking Construction: The Egan Report (1998), in *Construction Reports 1944–98*, (eds M. Murray and D. Langford) Blackwell, Oxford, pp. 178–195.

OST (1995) *Technology Foresight: Progress Through Partnership*, Office of Science and Technology, London.

Pickrell, S., Garnett, N. and Baldwin, J. (1997) *Measuring Up: A Practical Guide to Benchmarking in Construction*, Construction Research Communications Ltd. Garston.

Taylor, F.W. (1911) *Principles of Scientific Management*, Harper & Row, New York.

Wood, Sir Kenneth (1975) *The Public Client and the Construction Industries*, HMSO, London.

Womack, J., Jones, D. and Roos, D. (1990) *The Machine that Changed the World*, Rawson, NY.

6 Understanding Clients: Beyond the Machine Metaphor

6.1 Introduction

The Egan report undoubtedly had a significant influence on the debate about construction improvement. It was characterised throughout by an overriding obsession with customer responsiveness and efficiency. Yet at the same time it was rather illusive in terms of the substantive content of precisely what was being advocated. Customer responsiveness and efficiency are of course two themes that are difficult to argue against. Arguments in favour of efficiency always play out well in the construction sector, even if the solutions on offer leave much to be desired. In the context of the enterprise culture, there were few at the time who did not accept that the construction sector needed to be more responsive to the needs of their clients. But there are other perspectives on construction improvement which are equally important. If we focus ruthlessly on efficiency alone, there is a danger that we become very efficient at doing the wrong thing. And if we focus on the immediate needs of our clients then there is a danger that we neglect the needs of other stakeholders.

Given that *Rethinking Construction* was ultimately vague in terms of the substantive content of its recommendation, it is perhaps more enlightening to focus our attention on the underlying metaphors. The repeated emphasis on efficiency and the elimination of waste is immediately suggestive of an implicit machine metaphor. The construction sector was equated to a machine which needs to operate more efficiency. In was further taken for granted that the cause of efficiency serves the interests of all parties, which could be true in those instances where the benefits are shared equally among the participants. *Rethinking Construction* undoubtedly had a pivotal influence on the practices which were to become routinely accepted as 'best practice'. But such conceptualisations of best practice tend to be dominated by the same mechanistic quest for operational efficiency which characterised the original report.

Making Sense of Construction Improvement, First Edition. Stuart D. Green.
© 2011 Stuart D. Green. Published 2011 by Blackwell Publishing Ltd.

Ironically, despite its focus on customer responsiveness, *Rethinking Construction* adopted an impoverished view of client organisations. It was seemingly taken for granted that clients could specify their requirements in isolation of any interaction with the design team. It was further assumed that clients are unitary entities whose needs remain consistent over time. In other words, client organisations were assumed to operate as if they were machines. Hence, the storyline of customer responsiveness was underpinned by the same machine metaphor which shaped the consideration of efficiency.

The purpose of this chapter is to challenge *Rethinking Construction*'s default use of the machine metaphor by opening up a range of alternative ways of thinking about organisations. The concept of organisational metaphors was introduced originally by Morgan (2006; first published 1986) who suggests eight metaphors, each of which is seen to provide a different way of thinking about organisations. Each metaphor will be outlined in turn with an accompanying emphasis on the implications for best practice. Given the repeated emphasis on satisfying client requirements, particular attention will be given to the way metaphors can be used to make sense of client organisations. It will be demonstrated that the dominance of the machine metaphor was challenged repeatedly throughout the 1980s, only for it to be reinforced subsequently with monotonous regularity. Within the context of a single chapter, justice cannot be done to the richness of debate which has been provoked by Morgan's notion of organisational metaphors. Its coverage, however, will be sufficient to inform upon subsequent critiques of supposed best practice recipes such as business process re-engineering, partnering and lean construction.

6.2   Metaphorical perspectives on organisation

6.2.1   Machine metaphor

The first metaphor which Morgan (2006) describes is that of the machine. Viewed from this perspective, an organisation is perceived as a goal-seeking machine which consists of interchangeable parts. Organisations are further assumed to be unitary in nature with all parts working in harmony towards a predetermined set of objectives. The overriding task of management is the quest for efficiency. People are directly comparable to cogwheels within a machine. If a cogwheel in a machine were to break, we would simply remove it and replace it with an identical component. Likewise, if a person in an organisation were to break down we would similarly replace him (or her) with an identical component. This, in essence, sums up the construction sector's default recipe of human resource management (HRM). The machine perspective is further characterised by an assumption that the organisation is operating within a static environment. Management is therefore able to concentrate on internal efficiency without worrying too much about the direction of travel. The image is that of a steam engine pursuing a pre-determined path. Management is busy on the footplate keeping the train moving forward and innovation is limited to speed and efficiency. No one is especially interested in checking the direction of travel which is taken almost entirely for granted. This way of thinking is evocative of bureaucratic organisations of the 1950s which had the luxury of operating in a relatively stable business environment. However, external shocks to the business environment throughout the 1970s and beyond confined such certainties to a small minority of firms. Firms that focus continually on internal efficiency can be blind to the way their markets are changing; often with disastrous consequences. In the changing world of the 1980s construction firms soon learned that the key to competitive success did not lie solely in machine efficiency. Of greater importance was an ability to expand and contract in response to fluctuations in demand.

Morgan (2006) suggests that the machine metaphor is implicit within classical management theory as epitomised by 'scientific management' (Taylor, 1911). Keys (1991) has further suggested that the assumptions of the machine metaphor also underlie traditional operational research and the associated field of systems engineering. Such ideas have a modern representation in the guise of business process re-engineering (see Chapter 7). The perception of an organisation as a goal-seeking machine is also widespread within the field of micro-economics, and is especially true of the neoclassical 'theory of the firm'. Advocates whose thinking is dominated by the machine metaphor tend to describe organisational goals in terms of profit maximisation. Their natural response to declining profitability would be to make the 'machine more efficient'. The issue is not to argue that profitability and efficiency are not important, because clearly they are. The point is rather to argue that profitability and efficiency are not the only things which are important, and it is unnecessarily restrictive always to think only in these terms.

The machine metaphor is further closely associated with what has become known as 'closed systems thinking'. The stereotypical image is that of a mechanical engineer who tends not to think beyond the operational mechanics of the system for which he (or she) is responsible. Economists are similarly often accused of closed systems thinking, but in truth debates about whether or not externalities should be taken into account have longsince characterised the 'dismal science'. Within economics, externalities are defined as the impacts on other parties not directly involved in a given economic transaction. They occur when economic activity causes external costs or external benefits to third party stakeholders. The discourse of the enterprise culture of course held little truck with any notion of externalities. The concept of externalities was of central importance to the planned economy of the post-war era, with particular implications for the construction sector. Such ideas, however, were swept away during the 1980s by the remorseless focus on the 'customer'. However, the discourse of sustainability is currently bringing externalities to the fore once again, with significant implications for the construction sector improvement agenda.

6.2.2 Organic metaphor

The second metaphor described by Morgan (2006) sees organisations as bio-logical organisms which adapt continually to changes in their environment. The 'open systems' and 'contingency' approaches to organisations were central to the theory of organisations which dominated during the 1970s (cf. Kast and Rosenweig, 1985; first published 1970). Both were underpinned by a dominant organic metaphor in that they argued that different organisa-tional forms evolve in order to 'survive' in different environments. Organisations are therefore perceived to be analogous to species which evolve in different ways to survive in different natural environments. Such a perspective can be positioned directly against theories of 'intelligent design' whereby organisations are designed purposefully for a particular set of circumstances. The key argument of contingency theory is that organisa-tions are 'contingent' upon the environment within which they operate (Burns and Stalker, 1961; Lawrence and Lorsch, 1967). Such a view argues against generic notions of 'best practice' in favour of different practices being appropriate for different circumstances. For example, to use an analogy from the natural world, best practice in the Arctic might look like a Polar Bear, whereas best practice in South Asia might look like a Tiger. In other words, best practice becomes context specific. Yet bodies such as the Construction Best Practice Programme (see Chapter 5) advocate generic 'best practice' recipes which are supposedly valid across a huge diversity of contexts.

In contrast, the organic metaphor advocates the need for 'open-systems' thinking' whereby an ongoing interchange with the environment is crucial for sustaining the health of the organisation. The open nature of biological and social systems is often contrasted with the relatively closed nature of mechanical and physical systems (Morgan, 2006). Systems theorists also tend to refer to intra- and inter organisational relationships in terms of sub-systems which operate within an environmental supra-system. Thinking in this way can help escape the constraints of thinking only about the efficiency of organisations as if they were closed mechanical systems. Ultimately, the organic metaphor emphasises interconnectivity across per-meable system boundaries. The argument can also be applied to the level of individuals within organisations, who are recognised to be living entities in their own right. In contrast to the machine metaphor, individuals are held to have their own complex sets of needs which have to be held in balance with those of the organisation. Here lies the essential distinction between the machine perspective and the organic perspective. Whereas the former concentrates on efficiency and optimisation, the latter emphasises the need to maintain a balance through continuous re-adjustment. Initial inspiration for the organic metaphor was provided by the General Systems Theory (GST) developed by von Bertalanffy (1968).

Margaret Thatcher's famous declaration that 'there is no such thing as society' was indicative of the way the organic metaphor was squeezed out by the enterprise culture during the 1980s. There is of course a direct connection here with the preceding discussion of externalities. The advocates of the machine perspective routinely deny the importance of externalities. But for the advocates of the organic perspective they become an essential part of a manager's domain of interest. At the time of writing, organic metaphors are strongly evident in underpinning arguments in support of sustainability and corporate social responsibility (CSR) (Green, 2009). Advocates of the organic metaphor emphasise consistently the need to be responsive to the broader environment within which organisations operate. Economic, environmental and social externalities cannot therefore be ignored, they must be factored into the decision-making process.

Those who respond typically to difficult trading conditions by arguing that the organisation must 'adapt in order to survive' are evoking an organic metaphor. They therefore bring a very different emphasis in comparison to those who argue the need for ever-greater efficiency savings. The organic metaphor thus offers a means of breaking out of a recurring narrow fixation with narrowly-defined efficiency. But it should also be noted that the two responses are not necessarily mutually exclusive.

6.2.3 *Organisations as brains*

Morgan's (2006) third metaphor extends the organic allegory in that organisations are seen to possess a central intelligence which can predict change in the environment, rather than react to it passively. This quasi-cybernetic metaphor emphasises the need to collect information which can then be processed by the central brain (Green, 1996). Decisions are made on the basis of this information and instructions are communicated to the outlying 'limbs'. Instructions flow outwards, and information flows inwards. Simon (1947) was particularly influential in propagating this information-processing view of organisations. It is this avenue of thinking which has spawned the huge literature on information systems and decision-support systems. The subsequent development of the discipline of cybernetics saw organisations characterised increasingly as possessing 'brains' which can learn from experience. Perhaps even more important was the emphasis on the second-order capability of learning how to learn. Organisations of course cannot literally learn anything, but this has not prevented the development of a huge literature which rests on an underlying cybernetic metaphor.

Seminal proponents of the cybernetic metaphor within organisational theory include Argyris and Schön (1978) and Beer (1981). Senge (2007) has been further influential in promoting the cause of organisational learning, thereby softening the original emphasis on cybernetic control with a more humanistic and developmental storyline. However, of particular interest is the more recent concept of knowledge management. Nonaka and Takeuchi (1995) deserve much of the credit for popularising knowledge management, although Davenport and Prusak (2000) have argued that many of the ideas were still in their infancy as recently as 1998. Nevertheless, it is surprising that knowledge management did not receive even stronger recognition within the pages of *Rethinking Construction* (Egan, 1998). The proposal to turn the Construction Best Practice Programme into a 'knowledge centre for construction' is indicative of the prevailing popularity of improvement recipes which emphasised the key importance of 'knowledge'. It is also suggestive of a particular set of assumptions relating to the way that knowledge can supposedly be captured and commodified.

The conditions for the success of knowledge management were arguably created by the prevailing widespread disenchantment with business process re-engineering (Easterby-Smith and Lyles, 2005). Thus the limitations of one management fashion provide fertile ground for subsequent management fashions. Firms that have experienced extensive outsourcing and downsizing perhaps inevitably end up worrying about 'knowledge management'. The essential problem is how to maintain knowledge within the firm without having to carry people on the payroll. By extension, hollowed-out main contractors ultimately end up advocating participative planning in conjunction with their subcontractors. Part of the logic relates to the need to ensure 'buy-in' from subcontractors, but it relates further to the likelihood that the contractor is no longer likely to have the necessary expertise in-house. Lean organisations understandably become obsessed with ideas such as 'capturing knowledge' in direct recognition of the inherent weakness of 'leanness' as an organisational form.

As was the case with the two previous metaphors, it is useful to speculate how devotees of the cybernetic metaphor would respond to recession. Certainly advocates of the information-processing approach would emphasise the need for better 'intelligence', together with more effective control and feedback. Those who had bought into the knowledge management storyline would tend to focus on the benefits of knowledge sharing across organisational silos. But the overriding perspective beyond such arguments remains essentially unitarist. More often than not, there is seen to be one central 'brain of the firm' which collates information and issues instructions. The same is also true of knowledge management initiatives which emphasise the use of central repositories of knowledge facilitated by information technology (IT). In many respects the cybernetic metaphor can therefore be considered to be a 'high-tech' extension of the machine metaphor. But it must also be conceded that multi-perspective (and contested) concepts such as knowledge management are not so easily categorised in accordance with single metaphors. Indeed, this difficulty of classification is an essential condition of their success.

At this point it is worth making clear that metaphors rarely (if ever) offer 'complete' explanations of anything. Organisations are no more brains than they are machines or organisms. There is therefore a danger in advocating metaphors as yet another management cure-all. The argument being promoted here is rather more modest. It is contended that sensitivity to the use of metaphor can bring additional insights into the nature of organisations. And perhaps even more crucially, it can help bring an important degree of cynicism in response to the seemingly endless stream of managerial panaceas directed at the construction sector.

6.2.4   Culture metaphor

The culture metaphor focuses upon the importance of an organization possessing shared values and beliefs (Morgan, 2006). Definitions of organizational culture invariably emphasize the importance of shared values and beliefs and the way in which they affect individual behaviour (Kast and Rosenzweig, 1985; first published 1970). The following definition is typical of many:

> 'organisational culture is the unique configuration of norms, values, beliefs, ways of behaving and so on that characterise the manner in which groups and individuals combine to get things done.' (Edridge and Crombie, 1974)

However, it was Peters and Waterman (1982) in their search for excellence among American companies who cemented the assumption that organisational performance depends upon an alignment between employee values and managerial strategy. This ushered in the era whereby organisations seek routinely to establish a strong corporate culture by use of mission statements. Wilkins and Ouchi (1983) suggested subsequently that highly developed unique cultures tend to exist within organic organisations. They argued that in many cases it is the shared values, beliefs and accepted norms of behaviour which provide guidelines for action. In contrast, mechanistic organisations were rarely seen to develop unique cultures, the necessary guidelines for action supposedly being provided by written rules and regulations.

In the context of the 1980s, one of the reasons that the culture metaphor was popular was that it could be mobilised to justify the break-up of organisational bureaucracies. More responsive forms of organisation centered on notions of 'excellence' were much more compatible with the concept of enterprise. But for many the concept of excellence remained stubbornly illusive. Excellence – like best practice – is notoriously difficult to define. But it is equally difficult to argue that it is not a good thing. Peters and Waterman's (1982) focus on excellence coincided with the rise in popularity of total quality management (TQM) with an associated insistence on quality culture. Both concepts were adopted hungrily by those who were advocating change for the construction sector.

Every improvement initiative directed at the construction sector tends to be accompanied routinely by exhortations in favour of an associated cultural change. Indeed, more often than not, the required 'change in the industry's culture' is seen to be an essential pre-requisite of success. The culture metaphor is a particular favourite amongst the harbingers of change; but in truth it relies rather more on rhetorical appeal than it does on substantive meaning. Cultural change within organisations is difficult enough, let alone cultural change within an entire sector. And given the heterogeneous nature of the construction sector it would seem rather lazy to assume that it possesses a single homogeneous culture in the first place. This is especially true given the extensive structural changes which characterised the 1980s and early 1990s. Furthermore, supposed 'cultures' are rarely free of internal contradictions. Indeed, they are often best described in terms of their inherent contradictions and paradoxes. This is certainly true of the 'enterprise culture' (see Chapter 2). It has already been argued that it unrealistic to perceive a culture to be a variable which exists independently of the instructional structures within which it is constituted.

Unfortunately, the continued popularity of the culture metaphor amongst those who preach change in the construction sector belies the supporting evidence. There is a notable absence of any research to support the contention that cultural change can be achieved through the mechanisms described by management gurus such as Peters and Waterman (1984). Even on the level of organisations, several authors have questioned the extent to which culture is subject to top-down manipulation and control (e.g. Antony, 1994; Legge, 1994; Willmott, 1993). A particular interesting area for research is the possibility that managerial action may promote unforeseen counter-cultures that are dysfunctional. Hence the imposition of BPR creates 'adversarial attitudes' in response to subsequent improvement recipes.

Green (2006) discusses the prospects of cultural change in the construction sector, and draws a comparison with Ogbonna and Harris's (2002) review of cultural change in the food retailing sector. Central to Ogbonna and Harris's argument is the contention that managerial ambitions for cultural change are invariably over optimistic. They present two case studies of planned cultural change initiatives. In the first, they describe how the outcome for the vast majority of employers was a sense of resigned behavioural compliance rather than any shift in intrinsic values. Their second case study was more positive in that they found *some* evidence of change, albeit limited and unpredictable. But most crucially they observe that employees are less likely to be convinced by the espoused rhetoric of cultural change when it is in direct conflict with the material changes which they observe happening around them. This is a point to which we shall return several times throughout the remainder of this book.

6.2.5 Political metaphor

A further model of organisations is provided by the advocates of the political metaphor (Clegg, 1989; Pfeffer, 1981). Whereas the culture metaphor tends to emphasise cooperation amongst individuals, the political metaphor emphasises the role of power and conflict (Morgan, 2006). From this perspective, individuals are seen to use organisations as vehicles to achieve their own aspirations. Organisations become characterised by intrigue and internal 'politicking', which are held to be the norm rather the exception. Rather than perceiving organisations as unitary entities where all parties pull in the same direction, it is recognised that they comprise groups of individuals with their own agendas. Such groups invariably form temporary coalitions for the purposes of mutual self-interest. These alliances are always changing as new power groups emerge and individuals transfer their allegiance from one group to another. In this context, 'best practice' becomes a rhetorical device by means of which a given interest group seeks to progress its own agenda. For example, it is much more lucrative to be an advocate of BPR than it is to be a recipient (see Chapter 7).

The political perspective challenges the very notion that an organisation can ever possess objectives of its own. It is frequently held to be particularly important in that it legitimises the use of power, conflict and coalitions in order 'to get things done' (Morgan, 2006). Industrial relations issues are often best understood by means of the political metaphor. Indeed, much of the negotiation literature is directed towards ways of resolving conflict between conflicting interest groups. It should also be recognised that conflict and political action are not necessarily dysfunctional, but can often act as a useful catalyst for creativity. The long-term success of any organization will, however, depend upon an underlying willingness amongst its members to form temporary coalitions; otherwise the organisation will inevitably disintegrate. It is notable that construction project organisations are often conceptualised as temporary coalitions (Cherns and Bryant, 1984).

The extension of the political metaphor to the industry level readily exposes the myth that the construction sector possesses a unified set of interests. The construction industry has struggled consistently to maintain a unified front for the purposes of lobbying government. Instead, there exists a bewildering array of ever-changing representative bodies for every interest group in the sector. Attempts to reform the industry have also floundered on a consistent failure to harness the industry's multitude of interest groups towards a common purpose. Even within government, responsibility for the construction sector has been shifted periodically between departments; indeed responsibility for different aspects of the construction sector has on occasion been spilt between different government departments.

But there is another irony which cannot pass without comment. As has already been noted, numerous policy sources over the years have identified fragmentation as a major barrier to construction sector improvement. And yet continuous fragmentation is a blight that seemingly affects the reform movement. Indeed, the metaphor of continuous flux and transformation (see below) also characterises the endless reorganisation of the construction sector's reform infrastructure. Memories of a multitude of reform bodies dissolve gradually into acronym soup: the Construction Industry Board (CIB), the Movement for Innovation (M4), Rethinking Construction, the Construction Best Practice Programme, the Reading Construction Forum, and the Design Build Foundation have all at various times promoted 'best practice' only to disappear as the coalitions of change are reformed. Few would bet against a similar fate befalling Constructing Excellence and the Strategic Forum for Construction. The Construction Industry Council (CIC) is perhaps unusual in its relative longevity. Other bodies which still leave their shadow include the National Contractors' Group, which subsequently morphed into the Major Contractors Group; only to be more recently re-launched as the UK Contractors Group. On the clients' side we have seen the Construction Clients' Forum (CCF) followed by the Confederation of Construction Clients (CCC). Even more vaguely remembered is the Construction Round Table. Also deserving a mention is the Construction Research and Innovation Strategy Panel (CRISP), which probably deserves the prize for the best acronym. After having died a slow death it was subsequently reborn as the *new* Construction Research and Innovation Strategy Panel (nCRISP). Eventually it disappeared all together. Given the politics, intrigue and vested interests which continue to characterise this bewildering terrain of quangos, it is perhaps hardly surprising that its denizens have often struggled to find time to engage with the day-to-day challenges faced by construction firms.

The description above has done scant justice to the recent history of the UK construction reform movement. A rather more careful account has been provided by Adamson and Pollington (2006) in *Change in Construction: an Account of the UK Construction Industry Reform Movement*. It is easy to be dismissive of the role played by such bodies, but the constantly evolving cross-organisational networks sustained by such quangos cannot be dismissed as irrelevant. As we shall see in subsequent chapters, these are the networks through which ideas of 'best practice' are mobilised and legitimised. Indeed, it will be further argued that best practice has no meaning independent of these cross-organisational networks. Nevertheless, the innate instability of the organisations listed above raises questions about the industry's residual memory. No wonder the industry struggles to learn as a collective entity if the consultative structures are so inherently unstable. It is also useful to understand reports such as *Rethinking Construction* as the output of an undeniably politicised process. This is not to suggest that such reports consciously promote a particular set of interests, although the Egan report is much more openly biased towards the interests of clients than its predecessors. A more likely diagnosis is that the participants in the process ensure that nothing within the advocated agenda goes against their interests. The ultimate effect is pretty much the same. It should be added that such behaviour is by no means inevitable; this is simply the kind of behaviour which is emphasised when organisations are viewed through a political lens.

As a caveat to the above, the professional institutions which serve the construction sector have remained relatively stable. Professionals – and professional firms – have always been strong in construction, especially within the domains of architecture and engineering. However, from the late 1970s onwards the status of professionals came under sustained attack from the advocates of the enterprise culture (Green, 2009). Professional groups were cast repeatedly in the role of self-serving monopolies which were standing in the way of open competition (Flynn, 2002). The professions were in turn resistant to the contention that their members should simply compete in the market place in the same way as a trader of commodities. The self-identities of practising architects and engineers had long since been shaped by notions of serving society at large. They therefore routinely took account of externalities beyond the requirements of the immediate client who was paying the bill. But clients often by-passed such élitist structures to commission design-and-build architecture from the local contractor. The results of which continue to scar the urban built environment. Especially when viewed from the (privatised) trains as they enter and leave Britain's major conurbations.

In the main, the professional institutions maintained a sceptical distance from the excesses of the Egan agenda and continued to insist on defending the content of the educational programmes over which they resided. Had the professional institutions stood aside, degree courses would have been re-engineered radically in accordance with the demands of the Construction Industry Board (1996) report on *Educating the Professional Team*. Two years later they would probably have been revamped once again in the wake of *Rethinking Construction*. Many would now agree that the slow-to-change nature of the professions can act as an important brake on the more extreme ideas which are promoted by the harbingers of change. The industry has also belatedly introduced a degree of protection against tin-box architecture which serves only the narrow requirements of self-interested clients. We now have the Design Quality Indicators (DQIs) presided over by the Commission for Architecture in the Built Environment (CABE) and promoted by the Construction Industry Council (CIC). CABE was created in 1999 and is indicative, together with the advent of sustainability, that the enterprise culture is no longer in the ascendancy – at least when it comes to issues of building design. The bogusly self-employed seemingly lack effective lobby groups and at the time of writing there remains little interest within the CIC for the establishment of Employment Quality Indicators (EQIs).

6.2.6   Psychic prison – thinking outside the box

Another common metaphor which often characterises the debate about construction sector improvement is that of the psychic prison. Morgan (2006) contends that this metaphor stems from a perceived tendency of people, and organisations, to become trapped into the favoured way of thinking. Organisations can develop strong norms of behaviour which can act to stifle individual creativity. The phenomenon of 'groupthink' provides an extreme example of decision making which is constrained by the prevailing group norm (Janis, 1972). The tendency to persevere with tried and tested solutions is perhaps even stronger within bureaucratic organisations, where the normal methods of procedure are enshrined within rules and regulations. Although not emphasised especially by Morgan, the psychic prison metaphor can be seen to underpin implicitly the extensive literature on innovation and creative thinking (e.g. De Bono, 1990; Csikszentmihalyi, 1996; Dawson and Andriopoulos, 2008). Much of the literature on innovation and creativity starts from the perspective of breaking out of fixed and taken-for-granted patterns of thinking. But whether such patterns of thinking need to be disrupted will often be linked to political battles between interest groups within the organisation for power and control of resources. The need to overcome fixed ways of thinking through brainstorming techniques is often invoked within the context of value management (Kelly *et al.*, 2004). Creative approaches to problem-solving are also often advocated within the context of partnering (see Chapter 7).

Organisational metaphors are rarely invoked in isolation. Very often different metaphors are mobilised by different interest groups to sustain diametrically opposed positions. For example, the metaphor of a psychic prison is often mobilised as a counter-argument to those who argue in favour of strong corporate cultures. They both address the same issues, albeit from a different perspective. To possess a strong corporate culture is, apparently, a good thing; but at the same time to be trapped in fixed ways of thinking is a bad thing. So perhaps the more managers strive to implement cultural change programmes in pursuit of aligned values, the more they are obliged to encourage diversity of thought by promoting a counter-discourse of innovation. The psychic prison metaphor can undoubtedly be useful in encouraging managers to dig beneath the surface to discover the implicit processes and practices that trap organizations into unchallenged modes of working (Morgan, 2006). Here lies an important pointer towards the ways that metaphors are mobilised as rhetorical devices in support of different arguments.

The psychic prison metaphor can also be useful in identifying some of the barriers to change and innovation. Certainly, in recent years the discourse of innovation has become evermore pervasive in construction policy circles. In common with Peters and Waterman's (1984) concept of excellence, innovation is difficult to define … and equally difficult to argue against. This suggests that innovation is more meaningfully considered as a discourse which is mobilised as a legitimising devise at a political level (Davies, 2006). Certainly the word 'innovation' is used to describe a multitude of phenomena which are conceptually and practically distinct. Davies (2006) argues further that this flexibility in use can contribute more to confusion rather than to clarity.

The Movement for Innovation (M4I) has already been mentioned among the list of the construction sector's short-lived quangos (see Chapter 5). M4I resulted directly from a recommendation by the Egan Report, and its mission was to involve firms in showcasing the benefits of 'new ways of working' through the establishment of demonstration projects. At its peak in

2004, M4I resided over 440 projects with a combined capital value of £7 billion. Adamson and Pollington (2006) cite the demonstration projects as being of key significance in the post-Egan reform programme. The demonstration projects were especially notable for their focus on performance measurement, thereby cementing the industry's predilection for key performance indicators (KPIs). There is of course an obvious tension between advocating innovation on the one hand, and imposing a set of KPIs that serve to embed existing ways of thinking on the other. But of particular note is the lack of any objective process of verifying the claimed innovations on the demonstration projects (Gray and Davies, 2007). It is perhaps indeed easier to understand innovation as a legitimising discourse. 'Innovation' and 'excellence' can both be read as essential components of the discourse of enterprise, complete with all its paradoxes and ambiguities.

6.2.7   Flux and transformation – reality in flight

The seventh metaphor presented by Morgan (2006) is organisation as flux and transformation. On first reading, this seems somewhat abstract in that it perceives organisation as a verb, rather than a noun. The argument is that an organisation cannot be understood by means of a snapshot at a given point of time. This is true irrespective of how careful our analysis is, or how many different metaphorical standpoints are taken. To understand the characteristics of any individual organisation it is necessary to understand the path along which it has travelled. An organisation's history serves to shape and constrain the way it can respond to new opportunities. In this respect, it is popular to talk of *embedded practices* and to emphasise the importance of path dependency. But firms can change direction through the actions of individuals (see the innovation metaphor above). In high-technology sectors such as aerospace or ICT, change is often driven through technological innovation. Established company structures are often overturned by disruptive technologies. Construction is by no means immune to change through technological innovation, and parts of the sector are undoubtedly at the forefront of technology. However, many other niches remain reliant on technology which is relatively unchanged. Change is further constrained by embedded practices and professional structures. The professional institutions are often criticised for preserving out-dated structures and modes of working. But such constraints to change should not necessarily be construed as a negative. Had the UK financial sector been more resistant to 'innovation' the banking crisis of 2007–09 might have been averted. But arguments in favour of 'constraints to innovation' are not of course what the advocates of enterprise want to hear.

The shaping influence of path dependency means that firms frequently possess distinctive capabilities and embedded practices which cannot easily be replicated by others (Green *et al.*, 2008). Such a perspective further discredits the possibility of transferable generic best practice recipes. It also sidelines arguments which present 'best practice' as being contingent upon an organisation's environment. The metaphor of flux and transformation emphasises the way in which organisational practices have been shaped by a succession of adjustments to previous environments. It further challenges the possibility of any steady state condition. Business environments can change very rapidly; this has been especially true in recent decades where changes have been initiated by unanticipated shocks. Metaphors of continuous evolution therefore fail to capture the essence of these processes of continuous and rapid change. An alternative perspective sees organisational practices to be a continuous state of adaptation, and the key to success lies in the ability of firms to adapt quickly. Thus, if we are to understand an organisation, we must understand the processes through which the organisation continuously re-invents itself over time. Many construction companies exist in a continuous state of reorganisation in terms of whether they are organised against specific business sectors or in terms of geographical regions. The ongoing task of management is to strive for an appropriate balance between these different organisational forms. Hence the contention that 'organisation' is best understood as a verb rather than a noun. The day-to-day reality of many larger construction firms is characterised by the need to make sense of a seemingly endless succession of mergers and acquisitions. Companies are often striving continuously to bring a degree of consistency to the way they operate across a range of diverse business divisions. The ongoing pattern of mergers and acquisitions means that a steady state condition is never reached. Static notions of 'best practice' are therefore rendered obsolete.

The argument above is especially pertinent when extended to the construction sector at large. Different parts of the construction sector have very different path dependencies, and hence will have very different sets of embedded practices. Indeed, differences within the sector are often greater than the differences with other sectors. However, from the perspective of Morgan's seventh metaphor, it is argued that it is not possible to represent or describe reality using static terminology. Reality is essentially viewed as emergent, evanescent and fleeting; reality never stops in order to actually *be*, but is in a continuous state of *becoming* (Chia, 1995; Linehan and Kavanagh, 2006). Such a perspective thereby emphasises continuous processes of readjustment, rather than seeking to isolate static characteristics that can be supposedly measured and subjected to abstract statistical analysis.

In no small way, the advocated focus on flux and transformation is central to the rationale of this book. The need to understand the inter-relationship between the practice of construction management and the broader context is not of course an entirely new argument. Indeed, the dominant paradigm of construction management research has often been criticised for its tendency to extract issues outwith the context which makes them meaningful (Bresnen and Marshall, 2001; Fernie *et al.*, 2006). But within the broader field of organisation studies there is a long tradition of contextualist research which has to date received little recognition amongst construction researchers. Of particular importance is the recognition that contexts are invariably dynamic, and as such should be recognised as an active part of any analysis. Pettigrew (2003) emphasises the importance of studying 'reality in flight' and of locating present behaviour in the context of its historical antecedents. This was something which the advocates of construction sector improvement had done notably up until the late 1980s. But by the time *Rethinking Construction* was published in 1998 it had become the norm simply to advocate instrumental, and supposedly universal, improvement recipes. The shaping processes of the past were no longer judged to be relevant. The past became a dark and unexplored place, occupied only by dinosaurs and adversarial attitudes. The focus of attention shifted towards an alternative future which was to be created through the application of a range of modern management methods. Techniques such as benchmarking, partnering and lean construction were all that were required. If they were not successful first time around, then it was argued they needed to be implemented with greater enthusiasm. The end result is that instrumental management techniques have progressively become privileged over the need to understand industry development over time. History is placed in the dustbin and replaced with a set of key performance indicators (KPIs).

The argument in favour of contextualist research would be well understood by any student of history. For example, few would argue that the current situation in the Middle East could be understood without knowledge of events that have occurred over several generations. Whilst the perspective of flux and transformation is strangely neglected in the construction management literature, it does add a further obstacle to the possibility of ever obtaining a single objective interpretation of any organisational situation. Not only are we faced with different meanings attached to current events, we are also faced with different myths relating to past events. Such myths are clearly likely to develop and change over time. This may happen to such an extent that they lose any connection with the 'truth' of the initiating event. It is often myths of this nature which constitute an important part of a construction firm's culture. The development of these myths over several decades militate against any short-term efforts to 're-engineer' an organisation's culture.

6.2.8 *Instrument of domination*

Morgan's (2006) final metaphor is that of an organisation as an instrument of domination. From this point of view, organisations are the means by which powerful individuals impose their will on others. The notion of corporate culture is seen to be less about shared values and more about 'brainwashing' individuals into accepting the required model of behaviour. In some respects, the domination metaphor is an extreme extension of the political metaphor. However, the emphasis no longer lies on coalition-forming, but on the exploitation of the organisation's employees by the dominant power group. Morgan also argues that it is possible to perceive multi-national corporations as instruments of domination with respect to their host communities. The possibility that organisations exploit their environment gives an alternative perspective to that of the organic metaphor. It is perhaps worth mentioning that Karl Marx maintained that every capitalist firm, and indeed, the capitalist system as a whole, is an instrument of domination. For Marx, the basis of this domination is the quest for surplus value and the accumulation of capital (Morgan, 2006). Marxist ideologies and philosophy have of course been rendered passé by the collapse of the former Soviet Union. Even within the Labour Party there are relatively few politicians who still address fellow party members as 'comrades'. The singing of the 'Red Flag' has notably been ditched from Labour Party conferences. Such practices have been marginalised by New Labour's acceptance of the 'enterprise culture'.

Undoubtedly, there is little nostalgia for the communist utopia envisaged by Marx. The argument that the capitalist system contains 'inner contradictions that seed its ultimate destruction' seemed briefly prescient during the banking crisis of 2007–09. Yet capitalism comes in many forms, and is by no means synonymous with the enterprise culture. Capitalism has survived many previous crises and continues to possess a healthy capacity to continuously re-invent itself, and to absorb criticisms into the mainstream. In the 1980s, the advocates of unbridled enterprise were able continuously to cement their legitimacy by positioning themselves against the 'loony left'. Hence it can be argued that the demise of the political left is not just a problem for socialism; it is also a problem for the enterprise culture.

Nevertheless, the 'domination' perspective remains much loved by Marxist migrants into business schools from sociology departments. Indeed, recent years have seen a resurgence of interest in what has become known as critical management studies (CMS) (Alvesson and Willmott 1996; Alvesson and Deetz, 2000; Grey and Willmott, 2005). Such sources can provide a refreshing counter-balance to the manageralist rhetoric which too often dominates the conversation about construction sector improvement. The credit crunch of 2008–10 and the associated failures of the banking sector did much to undermine the iconic status of the financial services sector. CMS may similarly play an important role in sensitising us to some of the paradoxes and inconsistencies that underpin modern management improvement recipes. Despite the continuous mantra of 'wealth creation', the last two decades have seen sustained public hostility towards 'fat cat' salaries among CEOs. Of particular note within the current climate are the substantive bonus payments enjoyed by banking executives as rewards for their 'innovative' approach to risk management. Indeed, their approaches to risk management were so innovative they ultimately led to several high street banks being bailed out by the tax payer. Strangely, few were advocating that the construction sector should similarly be taken into quasi-public ownership. Nevertheless, it remains pertinent to remind ourselves that the financial rewards from productivity initiatives are very rarely shared equally with the workforce even in periods of economic growth. This is no less true in construction than it is in other industrial sectors. For example, according to US statistics during the 1990s executive pay jumped by 535% (before adjusting for inflation). The growth in worker pay during the same period was 32%, which barely outpaced inflation at 27.5% (Anderson *et al.*, 2000). The figures for the United Kingdom display a similar rapid increase in wage inequality since 1978 (Machin, 1996).

More recent figures confirm that little has since been done to reverse the large inequality in growth which occurred between the late 1970s and early 1990s (National Equality Panel, 2010). The available statistics are therefore in direct conflict with the prevailing assumption that the rewards of industrial innovation are shared equally. Such background statistics provide an important reminder of the dangers of assuming that the benefits of improved performance are shared necessarily equally amongst all parties. And this remains the case regardless of the extent to which the advocated initiative is dressed up in the rhetoric of 'collaborative working'. However, managers who are only able to mobilise arguments shaped by a single metaphor inevitably become trapped in a fixed way of thinking. This applies equally to all the metaphorical perspectives rehearsed above, but is especially true of the instrument of domination. Metaphors can in themselves become psychic prisons. Or, in other words, every way of seeing also becomes a way of not seeing (Morgan, 2006).

It should be emphasised that this book is aimed primarily at helping construction professionals make sense of the sector's unfolding 'improvement agenda'; the extent to which it is appropriate to explore the complex theoretical nuances of critical sociology is thus strictly limited. The metaphor of organisation as an instrument of domination can undoubtedly bring useful insights, but judgement must always be exercised in terms of when such insights are appropriate. Those who insist on seeing the world in terms of exploitation and domination inevitably marginalise themselves from mainstream conversation. In consequence, they run the risk of ending up standing on street corners selling copies of the *Socialist Worker* (metaphorically, if not literally) which nobody wants to buy. The present author has certainly enjoyed himself on occasion by invoking critical perspectives of 'new' management practices, and many of these arguments are rehearsed in the course of this book. But condemning every new initiative as a means of technocratic totalitarianism ultimately becomes sterile and threatens to replace one set of supposed dogma with another.

It should be added that traditional critical arguments based on materialist interpretations of power have in recent years have been challenged by postmodernist interpretations that advocate the need for a discursive understanding of power (cf. Fournier and Grey, 2000). There is therefore a danger in assuming that the advocates of CMS share any basic position other than being self-consciously against the mainstream (even then, they would be unlikely to agree on any definition of the 'mainstream'). In truth, CMS is a loose umbrella term which includes a multitude of very different 'critical' perspectives. The result is a community of critical academics who spend most of their time attacking the theoretical positions of other critical academics, seemingly with little interest in engaging with the empirical realities experienced by practitioners. Monty Python's film *Life of Brian* comes to mind here, with the internecine rivalries between the 'People's Front of Judea' and their various rivals, including the 'Judean People's Front', the 'Judean Popular People's Front' and the 'Popular Front of Judea'. Needless to say, these various factions did not spend a great amount of time fighting the Romans. But this of course is a political metaphor which can bring useful insights, whilst also denying the legitimacy of insights from other perspectives.

6.3   Metaphors as a process

6.3.1   Limitations of metaphor

In the cause of reflective practice, it is also necessary to concede some of criticisms that have been directed at the concept of organisational metaphors. Morgan makes great claims for the new understandings which can be liberated through metaphorical thinking; even going so far as to suggest that the resulting insights can lead to creative action (see especially Morgan, 1993). Certainly it is wise to be cautious of any claims that thinking in terms of metaphor will produce better instrumental outcomes; no such claim is made in this book. With reference to the less contentious claim that metaphorical thinking can bring new insights, it is perhaps ultimately best left to the reader to be the judge of this. Certainly metaphorical thinking has its limitations. If the idea of metaphor is taken too far, there is a risk of losing any perspective on what is real and what is metaphorical. Morgan (2006) is especially clear that in creating new ways of seeing, metaphors also create ways of *not* seeing. For example, there is a danger of becoming overly focused on particular favoured metaphors when other less favoured ones are more readily apparent. If we allow ourselves continuously to look for those metaphors which we find most appealing we undoubtedly run the risk of realising a distorted perspective. An over-reliance on metaphor can therefore be counter-productive; but sensitivity to metaphor can also undoubtedly realise important critical insights into the persuasiveness of managerial storylines.

A related potential problem is that an over-enthusiastic reliance on metaphor runs the risk of failing to take any organisational theory seriously on the basis that 'it's all metaphorical'. Certainly there is a danger of equating theories and metaphors rather too easily. A metaphor is decidedly not the same thing as a theory, and there is more to theory than metaphor. This is most decidedly true within the natural sciences where Nature acts as the final arbiter. Nevertheless, precisely what constitutes a 'theory' within organisational studies is in itself a subject of much debate, and the context of the built environment brings particular challenges to the role of theory (cf. Cairns, 2008; Koskela, 2008; Raberneck, 2008). And even if we could agree on what a theory is, there is little agreement on what we should do with it once we have it. Certainly there are those who are rather too pompous about the role of theory, and there are others who mobilise theories in ways that can only be admired. The domain of construction management undoubtedly needs more theoretically-informed research. But it is important to re-emphasise that this book is not about the development of predictive theoretical models; rather it is a book about making sense of managerial improvement recipes. As shall be demonstrated throughout the subsequent chapters, exhortations in support of industry change rarely have much to do with theory and often have much more to do with management fashion. And sensitivity to metaphor becomes central in understanding the recurring nature of the arguments being advocated.

Without wanting to bore the reader with too much about 'theory', it is possible to ascribe a useful role to theory without losing the all important connection to practice. Indeed, in an applied discipline such as construction management if a theory cannot usefully engage with practice then it is pertinent to ask what it is for. On this note, Green and Schweber (2008) highlight the important role of middle range theories which are seen to provide a form of theorising which lies between abstract grand theorising and atheoretical local descriptions. Such theories are characterised by the way in which they engage directly with the concerns of practitioners. In a practice-orientated discipline such as construction management it is important that theories support a model of reflective practice. Schön's (1983) seminal book *The Reflective Practitioner: How Professionals Think in Action* is important here in articulating a mode of practice whereby practitioners engage in a form of theorising which enables them to reflect on their underlying assumptions before committing to a particular understanding or course of action.

6.3.2   De-valuing of literal language

McCourt (1997) rehearses many of the arguments in favour and against the use of metaphor and argues that it is necessary initially to distinguish between two epistemological positions which he characterises as 'non-constructivist' and 'constructivist' (after Ortony, 1993). Readers with a low tolerance for long words should feel free to skip to the next section. But for the benefits of the two people still reading, non-constructivists – otherwise known as positivists – believe there is an objective world 'out there' which can be known unproblematically through the direct evidence of our senses. In contrast, constructivists believe that there is an objective world, but that we can only know it though our senses which inevitably filter the data they receive and impose their own structure. McCourt places Morgan firmly within the constructivist position, but also suggests that he verges towards solipsism (an accusation which Morgan rebuffs). Notwithstanding this latter point, it has probably been Morgan's greatest achievement to inform students of organisation of the weakness of positivist epistemology. Indeed, Morgan's ideas of metaphor were influential in helping the current author to escape from the positivism of his engineering training.

McCourt (1997) further distinguishes between 'strong' and 'weak' versions of the constructivist position. The strong version asserts that all scientific language is essentially metaphorical and hence discredits the possibility of literal scientific language. The weak version claims only that scientific language may be metaphorical; but still concedes the possibility of literal scientific language. It is the weak version which seems most sensible; to argue that language is *sometimes* metaphorical is not the same as arguing that it is *always* metaphorical. As a rule of thumb we should only revert to metaphorical interpretation once we have been convinced that the literal meaning is flawed. This would seem a sensible guiding rule to apply throughout the remaining chapters of this book. But it should also be conceded that popularised construction improvement recipes are something of a soft target. Hence it should perhaps come as no surprise that they are weak on literal scientific language and strong on metaphor. But it is remarkable how often the advocated recipes are taken entirely at face value.

6.3.3   From knowledge creation to sense making

Following on from the above, it is possible to argue that an emphasis on theory as an end in itself is somewhat misplaced in the context of professional practice. Rather than concentrate on theory *per se*, it is perhaps more appropriate to emphasise the need for continuous knowledge creation. For those who remain unconvinced, the real clincher here is to point towards the dynamic context within which construction takes place. Even if we were able to develop the most perfectly valid theory imaginable, it would rapidly become sidelined by the ever-changing policy environment within which construction takes place (cf. Raberneck, 2008). However, not only are our aspirations for the built environment changing continuously, so are the prevailing market conditions. A further source of disruption to any static theory is the continuous flow of technological innovations which characterise the ever-changing construction sector.

Thus, it is understandable that many construction professionals appear ambivalent to the role of theory. Nevertheless, they are often persuaded that innovation and continuous knowledge creation are important in maintaining their competitive position. It is currently fashionable to talk of 'Mode 2' knowledge production as popularised by Gibbons *et al.* (1994). The underlying argument is that a new form of knowledge production started to emerge in the mid-twentieth century which Gibbons *et al.* (1994)

characterise as context-driven, problem-focused and interdisciplinary. Mode 2 is seen to involve multidisciplinary teams coming together for short periods of time to work on specific problems in the 'real world'. This is distinguished from so-called traditional patterns of knowledge production (i.e. 'Mode 1') which Gibbons *et al.* (1994) characterise as being academic, investigator-initiated and discipline-based knowledge production.

Mode 2 knowledge production sits comfortably with the aspirations of a research agenda informed by the enterprise culture, as exemplified by *Realising Our Potential* (see Chapter 4). The emphasis on the need for 'real world' research implies a degree of discomfort that too much academic research is disconnected from the 'real world'. The idea of Mode 2 knowledge production also challenges widely help assumptions relating to the supposed linear relationship between research and innovation. An alternative perspective is advocated whereby theorising and innovating are wrapped around each other in a continuous spiral of activities. Perkman and Walsh (2007) are persuasive in emphasising the non-linear, iterative and multi-agent nature of the innovation process. Weick's (1995) notion of sense making has been hugely influential in shifting the emphasis from static ideas of theory towards dynamic, multi-participant notions of sense making. And it is within the latter context that it becomes useful to think of the way in which practising managers mobilise metaphors continuously as sense making mechanisms.

6.4   Practical implications

6.4.1   *Metaphorical lenses*

For present purposes, the importance of Morgan's contribution is that it moves the debate away from asking which organisational theory is more 'correct' towards an understanding of the metaphors which are implicit within different theories. This arguably enables practitioners to adopt the perspective which provides the best insight into the particular problem in hand, rather than being limited to one specific way of thinking. Although quite how a particular insight is judged to be 'best' remains problematic. 'Best' is perhaps usefully translated as 'most persuasive', and this will certainly depend upon the skill with which it is mobilised and the extent to which is resonates with the broader accepted discourses. Hence a metaphor might be judged more persuasive if it were to resonate with the discourse of the enterprise culture. Sense making of course is not only an individual activity, but something which is carried out collectively within groups. In some situations the machine metaphor may be accepted as the most useful guide to action, in others the cybernetic metaphor might be more judged more persuasive. In most situations, managers would be able to gain different insights from the use of different metaphors. The agreed narrative of action would therefore be likely to reflect a hybrid combination of metaphors. Hence reflective practitioners can be seen to be mobilising different theories as short-term sense making mechanisms. Such an interpretation would be broadly consistent with Schön's (1983) notion of reflection-in-practice. However, it would be misleading to suggest that metaphors can be selected in the same way we select goods from the supermarket shelf. Metaphors are everywhere and whenever we write or speak we always leave a metaphorical trail. And we are always presented with the metaphorical trail left by others.

Flood and Jackson (1991) find it useful to perceive metaphors (and theories) as being lenses through which we can view organisations. The use of different lenses will accentuate different aspects of organisational reality. Such an argument does not necessarily depend upon an individual having any particular knowledge of organisational theory. The lenses used by practising managers are often implicit to their 'view of the world'. Practitioners further tend to value theories which accord with, and reinforce, their pre-existing assumptions. Hence BPR appeals to practitioners with a predilection for machine metaphors. Every manager views organisations from a particular viewpoint; and makes simplifications in order to make their own sense of organisational life. An explicit recognition of Morgan's (2006) notion of organisational metaphors purports to bring a degree of 'liberating value' in that it helps managers develop a self-awareness of the metaphors they are using.

6.4.2 Metaphors as self-fulfilling prophecy

The concept of organisational metaphors rests within broader constructivist arguments that the 'reality' of an organisation cannot be accessed without some sort of structure being imposed by the adopted theoretical lens. It has already been argued that organisations are not only characterised by 'objective facts', but also by myths, meanings and interpretations. If we accept that the meanings ascribed to events are dependent on which metaphorical lens we use in order to understand events, then it follows that the nature of organisational reality is influenced by the sense making mechanisms adopted by practitioners. Hence sense making is not only about reading, but is also about writing (Weick, 1995). For example, if an organisation's management acted consistently in accordance with the narrow interpretation of the machine metaphor, then the organisation would become increasingly machine-like in nature. Those employees who objected to being treated as mindless components would seek employment elsewhere; management would then seek to replace them with people who were comfortable with being treated in this way. The machine metaphor would therefore tend to become a self-fulfilling prophecy. In retrospect, McGregor (1960) was undoubtedly before his time in recognising that the adopted management theory will influence 'reality'. The assumptions of the machine metaphor regarding human behaviour are encapsulated neatly in McGregor's well-known 'Theory X'. McGregor argued that if management acted as if these assumptions were true, there was a likelihood that they would indeed become true. In other words, McGregor recognised that the metaphor which we use to make sense of reality is likely to influence the behaviour of others, and therefore the nature of 'reality' itself.

It is easy to see how the argument of the self-fulfilling prophecy could also apply to the use of other metaphors. If a manager acted consistently in accordance with the political interpretation of organisational life, then others would be likely to respond in a similar way. The organisation would therefore become characterised increasingly by intrigue and conflict. On a higher level of abstraction, the notion of a shared culture can be equated with a 'shared view of the world', or, in other words, a shared metaphorical lens. Purposeful action within an organisation would clearly be hampered if key personnel viewed events consistently through different lenses. A strong corporate culture thereby becomes something which is to be aspired to, but at the same time a strong sense of 'how we do things around here' can become a psychic prison with the result of stifling individual creativity and thereby making the organisation less responsive to change.

6.4.3   *Building credibility*

In consultancy situations, an appreciation of organisational metaphors could arguably provide an enhanced understanding in two ways. Firstly, the use of different metaphors can help provide different insights into client organisations. Secondly, important insights can be gained by detecting the dominant metaphors which govern the behaviour of individual interest groups. It is also useful to recognise that the initial acceptability of a consultant is likely to be enhanced significantly if he (or she) is able to reflect the favoured metaphors of their sponsor. There is a persuasive argument that clients ultimately rate consultants in accordance with the perceived success of their relationship. There is perhaps some truth in the old adage which says '*I like this guy, he talks my language*'.

An understanding of metaphor may also conceivably be useful in helping consultants prepare for an interview with a client organisation. An examination of public statements by senior client representatives may help to identify the metaphors which characterise the preferred discourse of improvement. For example, the client may be shown to have a predilection for improvement storylines which combine machine and culture metaphors. It should also be recognised that the dominant preferred discourse of improvement is unlikely ever to be homogeneous and will almost certainly be in a constant state of flux. Different metaphors are also likely to be favoured by different interest groups, some of which may well be represented on the interview panel. But in truth, what is being advocated here is nothing which is new. We are all already engaged in deciphering metaphors as part-and-parcel of our day-to-day activities.

Most would readily agree that an ability to 'talk the same language' as the sponsor is likely to be crucial in determining the success of the sponsor-consultant relationship. But few people appreciate a 'nodding donkey' for long. Once a consultant has established his (or her) initial credibility he (or she) will then be able to encourage the development of different ways of thinking by invoking different metaphors. Indeed, over a period of time the consultant may well be able to change the dominant metaphors which are used within the client organisation (or at least parts of it). As a caveat to this argument, it is also important for consultants to recognise which metaphors are likely to be politically unacceptable. For example, clients may not always appreciate outsiders interpreting their organisation as an 'instrument of domination'.

6.4.4   Metaphors in client briefing

By way of example, it is revealing to consider the organisational metaphors which lie behind different approaches to client briefing. Blyth and Worthington (2001) describe how in the 1970s briefing was conceived as a linear process which preceded design activity. The brief comprised essentially a schedule of technical requirements and success was achieved through adhering to standard checklists (cf. Salisbury, 1998). The underlying mentality of this approach is characterised nicely by Bennett (1985), who draws from Townsend's (1970) 'tongue-in-cheek' advice on how an organisation should move head office. Townsend advocates that one person should be appointed to take responsibility for the entire process. The appointed person should then seek to provide standard accommodation for everybody and on no account should the building users be consulted. If consultation were allowed, Townsend suggests that the process would take 'twice as long and cost three times as much and all key people in the firm will be completely preoccupied with status symbols and would have no time for their work'. A successful outcome is seen to have been achieved provided that the building works 'reasonably well' and is completed on time. It is further contended that the cries of outrage will die out after 30 days.

Townsend's approach to briefing would seem to relate to the machine-bureaucracy of classical management theory. The dominant implicit metaphor is that of the machine, with undertones of the instrument of domination. It is entirely taken for granted that the client's objectives are clear and predetermined and that they remain static over time. It is further assumed that the organisation's objectives are entirely independent of the aspirations of the workforce. Such organisations may have been common in the 1950s, but they increasingly are few-and-between. And yet strangely, it is the machine metaphor which characterises the references to briefing within *Rethinking*

Construction, where the primary concern would seem to relate to the use of management techniques for the purpose of eliminating waste from the brief:

> 'Value management is a structured method of eliminating waste from the brief and from the design before binding commitments are made.' (Egan, 1998; 10)

Blyth and Worthington (2001) refer to the way in which mechanistic approaches to client briefing are often a joy for project managers, with their focus on untested quantifiable requirements and checklists. The associated machine metaphor remains deeply rooted in the psyche of many project managers within the construction sector. Whether or not the finished building fulfils the aspirations of those who will occupy it is relatively unimportant. The important thing is to make sure that it is completed on time and to cost; and, as Townsend suggests, the cries of outrage will die out relatively quickly. In any case, neither the project manager nor the architect is likely to hang around long enough to hear the grumbles.

But there is another model of briefing with an almost equally long heritage that can be traced back to Goodacre *et al.* (1982). This alternative view suggests that briefs are best produced by means of extensive collaboration between clients and designers over a period of time. It is further suggested that clients are often incapable of producing their own briefs and that their needs and objectives need to be probed in depth. Goodacre *et al.* (1982) perceive the processes of briefing and designing to be iterative and interdependent, thereby questioning the assumption that the client's objectives can be pre-determined and that they remain static over time. Furthermore, design is no longer perceived to commence only when the brief has been completed.

Bringing the discussion back up to date, the recommendations of Goodacre and his colleagues are now broadly representative of accepted mainstream thinking. Blyth and Worthington (2001) emphasise that the process of briefing is often more important than the end product. Murray *et al.* (1993) provide further clarification by observing that inexperienced clients are often unable to articulate their requirements until they have been exposed to a range of initial design concepts. Barrett and Stanley (1999) likewise emphasise the importance of empowering the client and the need to manage the project dynamics. They also emphasise the importance of appropriate user engagement. Kao and Green (2002) take such arguments further to conceptualise briefing as a collective knowledge creation process involving both client representatives and designers. From this perspective, the challenge is not to have the right checklist, but to manage the appropriate form of engagement. Latham (1994) notably recognised specifically that briefing depends upon a dialogue between clients and designers, and that clients often need training to perform their role adequately. Within this second model of briefing there is also a recognition that freezing the brief as early as possible (*a lá* Townsend) is not always appropriate. Blyth and Worthington (2001) have championed the cause of freezing (or at least 'lightly chilling') the brief at the last responsible moment. The recognition here is that false precision too early is likely to be counter-productive. Of key importance is the recognition that client organisations are not unitary entities, but that they comprise multiple interest groups which are in continuous competition for power and resources. As an aside, it is worth mentioning that different interest groups are likely to mobilise different organisational metaphors and are thereby competing in a process of reality construction.

There is therefore a considerable body of opinion which contradicts the long-standing assumption that client requirements can be pre-determined. The organisational metaphors which underlie this second approach to client briefing are clearly more complex than that of the machine. The previously dominant metaphors of machine and domination have given way to the organic and political metaphors. The contention that clients can learn as a result of their interaction with designers is also suggestive of an underlying cybernetic metaphor.

The third approach to client briefing characterised by Bennett (1985) is that which was adopted by Greycoat Estates during the property boom of the 1980s. It was a continuation of this approach which was later employed by Rosehaugh Stanhope on the prestigious Broadgate development in central London (see Chapters 2 and 4). As repetitive developers, these clients were able to develop standard briefs which could be presented to designers at the time of their appointment. The tendency towards standard briefs was further recognised by Latham (1994). Such standard briefs represented the lessons which had been learnt on previous projects. However, designers were encouraged to suggest improvements to the brief on the basis of their own experience. The underlying philosophy was one of continuous improvement by means of 'responsible innovation'. Not only were these clients notable for their sharp commercial acumen, they were also distinguished by their commitment to research which kept them up to date with the best international construction practice. Their philosophy required designers to challenge constantly the accepted UK methods of construction. The practice of developing standard briefs, together with a commitment to continuous development, has also been widely adopted by supermarket and other retail developers (NEDO, 1988).

Whilst the Greycoat/Stanhope approach has some similarities with that advocated by Townsend, there are also a number of important differences. The commitment to progressive improvement by learning from experience is strongly reflective of an underlying cybernetic metaphor. The emphasis given to innovation, especially in terms of challenging the accepted methods of construction, is also suggestive of an underlying psychic prison metaphor. A consultant who limited himself (or herself) to machine metaphors would therefore be unlikely to find favour.

6.4.5 *Trapped in fixed ways of thinking*

Blyth and Worthington (2001) were not the first authors to suggest that designers sometimes adopt an inappropriate approach to client briefing. Cherns and Bryant (1984) had previously observed that most client organisations are more complex than project team members often acknowledge. They also observed that construction professionals sometimes seem impatient with this complexity, or are even embarrassed by it. Allen (1984) has also contended that professionals tend to assume erroneously that the client already knows what his (or her) requirements are. Peña *et al.* (1987) have further suggested that designers give far too little time and attention to 'exploring the problem'. The implication which lies behind these various observations is that construction professionals often adopt a machine metaphor in inappropriate circumstances. It is notable that all of the views above were expressed in the 1980s, and the policy agenda ever since has continued remorselessly to promote machine metaphors and unitary conceptualisations of client organisations. Rather than promote radical change, there is another possibility that *Rethinking Construction* merely reflected and reinforced pre-existing regressive ways of thinking. Hence the Egan report becomes part of the problem rather than part of the solution.

On the basis of the discussion above it becomes possible to suggest that different organisational metaphors are appropriate for different situations. If the client organisation is unitary, and operates within a static environment, then the machine metaphor may well provide the most efficient means of operation. In some cases the existence of a pre-determined set of objectives can indeed be taken for granted. However, if the client organisation is multi-faceted, with no broad agreement on objectives, then it becomes necessary to construct a consensus understanding among the key stakeholders. The briefing process advocated by Goodacre *et al.* (1982) achieves this social construct by means of a dialectic debate amongst client members and designers over a period of time. Once a clear and accepted statement of the client's strategic objectives has been negotiated successfully, it then becomes possible to draft the outline briefing document. If the client's business environment is stable, then the requirements stated in the briefing document are likely to remain constant over time. But if the business environment is dynamic, then the requirements may well change as the design evolves. The Goodacre *et al.* (1982) model of briefing – as dated as it is – recognises clearly the second of these possibilities in recommending that the outline brief should not be 'frozen' until the end of Stage C of the RIBA Plan of Work. The expectation that the brief needs to evolve in accordance with a changing business environment is reflective of an organic metaphor. The idea of a socially negotiated consensus is immediately evocative of a political metaphor. There is also a suggestion of collective learning over time, which rests on an implicit cybernetic metaphor.

Unfortunately, such arguments were progressively squeezed out of the improvement agenda by the Eganites, who preferred to focus upon mechanistic approaches to the 'elimination of waste'.

6.5   Summary

The discussion in this chapter has opened up new ways of thinking about construction improvement. It has also served to highlight the way in which the machine metaphor is adopted repeatedly as the basis for any discussion of construction best practice. *Rethinking Construction* has been seen to be especially reliant on machine metaphors thereby denying the legitimacy of insights from other perspectives. Attention has been drawn to the dangers of becoming trapped into fixed ways of thinking.

The concept of organisational metaphors is useful in that it offers a range of alternative perspectives from which to make sense of organisations. It also serves to sensitise us to the dominant metaphors which lie behind particular improvement recipes such as business process re-engineering and partnering – both of which will be addressed in the following chapter. It is also important to emphasise that the eight metaphors described in this chapter are by no means exhaustive. There are also other metaphors which are mobilised routinely in discussions about construction improvement. However, most can be linked back to those which have been discussed in the course of this chapter. For example, the image of a dinosaur is commonly used to depict those who adhere stubbornly to outdated working practices, but the underlying root metaphor is that of the organisation as an organism. Dinosaurs are seen to have failed because they were unable to adapt to changing circumstances. A further common metaphor which is mobilised repeatedly is that of 'teamwork'. However, all too often a 'good team player' is seen to be someone who is willing to follow instructions blindly, rather than offer their own informed opinions. Metaphors based on teamwork are thus often uncomfortably close to metaphors based on machine efficiency. The next chapter will address some of the paradoxes and contradictions which routinely undermine the teamwork storyline within the partnering discourse.

It has further been argued that there are dangers of becoming too reliant on metaphorical thinking. It is clearly important that we should not lose our ability to distinguish between what is metaphorical and what is real. For example, it would be possible to analyse the metaphors which shape the health and safety agenda in the construction sector. However, such an analysis cannot be allowed to shift attention away from the substantive issues which contribute to the construction sector's poor track record, such as the widespread reliance on non-standard forms of employment. It is wise therefore only to revert to metaphorical interpretation once we are convinced that that literal meaning of what is being advocated is flawed.

Metaphors have been held to be particularly useful in highlighting the complex nature of the client briefing process. Construction professionals too often expect clients to behave as if they were machines. Such default assumptions should be challenged continuously in accordance with the principles of the reflective practitioner. Unfortunately, none of these arguments was recognised by the authors of *Rethinking Construction*, whose reliance on simplistic machine metaphors arguably set back the construction industry improvement debate by at least ten years. But the Egan report drew its legitimacy from changes that had already happened, especially in terms of the industry's reliance on leanness and agility. In truth, by 1998 relatively few contracting firms were interested in efficiency. Responsibility for efficiency in production had long since been delegated to the supply chain, along with responsibility for skills and capability development. *Rethinking Construction* may indeed have succeeded in making lean thinking fashionable, but the reality was that the UK construction sector needed little encouragement to adopt lean ways of working. This had already happened over the preceding 20 years.

References

Adamson, D.M. and Pollington, T. (2006) *Change in the Construction Industry: An Account of the UK Construction Industry Reform Movement 1993–2003*, Routledge, Abingdon.

Allen, D. (1984) Towards the client's objectives, in Quality and Profit in Building Design, Brandon, P.S. and Powell, P.A. (eds), Spon, London, pp. 327–338.

Alvesson, M. and Deetz, S. (2000) *Doing Critical Management Research*, Sage, London.

Alvesson, M. and Willmott, H. (eds) (1992) *Critical Management Studies*, Sage, London.

Anderson, S., Cavanagh, J., Collins, C, Hartman, C. and Yeskel, F. (2000) *Executive Excess 2000*, Institute for Policy Studies, Washington, USA.

Antony, P.D. (1994) *Managing Culture*, Open University Press, Milton Keynes.

Argyris, C. and Schön, D. (1978) *Organisational Learning: A Theory of Action Perspective*, Addison-Wesley, Reading, Mass.

Barrett, P. and Stanley, C. (1999) *Better Construction Briefing*, Blackwell Science, Oxford.

Beer, S. (1981) *Brain of the Firm*, John Wiley & Sons, Chichester.

Bennett, J. (1985) *Construction Project Management*, Butterworths, London.

Blyth and Worthington (2001) *Managing the Brief for Better Design*, Spon Press, London.

Bresnen, M. and Marshall, N. (2001) Understanding the diffusion and application of new management ideas, *Engineering, Construction and Architectural Management*, **8**(5/6), 335–345.

Burns, T. and Stalker, G.M. (1961) *The Management of Innovation*, Tavistock, London.

Cairns, G. (2008) Advocating an ambivalent approach to theorizing the built environment. *Building Research & Information*, **36**(3), 280–289.

Cherns, A.B. and Bryant, D.T. (1984) Studying the client's role in construction, *Construction Management and Economics*, **2**, 177–184.

Chia, R. (1995) From modern to postmodern organisational analysis *Organization Studies*, **16**: 4, 579–604.

Clegg, S. (1989) *Frameworks of Power*, Sage, London.

Construction Industry Board (1996) *Educating the Professional Team*, Thomas Telford, London.

Csikszentmihalyi, M (1996) *Creativity*, HarperCollins, UK

Davenport, T. and Prusak, L. (2000) Working Knowledge, Harvard Business School Press, Boston, Mass.

Davies, R. (2006) Talking about construction innovation. *In:* Boyd, D. (Ed.), *22nd Annual ARCOM Conference*, 4–6 September 2006 Birmingham, UK. Association of Researchers in Construction Management, Vol. 2, 771–80.

Dawson, P. and Andriopoulos, C. (2008) *Managing Change, Creativity and Innovation*, Sage, UK

De Bono, E. (1990) *Lateral Thinking: a Textbook of Creativity*, Penguin, UK.

Easterby-Smith, M. and Lyles, M.A. (eds) (2005) *Handbook of Organizational Learning and Knowledge Management*, Blackwell, Oxford.

Egan, Sir John. (1998) *Rethinking Construction*. Report of the Construction Task Force to the Deputy Prime Minister, John Prescott, on the scope for improving the quality and efficiency of UK construction. Department of the Environment, Transport and the Regions, London.

Eldridge, J.E.T. and Crombie, A.D. (1974) *A Sociology of Organizations*, George Allen and Unwin, London.

Fernie, S., Leiringer, R. and Thorpe, T. (2006) Change in construction: a critical perspective. *Building Research and Information*, **34**(2), 91–103.

Flood, R.L. and Jackson, M.C. (1991) *Creative Problem Solving: Total Systems Intervention*, John Wiley & Sons, Chichester.

Flynn, R. (2002) Managerialism, professionalism and quasi-markets, in M. Exworthy and S. Halford (eds.) *Professionals and the New Managerialism in the Public Sector*, Open University Press, Buckingham, pp. 18–36.

Fournier, V. and Grey, C. (2000) At the critical moment: conditions and prospects for critical management studies, *Human Relations*, **53**(1), 7–32.

Gibbons, M., Limoges, C., Nowotny, H., Schwartzmann, S., Scott, P. and Trow, M. (1994). *The New Production of Knowledge: The Dynamics of Science and Research in Contemporary Society*. London: Sage.

Goodacre, P., Pain J., Murray, J. and Noble, M. (1982) *Research in Building Design*, Occasional Paper No. 7, Department of Construction Management, University of Reading.

Gray, C. and Davies, R.J. (2007) Perspectives on experiences of innovation: the development of an assessment methodology appropriate to construction project organizations *Construction Management and Economics*, **25**(12), 1251–68.

Green, S.D. (2006) Discourse and fashion in supply chain management, in *The Management of Complex Projects: A Relationship Approach*, (eds S. Pryke and H. Smyth), Blackwell, Oxford, pp. 236–250.

Green, S.D. (2009) The evolution of corporate social responsibility in construction: defining the parameters, in *Corporate Social Responsibility in Construction*, (eds. M. Murray and A. Dainty), Taylor & Francis, Abingdon, pp. 24–53.

Green, S.D. (1996) A metaphorical analysis of client organisations and the briefing process, *Construction Management and Economics*, **14**(2), 155–164.

Green, S.D., Larsen, G.D. and Kao, C.C. (2008) Competitive strategy revisited: contested concepts and dynamic capabilities, *Construction Management and Economics*, **26**(1), 63–78.

Green, S.D. and Schweber, L. (2008) Theorising in the context of professional practice: the case for middle range theories, *Building Research & Information*, **36**(6), 649–654.

Grey, C. and Willmott, H. (2005) *Critical Management Studies*, Oxford University Press, Oxford.

Janis, I.L. (1972) *Victims of Groupthink*, Houghton Mifflin, Boston.

Kao, C. and Green, S.D. (2002) The briefing process: a knowledge management perspective, in *Value Through Design* (eds C. Gray and M. Prins), CIB, Rotterdam, pp. 81–92.

Kast, F.E. and Rosenzweig, J.E. (1985) *Organization and Management: a Systems and Contingency Approach*, 4th edn., McGraw-Hill, New York.

Kelly, J., Male, S. and Drummond, G. (2004) *Value Management of Construction Projects*, Blackwell, Oxford.

Keys, P. (1991) Operational research in organisations: a metaphorical analysis, *Journal of the Operational Research Society*, **42**(6), 435–446.

Koskela, L. (2008) Editorial: Is a theory of the built environment needed? *Building Research & Information*, **36**(3), 211–215.

Latham, Sir Michael (1994) *Constructing the Team*. Final report of the Government/ industry review of procurement and contractual arrangements in the UK construction industry. HMSO, London.

Lawrence, P.R. and Lorsch, J.W. (1967) *Organization and Environment*, Harvard Press, Cambridge, Mass.

Legge, K. (1994) Managing culture: fact or fiction, in Sisson, K. (ed.) *Personnel Management: A Comprehensive Guide to Theory and Practice in Britain*, Oxford, Blackwell, pp. 397–433.

Linehan, C. and Kavanagh, D. (2006) From project ontologies to communities of virtue, in *Making Projects Critical*, (eds. D. Hodgson and S. Cicmil), Palgrave Macmillan, pp. 51–67.

Machin, S. (1996) Wage inequality in the UK, *Oxford Review of Economic Policy*, **12**(1) 47–64.

McCourt, W. (1997) Using metaphors to understanding and to change organizations: a critique of Gareth Morgan's approach, *Organisation Studies*, **18**(3), 511–522.

McGregor, D. (1960) *The Human Side of Enterprise*, McGraw-Hill, New York.

Morgan, G. (2006) *Images of Organization* (updated edn), Sage, Thousand Oaks, CA.

Morgan, G. (1993) Imaginazation: The Art of Creative Management, Sage, Newbury Park, CA.

Murray, J.P., Gameson, R. and Hudson J. (1993) Creating decision support systems, in *Professional Practice in Facility Programming*, Preiser, W.F.E. (ed), Van Nostrand Reinhold, New York, pp. 427–441.

National Equality Panel (2010) *An Anatomy of Economic Inequality in the UK*, Government Equalities Office, London.

NEDO (1988) *Faster Building for Commerce*, HMSO, London.

Nonaka, I. and Takeuchi, H. (1995) *The Knowledge-Creating Company*, OUP, New York.

Ogbonna, E. and Harris, L.C. (2002). Organizational Culture: A Ten Year, Two-phase study of change in the UK Food Retailing Sector. *Journal of Management Studies*, **39**(5), 673–706.

Ortony, A. (1993) Metaphor, language and thought, in A. Ortony (ed.) *Metaphor and Thought*, 2nd edn., Cambridge University Press, Cambridge, pp. 1–17.

Peña, W., Parshall, S. and Kelly, K. (1987) *Problem Seeking: an Architectural Primer*, AIA Press, New York.

Perkman, M. and Walsh, K. (2007) University-industry relationships and open innovation: towards a research agenda, *International Journal of Management Reviews*, **9**(4), 259–280.

Peters, T.J. and Waterman, R.H. (1982) *In Search of Excellence: Lessons from America's Best-Run Companies*, Harper & Row, New York.

Pettigrew, A.M. (2003) Strategy as process, power, and change, in Cummings, S. and Wilson, D. (eds) *Images of Strategy*, Blackwell Publishing, **pp.** 301–330.

Pfeffer, J. (1981) *Power in Organizations*, Pitman, Marshfield, Mass.

Raberneck, A. (2008) A sketch-plan for construction of built environment theory. *Building Research & Information*, **36**(3), 269–279.

Salisbury F. (1998) *Briefing Your Architect*, 2nd edn., Architectural Press, Oxford.

Schön, D.A. (1983) The Reflective Practitioner: How Professionals Think in Action, Temple Smith, London.

Senge, P. (1990) *The Fifth Discipline: The Art and Practice of the Learning Organization*, Doubleday, New York.

Simon, H. (1947) *Administrative Behavior*, Macmillan, New York.

Taylor, F.W. (1911) *Principles of Scientific Management*, Harper & Row, New York.

Townsend, R. (1970) *Up the Organization*, Michael Joseph, London.

von Bertalanffy, L. (1968) *General Systems Theory: Foundations, Development, Applications*, Braziller, New York.

Weick, K.E. (1995) *Sensemaking in Organizations*, Sage, Thousand Oaks, CA.

Wilkins, A.L. and Ouchi, W.G. (1983) Efficient cultures: exploring the relationship between culture and organisational performance, *Administrative Science Quarterly*, 468–481.

Willmott, H. (1993) "Strength is ignorance: slavery is freedom": managing culture in modern organisations, *Journal of Management Studies*, **30**(4), 515–552.

7 From Business Process Re-Engineering to Partnering

7.1 Introduction

The preceding chapter highlighted the benefits of understanding client organisations from a range of different metaphorical perspectives. Particular note was made of the recurring popularity of the machine metaphor among the advocates of construction sector improvement. It was also emphasised that a metaphorical interpretation was only justified once the literal meaning has been shown to be flawed. This chapter sets out to evaluate the merits of two of the most influential improvement recipes of the 1990s: business process re-engineering (BPR) and partnering. Each of these will be critiqued in terms of their substantive content prior to exploring other explanations of why they became so persuasive. *Constructing the Team* (Latham, 1994) and *Rethinking Construction* (Egan, 1998) had both recommended the wider use of partnering, and the latter drew much of its language and imagery from the rhetoric of BPR. One of the core arguments of *Rethinking Construction* was that activities which did not add value to the client's businesses processes comprised waste, and should hence be eliminated. This was an argument shaped inexorably by the prevailing popularity of BPR.

At first sight, BPR and partnering would seem to offer very different approaches to construction sector improvement. While re-engineering is commonly associated with a ruthless focus on process efficiency, partnering is more often associated with an emphasis on relationships. The importance of relationships had of course been emphasised previously by Emmerson (1962) and Banwell (1964), but partnering entered the discourse of industry improvement primarily as a result of its endorsement in *Constructing the Team* (Latham, 1994). Despite their superficial differences it will be argued that BPR and partnering are underpinned by the same metaphors, and that they share similar paradoxes and contradictions. The reliance of BPR on an underlying machine metaphor is apparent from the very idea that organisations can be 're-engineered'. The accompanying storyline places great

Making Sense of Construction Improvement, First Edition. Stuart D. Green.
© 2011 Stuart D. Green. Published 2011 by Blackwell Publishing Ltd.

emphasis on the need to implement efficiency improvement irrespective of the opposition. The language of BPR invariably emphasises the need for radical change. In contrast, the language of partnering is much softer – at least on the surface. The overriding emphasis lies on working together to achieve mutual objectives. By definition, partnering was championed by 'enlightened' clients who had grown tired of adversarial relationships. However, as will become apparent, the industry's enlightened clients were not pursuing partnering for its own sake; they were interested in partnering because of what it was perceived to offer in terms of improved efficiency.

7.2   Business process re-engineering in construction

7.2.1   Enthusiastic endorsement

The enthusiasm for BPR was in part born from a widespread disenchantment with TQM and its focus on continuous incremental improvement (Fulop and Linstead, 1999). Certainly BPR replaced 'Total Quality Management' (TQM) as the hottest topic in the business press from 1993 onwards (De Cock and Hipkin, 1997). BPR advocated a much more radical approach to job re-design which clearly appealed to the mood of the time. Having decided that BPR was a 'good thing', its advocates seemingly limited their reading to those sources that justified their pre-assumption.

BPR was popularised originally by Hammer and Champy's (1993) phenomenally successful book *Re-engineering the Corporation*. In common with all the best management fashions, BPR was born in the United States. Hammer and Champy either judged the timing of their book very carefully, or they were very lucky. In any case, the rhetoric of BPR resonated perfectly with the mood of the time. Certainly there was no shortage of willing recipients within the UK construction industry; BPR was widely embraced as the solution to the industry's problems by practitioners, academics and policy makers alike. Here was seemingly a proven 'advanced management technique' which offered a radical solution to under-performing industries such as construction.

It has already been argued that the imagery and rhetoric of BPR influenced directly both *Progress through Partnership* and *Rethinking Construction*. Within the former, the endorsement of BPR could hardly have been more enthusiastic:

> '....re-engineering of basic business processes to provide "lean", rapid and effective performance is now commonplace throughout other industries but rarely occurs in construction' (OST, 1995)

The introduction of such 'advanced business techniques' in construction was further exhorted as a matter of urgency. At least within *Progress through*

Partnership, the emphasis on re-engineering was just one theme amongst several. However, the dominant argument within *Rethinking Construction* was entirely consistent with the construct of BPR, albeit freshened up with a sprinkling of lean thinking and sweetened with warm words about partnering. In short, the overriding message of *Rethinking Construction* was the need to eliminate waste through the elimination of activities which do not add value to the client. Of particular note was the unitary conceptualisation of the 'client' and the repeated reliance on simplistic machine metaphors. The advocated technical requirements for achieving the required improvements included a plea for a much greater application of 'business processes'. The storyline of BPR was further reflected in the call for a 'redefinition' of information needs and a 're-design' of organisational relationships. Even more striking was the specific recommendation that the UK research councils should fund multidisciplinary research into the application of improved business processes. But the outcome of any such research seemed already to have been predetermined:

> 'Business process analysis should be applied more vigorously to improve the efficiency and effectiveness of the construction industry.'
> (Egan, 1998)

The research was seemingly required to provide support for the approach which had already been decided upon. The UK Engineering and Physical Sciences Research Council (EPSRC) had in any case already committed itself to a 'business process' approach through its Innovative Manufacturing Initiative (IMI):

> 'Business processes may be viewed as those procedures, practices and methodologies that companies use in employing their assets to gain competitive edge in the translation of raw materials into finished products which satisfy the customer/consumer demand in the market place. This approach is used within the IMI to identify the manufacturing challenges and research priorities.... to improve industrial competitiveness.'
> (EPSRC, 1994)

The IMI programme accounted for a significant proportion of UK construction management research spending during the mid-1990s. The adoption of a 'business process approach' therefore became an essential requirement for university-based researchers if they were to be successful in their applications for funding. Ever resourceful, they quickly became skilled in presenting their research applications in the required rhetoric.

Not only was the efficacy of BPR taken for granted by those responsible for funding research, the same was also true for the harbingers of change who populated the networks of industry improvement. The 'best practice' reports published by Construct IT (1996, 1998) drew expertise from both industry and academia, but the published output was striking in its uncritical acceptance of BPR. Likewise, the sober façade of the Construction Industry Board (1996) did not prevent it from recommending that BPR should form part of the industry's ongoing programme of continuing professional development. Such tendencies were by no means limited to the United Kingdom, but were widespread throughout the English-speaking world. For example, the *Building for Growth* report (Commonwealth of Australia, 1999) cited process re-engineering as an essential element of process innovation. In the late 1990s it seemed that everyone was in favour of BPR. Only the un-reconstructed 'dinosaurs' were against.

7.2.2   Lessons from other sectors

The literature advocating BPR for the construction sector adhered to a recognised pattern. Almost without exception the starting assumption was that BPR had already been successful in other sectors. Mohamed and Tucker (1996) were typical in commencing their discussion with the claim that:

> '[m]any industries worldwide have found...BPR to be an effective approach in achieving dramatic improvements in production time and cost.'

Betts and Wood-Harper (1994) also looked towards BPR for new ideas without seeming to take care to question the claims made on its behalf. The overriding concern was how to apply Hammer and Champy's (1993) recipe of 'business process re-engineering' to the construction industry. It was taken for granted that BPR had already been successful in other industrial sectors that were supposedly more advanced in terms of their management thinking. Given the adopted frame of reference, the task was reduced to implementation. The challenge was to overcome the 'cultural barriers' to new ways of working; industry fragmentation was similarly seen to be a barrier which had to be overcome.

Other indicative sources within the construction management literature included *Construction Management: New Directions* (McGeorge and Palmer, 1997) which suggested that 're-engineering has the power to change the very structure and culture of the industry'. In fairness, McGeorge and Palmer (2002) were notably much more circumspect in the book's second edition, once their initial enthusiasm had abated. The early enthusiasm for BPR amongst construction management academics is well illustrated by the proceedings of the 1997 conference on *Construction Process Re-Engineering* (Mohamed, 1997). A review of the references used in the 60 papers is especially revealing; sources such as Davenport and Short (1990), Hammer (1990) and Hammer and Champy (1993) were all cited extensively. In contrast, there was an almost total absence of reference to the extensive critical literature which challenged the assumption that BPR had already been successful in other sectors. For example, sources such Grint (1994) and Micklethwait and Woodridge (1997) were already criticising BPR for its regressive tendencies; both sources also cite a failure rate of 70% for BPR implementation projects when measured against their own objectives. But this was not a message which the advocates of BPR for the construction sector wanted to hear. That others elsewhere were struggling to implement re-engineering did not fit with the required rhetorical argument that the construction industry was backward and behind the times in its management thinking.

7.2.3 *Vagueness of definition*

Despite its widespread rhetorical appeal, the essence of BPR remains notoriously difficult to define. The most commonly quoted definition of BPR is probably that of Hammer and Champy (1993):

> '... the fundamental rethinking and radical redesign of business processes to achieve dramatic improvements in critical contemporary measures of performance, such as cost quality and speed.'

It is therefore not difficult to see where the authors of *Rethinking Construction* drew their inspiration from. The quote above fits in perfectly with the core arguments mobilised by Sir John Egan's Construction Task Force. Enthusiastic support was also pledged for partnering and lean thinking, but as shall by demonstrated these ideas were tailored and shaped to fit the required agenda of meeting clients' requirements with maximum efficiency. There was little recognition of the needs of broader interest groups; certainly there was no mention of sustainability or of the needs of an increasingly casualised workforce. Sustainability of course was not recognised because this did not fit with the remorseless focus on satisfying narrowly-construed 'client requirements'.

The plea within *Rethinking Construction* to apply 'business process analysis' evermore vigorously is indicative of a recurring storyline within the BPR literature; if the medicine is not working it is because it is not being applied with sufficient enthusiasm. Conveniently, the advocates of BPR have an extensive list of reasons to explain why particular implementations have not been successful. Typical examples include: (i) poor implementation; (ii) lack of top management commitment and support; and (iii) resistance to change by subordinates. Hammer and Champy (1993; 107) were themselves especially clear that 'most re-engineering failures stem from breakdowns in leadership'. So if a BPR initiative in the construction sector ends in failure, it is no use blaming Hammer and Champey; it is the fault of weak management within the construction sector for not implementing BPR with sufficient conviction. And if all else fails, it can always be claimed that the 'so-called BPR failure were not "real" BPR programmes' (cf. Miers,

1994). Such claims were to become subsequently very common in respect of partnering; if the adopted approach did not work it could not have been 'real' partnering. But in the case of BPR, the literature contains so many inconsistencies and contradictions it is difficult to be sure what the key principles are.

The preceding argument points towards a deep-seated paradox which lies at the heart of BPR. Despite the prescriptive nature of the rhetoric, there remains an essential vagueness over precisely what BPR comprises and how it should be implemented. Such a characteristic is by no means unusual amongst 'new' management approaches, which often lack proper codification in terms of a commonly held set of principles and practices (Bresnen and Marshall, 2001). In the case of BPR, Jones (1995) argues that the terminology is so vague and imprecise that it is impossible to distinguish BPR from other management improvement recipes. The conflation of BPR, lean thinking and partnering, within the pages of *Rethinking Construction* suggests that they comprise constituent parts of a broader discourse. Empirical studies of the implementation of supposedly different management techniques often highlight a similar conflation of ideas. For example, De Cock and Hipkin (1997) compare the implementation of BPR to TQM and conclude that the concepts can only really be distinguished in terms of the rhetoric in which there are presented. This shifts the emphasis away from seeking to understand such management techniques in terms of their substantive content towards understanding them in terms of their implicit metaphors. Whereas TQM tends to be dressed up in the humanistic language of teamwork, BPR more often relies on violent metaphors such as 'breaking the china'. Despite its elusiveness of definition, BPR rapidly earned a reputation for its 'slash and burn' approach (Buchanan, 2000). An extensive critical literature very quickly emerged which associated BPR with regressive approaches to human resource management (HRM) (e.g. Grey and Mitev, 1995; Grint and Willcocks, 1995; Willmott, 1995; Mumford and Hendricks, 1996). Indeed, beyond the frequently violent rhetoric, the only aspect of BPR that appeared to be constant was the quest to secure greater output from fewer employees. This of course was of no great concern to the Eganites in the construction sector who were able to point towards the *Respect for People* initiative as proof of their commitment to enlightened HRM practices.

Hammer and Champy's (1993) claim to provide 'a radical new beginning' has also been challenged repeatedly in the literature. Grint (1994) examines the central arguments of BPR and concludes that few, if any, are innovations, let alone radical innovations. Several other authors have suggested that BPR represents nothing other than a return to the simplistic metaphors of Taylorism (Conti and Warner, 1994; Wood *et al.*, 1995; Broekstra, 1996). From this point of view, the supposedly-defining principles of BPR are no different from Taylor's (1911) concept of scientific management. Even some of the consultants who enacted BPR were not convinced by Hammer and Champy's (1993) claim to provide a 'radical new beginning'. For example, Johansson *et al.* (1993) readily conceded that the origins of the techniques of process mapping lay within the methods of work study developed initially by Frederick Taylor (1911). Rather than represent a radically new way of thinking, BPR arguably represented a return to a previous era dominated by scientific management. But of course many construction industry leaders retain a deep-rooted commitment to mechanistic management recipes. This perhaps is why BPR-type storylines continue to be popular amongst managers and policy makers in the construction sector; they are popular not because they offer something new, but because they reflect and reinforce what industry leaders already think. The expression 'old wine in new bottles' is something of a cliché, but it nevertheless conveys effectively the idea that BPR comprised essentially old ideas dressed up as something which was new.

7.2.4 Shoot the dissenters

There is much reason to suppose that the 'radical and dramatic' nature of the Egan agenda drew its inspiration from the rhetoric of BPR. Hammer and Champy (1993) were especially notable for their use of emotive terms such as 'fundamental', 'radical' and 'dramatic'. Certainly the advocates of BPR promoted a much more radical approach to organisational change than the advocates of TQM. Part of the remit of BPR was to challenge the very need for particular processes which, at least according to Hammer and Champy (1993), may exist for no other reason than historical precedent. Gentle notions of incremental improvement, as advocated previously by TQM, seemingly belonged to a previous age. In the words of Davenport (1993):

> 'Today firms must seek not fractional, but multiplicative levels of improvement – 10x rather 10.'

Not only was the language and imagery of BPR often emotive, it was also frequently quite violent. An infamous interview with *Forbes Magazine* (Kalgaard, 1993) accredits Michael Hammer with the expression 'shoot the dissenters'. Hammer and Champy (1993) were consistently clear that change 'must come from the top' and that BPR is something that must be imposed on a reluctant workforce. They further recognised that middle managers may actively resist change, and that:

> '[o]nly strong leadership from above will induce these people to accept the transformations that reengineering brings.' (Hammer and Champy, 1993; pp 207–8)

Throughout the BPR literature, there is a strong recurrent theme that senior management must make an unwavering commitment to radical change.

Furthermore, senior management must set ambitious goals and the re-engineering process must be initiated from the top. Frequent references are also made in favour of 'culture change', although it would seem to be behavioural compliance which is valued above innovation. Hammer and Champy (1993) also refer to re-designing work into self-managed teams and shifting managerial accountability to the front line. But the teamwork on offer would seem to be constrained within parameters which are dictated from above. Teams are allowed to be 'self-managed' provided that they display the appropriate attitude in terms of supporting the top-down prescription

Such arguments understandably played out well in the construction sector where middle-managers have long since been accredited with 'adversarial attitudes'. The repeated calls for 'culture change' also directly reflect the lexicon of BPR. Elsewhere, Champy (1996) was remarkably candid:

> '[c]apitalism is a system that quite literally works on fear ... the only way to persuade many folks to undertake a painful therapy like reengineering ... is to persuade them that the alternative will be even more painful.' (Champy, 1996; 49)

It is of course possible to share Champy's (1996) faith in capitalism without sharing his belief that it relies on fear to make it work. There are many variants of capitalism, but most enlightened thinkers would argue that any successful system should rest on aspiration rather than fear. Unfortunately, re-engineering programmes all too often tended to be advocated without regard for personal or social costs (Grey and Mitrev, 1995). In this respect they were the antithesis of enlightened HRM. Certainly there is no room for satisfying the needs of multiple stakeholders, neither is there any room for predicating notions of sustainability over the short-term imperatives of economic efficiency. This was undoubtedly the enterprise culture in full flight; this was Gordon Gekko incarnate. The use of the term 're-engineering' in itself invokes an image that organisations can be treated as machines in need of a 'technical fix'. In essence, BPR encapsulates a top-down approach to re-structuring organisations that concentrates on technical efficiency in isolation from the social aspects of organisations (Tinaiker et al., 1995). The overriding assumption is that organisations are unitary and that corporate objectives can be determined in isolation from the aspirations of the employees. In others words, top management knows what is best and everybody else should just do as they are told. Certainly Hammer and Champy (1993) were clear that BPR did not make people happy and that some employees would inevitably lose their jobs. Despite rather half-hearted attempts to distance themselves from 'downsizing', it remains clear that a key characteristic of BPR involves the rationalisation of jobs (Fulop and Linstead, 1999). The hierarchy of the organisation is flattened as middle-mangers give way to self-managed teams. This was a storyline which was especially persuasive in the construction sector of the mid-1990s. It was also a storyline which resonated with the structural changes which managers observed taking place around them.

Alvesson and Willmott (1996) did much to pioneer the field of critical management studies (CMS), and from their perspective re-engineering equates to technocratic totalitarianism. They argue that BPR provides one of the means by which top management seek to optimise organisational performance by the imposition of supposed 'scientific' solutions. This is once again reminiscent of exhortations in support of the Egan agenda in the construction sector, whereby no space was preserved for any alternative prescription. Such an absolute insistence on conformity, coupled with the emotive imagery of 'shooting the dissenters', certainly sits ill at ease with the *Respect for People* initiative. It is of course not necessary to be a devotee of CMS to feel uncomfortable with the way in which the Egan agenda was imposed on the construction sector. For the more critically aware observer, there would seem to be an inherent contradiction between notions of 'innovation', on the one hand, and 'command and control', on the other. All too often employees are empowered only to adopt courses of action that fit in with the 'one best way' advocated imposed by top management. Given the dominant competitive recipe of structural flexibility, the construction firms of the late 1990s had no real need for employees teeming with innovative ideas; this was not the basis upon which they sought to achieve competitive advantage. But as we shall see, the assumption that BPR leads necessarily to bad outcomes is ultimately as fallacious as the assumption that it leads inevitably to good outcomes.

7.3   The persuasive appeal of re-engineering

7.3.1   *Management gurus*

Given the doubts expressed above regarding the originality and coherence of BPR, it is pertinent to ask why it became so influential amongst practitioners. Part of the answer is to be found within the literature on management fashion, which will provide a recurring touchstone throughout the remainder of this book. What seems especially clear is that the core messages of BPR captured the attention of practising managers during the mid-1990s for reasons over and above the alleged originality of the message. *Re-engineering the Corporation* (Hammer and Champy, 1993) was reputedly the best selling management book of the 1990s; Michael Hammer is reported to have charged up to US$50 000 for a one-day seminar. The track record of BPR as an improvement recipe may be questionable, but its success as a management fashion cannot be denied. BPR is frequently held up as an exemplar of a management fashion, and Michael Hammer is similarly often cited as the epitome of a management guru. It is clear than such gurus undoubtedly play an important role in the propagation of fashionable management recipes. In seeking to understand why BPR was so persuasive during the mid-to-latter 1990s, it is useful to focus on the rhetorical persuasiveness of the arguments which were mobilised. It has already been demonstrated that the popularity of BPR cannot be explained by the originality or coherence of the constituent ideas.

7.3.2 Playing on insecurity

What cannot be denied is that the storylines of BPR found many receptive ears in the mid-1990s. Part of the explanation lies in the way the propagated storylines providing practising managers with a justifying narrative for the changes that they observed taking place around them. Of particular signifi-cance was the way in which BPR offered managers a role other than that of passive victim. Huczynski (1993) was one of the earlier writers to investigate the phenomenon of the management guru, who he saw as playing deliber-ately on the uncertainties and self-doubts of practising managers. In Huczynski's view, management gurus offer simplistic and easily digestible models which are seized upon by insecure managers who feel increasingly overwhelmed by complexity. While managers may be cynical of what they sometimes perceive to be the latest fad, they can feel obliged to pursue them just in case they might work. Jackson (1996) went further in comparing the rhetorical devices employed by management gurus such as Michael Hammer, James Champy and Tom Peters to those used by evangelist preachers. The gurus are seen to use fear to attract the attention of managers in the same way that an evangelist preacher might typically begin a sermon by invoking images of hellfire and eternal damnation. The evangelists deliberately create an initial sense of anxiety so that their audience will then be more susceptible to their subsequent message which offers a 'path to salvation'. According to Jackson (1996) the gurus of BPR in the mid-1990s used the same approach. They gained the attention of their audience by claiming that re-engineering is a survival issue. Re-engineering is therefore presented as an economic imperative. The only choice offered is between BPR and corporate extinc-tion; in other words, no choice at all. The prime candidates for downsizing were invariably middle managers. The underlying metaphor was one of 'cutting out the fat' to make the organisation more lean. The implied threat was that if middle managers did not themselves become the proponents of BPR, then they would become its victims. In simple terms the message was 'do it to others before they do it to you'. The ambitious young middle man-ager hence volunteered to become the BPR champion, and the worldly-wise and cynical very quickly became the barriers that had to be overcome.

7.3.3 Playing on patriotism

Jackson (1996) further observed the way in which the rhetoric of Hammer and Champy appealed to the patriotism of American managers. During the early 1990s the United States was finally emerging from its long post-Vietnam hangover, and re-engineering was presented as part of the American dream. Rather than seeking to change the intuitive practices of American managers, the power of BPR was seen to reside it the way it 'takes advantage of American talents and unleashes American ingenuity' (Hammer and Champy, 1993). The subtext was that American productivity was in some way hindered by external parties who are by inference 'un-American'. Therefore BPR was not only a survival issue for individual firms, it was also a survival issue for American industry as a whole. The promotion of BPR thus became a patriotic duty, and the dissenters risked being labelled as wreckers who were holding back the emergence of an espoused new and vibrant America.

The increasingly globalised nature of business in the 1990s meant that once management ideas became established in the United States, they very quickly became common currency internationally. Countries such as Germany and France were arguably more resistant to American management gurus than the Anglo-Saxon world, but their resistance was at best a rearguard action. Airport bookshops throughout the world display increasingly the same prescriptive textbooks preaching identical recipes, irrespective of local customs and culture. And the vast majority of these managerial improvement recipes continue to be sourced from within the United States. The ideas of lean production may have been derived superficially from the Toyota manufacturing system as initially developed in Japan, but the popularised version promoted by Womack *et al.* (1990) was in essence an Americanised re-interpretation (see the discussion in Chapter 8).

7.3.4 *Playing on historical significance*

A further persuasive technique identified by Jackson (1996) was the way in which the advocates of BPR dramatised its historical significance. BPR was equated directly with modernity on the basis that it represented a decisive break with previous management theory. As alluded to above, particular care was taken to explain why BPR was radically different from TQM. The language used by Hammer and Champy (1993) was the language of revolution; BPR was widely promoted as a 'paradigm shift' in management thinking. Johansson *et al.* (1993) invoked a similar metaphor by referring to the need to 'break the china' before embarking upon new ways of working. A recurring theme was the way in which BPR sought to overturn the outdated functional divisions that allegedly emerged from the Industrial Revolution. Much blame was laid repeatedly at the door of Adam Smith. Continued adherence to 'old thinking' was promoted as the route to mediocrity and economic obsolescence. To resist BPR was to be out of date, and those unwilling to sign up to its implementation were labelled as 'dinosaurs'. The metaphor portrays those who are unable to adapt to changing circumstances as being doomed to extinction.

7.3.5 *Playing on political resonance*

Perhaps the most convincing explanation for the popularity of BPR in the mid-1990s was the way in which the core messages reflected the social and political ethos of the day (cf. Grint, 1994). Few would now deny that the legitimacy of BPR amongst practising managers was derived at least in part from the broader discourse of the enterprise culture, as championed by Thatcher's Conservative government in the United Kingdom and the Reagan administration in the United States. Political discourse throughout this time emphasised the need to be competitive in the global economy. Such narratives provided both a supportive context and a source of legitimisation for management improvement recipes based on 'cutting out the fat'. The espoused policies sought to extend the domain of the free market throughout the economy in the cause of competition. The shift to the political right denied legitimacy for the continued state support of 'lame ducks'. National economies and individual firms were increasingly obliged to compete in the marketplace. Factory closures and rising unemployment were considered a price worth paying for improved competitiveness. Popular management discourse on both sides of the Atlantic reflected and reinforced this new emphasis on the 'survival of the fittest'. The diktats of BPR therefore resonated with the mood of the times. Fulop and Linstead (1999) contend that three key principles drove BPR: customers, competition and change; all of which were central to the broader discourse of the 'enterprise culture', and none of which were unique to BPR. There are therefore strong arguments in support of the claim that BPR is inseparable from the discourse of the enterprise culture. Such an argument stands in sharp contrast to the way in which BPR was presented as a 'cunningly clever management technique' in construction management textbooks such as McGeorge and Palmer (1997).

7.4 Information technology and process improvement

7.4.1 *Enabling technologies*

As persuasive as the preceding diagnosis is, it does not constitute a complete explanation of the popularity of BPR. However, it does emphasise the need to understand the evolution and diffusion of management ideas in a broader context. The focus on the persuasiveness of the rhetoric of BPR provides an antidote to the instrumental rationality which all too often frames the discussion of management techniques. BPR must undoubtedly be understood as a part of broader discourses of economic and political change. However, BPR also rode on the wave of the technological change. During the 1990s developments in *information technology* (IT), latterly christened *communications and information technology* (CIT), undeniably provided firms with ways of working in ways that previously did not exist. Indeed, so-called *enabling technologies* did indeed provide the basis for a radical re-think of working practices that few could afford to resist. Hammer and Champy (1993) referred to *disruptive technologies* that were capable of challenging fixed pre-conceptions about the way processes were performed. Repeated references are made to IT as a 'critical enabler' that allows organisations to do work in 'radically' different ways. Examples of the so-called disruptive technologies included shared databases, telecommunications, wireless data communications and portable computers. Few could argue that such technologies have not had a fundamental impact on the workplace, arguably beyond that envisaged by Hammer and Champy (1993). Perhaps of even greater impact on our day-to-day working lives have been e-mail and the internet. Such material changes in working practices cannot simply be dismissed as management hype. They have had a dramatic impact on the workplace and few firms (or individuals) have been able to resist their encroachment. The implementation of these technologies may have disadvantaged some in the short term, and they may even have caused a number of redundancies. But few would argue that their overall impact has been negative. In many respects, such technologies have served to democratise organisations and de-centralise decision-making. In directing critical attention at the over-hyped rhetoric of BPR, it is therefore important not to adopt a Luddite perspective regarding the introduction of new technology. But to advocate the benefits of new technology is one thing; to adopt the heady rhetoric of BPR is another.

7.4.2 Focus on processes

As already noted, the literature on BPR is riddled with contradictions, and for every source that emphasises the importance of IT there is another that argues that it is the focus on *processes* which is of primary importance. Of particular importance is the focus on *core* processes, although there is little recognition that there may be disagreement about which processes are core and which are non-core. Johanssson *et al.* (1993) emphasise that a core business process creates value by the capabilities it gives the company for competitiveness. They further suggest that there are usually between five and eight core business processes in any industry group, although the evidence for this diagnosis is at best unclear. BPR clearly rests on a unitary model of organisations. It is taken for granted that the objectives of the organisation are clear and shared by everybody, or at least if such a situation does not exist, that it can be brought about through a 'cultural change programme'. According to Hammer and Champy (1993) a process can be defined as 'a set of activities, that taken together, produce a result of value to the customer'. Processes are therefore evaluated on the extent to which they 'add value' for customers, thereby reflecting and reinforcing the 'cult of the customer'. Unfortunately, in the context of construction projects, the notion of customer responsiveness is not quite as straightforward as it may be for manufacturing firms. The advocates of BPR invariably adopt a much narrower conceptualisation of the 'customer' than those which are normally mobilised in the construction sector. Construction professionals have long held themselves to have broader responsibilities over and above an allegiance to those who pay their fees. In consequence, the advocates of BPR tend to equate professionals with old-fashioned vested interests which have to be overcome (see below).

Even if the conceptualisation of the 'customer' is limited to a single client organisation, there is still little reason to assume that there is any clear understanding on what constitutes 'added value'. The briefing literature makes it quite clear that the process of identifying client requirements is by no means straightforward. Client organisations very often comprise several different interest groups, each of which may have a different interpretation of what constitutes 'good value'. BPR reserves little space for such political metaphors of organisation, neither does it recognise the necessity for organisations to evolve over time in response to changing circumstances. The dominant metaphor within the BPR literature is that of the machine, and the prescribed task is to improve efficiency through the elimination of waste.

The unitary model of organisations which underpins BPR has significant implications for the advice on how processes should be mapped. The advice which is offered invariably follows a prescriptive number of stages, and the resultant diagrams are often reminiscent of the work-flow diagrams of scientific management (cf. Taylor, 1911). Certainly neither the focus on process, nor the associated process mapping techniques, would seem to constitute little that is new once stripped of the associated rhetoric. The overall sense is that analysts are required to 'drill down' to identify the really important processes that are of key importance to maintaining competitiveness. A popular alternative metaphor is the need to 'cut through the fog' in order to see what is really happening. There is of course something of a circular argument here. BPR depends upon 'clear analytical skills' for its implementation, and yet it is the techniques of BPR that supposedly provide the required analytical insight. What is clear is that BPR cannot be implemented by anyone. It plainly cannot be enacted by wreckers, dissenters or dinosaurs. The implementation of BPR requires somebody who is in tune with the diktats of the enterprise culture. In the UK construction sector of the late 1990s the most important qualification for those wishing to implement BPR was to be aligned with the Egan agenda. Apparently, only the Eganites had the rhetorical legitimacy to implement change. But in truth, the most significant changes had already happened as the industry became increasingly lean and agile in the cause of structural flexibility. Perhaps what BPR offered was a means of making sense of changes that were already taken place. BPR was but one part of the broader discourse of enterprise which included the re-structuring of the construction sector coupled with the emergence of the hollowed-out firm. The rousing arguments of *Rethinking Construction* are best understood as a constituent part of the same discourse. Thus, it becomes meaningless to debate the 'impact' of the Egan report in isolation from the pre-existing dynamics of industry change.

7.4.3   Generic design and construction process protocol

The popularity of the 'process approach' within the research community during the late 1990s is exemplified by the so-called *Generic Design and Construction Process Protocol* (GDCPP) (Kagioglou *et al.*, 2000). The development of the GDCPP was funded through the EPSRC *Innovative Manufacturing Initiative* (IMI) industry-led research initiative. Hence the adoption of a business process approach was an essential prerequisite to the receipt of funding. The justification for the advocated approach commences by rehearsing the oft repeated arguments that the UK construction industry is 'plagued by a number of problems, which have not disappeared in the last few decades'. Strangely, no mention is made of the extensive structural changes which re-shaped the construction sector over the preceding 20 years (see Chapter 3). The underlying assumption would seem to be that the construction sector is a static and fragmented entity which is waiting to be improved through the implementation of modern management techniques. Great emphasis is placed on the lessons that could be learned from manufacturing:

> '…a number of very effective philosophies and practices such as Just in Time (JIT), lean production and others have a legacy of optimised production in the manufacturing sector.' (Kagioglou *et al.*, 2000; 142)

Characteristically, little is offered in the way of evidence to support the claim above. That construction had much to learn from manufacturing was at the time axiomatic. The principle of producing models of the design and construction process was already well established and dates back at least as far as the RIBA Plan of Work (RIBA, 1963). Other influences which were cited include Walker (1989) and Hughes, (1991). The adopted approach to process mapping within GDCPP was IDEF-0, as developed originally by the United States Air Force in the 1970s. Process mapping was at the time hugely popular amongst IT consultants, and had previously been cited as an 'indispensible tool in business process reengineering' (Johansson *et al.*, 1993; 209). The origins of IDEF-0 can be traced back to the tradition of work study as developed originally by Taylor (1911).

Work study had previously been exceedingly popular within the British nationalised industries of the 1950s. In this respect IDEF-0 is a strange companion of the enterprise culture. But as has already been outlined, there was little substantive content within BPR that was new. The developers of the

GDCPP found IDEF-0 to be effective for modelling 'as-is' processes, but it was apparently less useful for the modelling of 'will-be' processes. Key principles of the adopted approach included the need to take a 'whole project view' together with the need for a consistent process. Such principles are by no means unreasonable, and in many respects would seem to accord with common sense. More intriguing was the notion of 'progressive design fixity' whereby phase reviews are conducted at the end of each stage. The classification of 'soft gates' and 'hard gates' was applied to enable the progressive fixing of decision decisions throughout the process. This is a useful way of thinking which can be of practical benefit in imposing a degree of control on complex design processes. However, it is notable that the overriding aim is to achieve client sign-off on design elements as early as possible (Cooper *et al.*, 2005). From a production perspective, this of course makes perfect sense. But others with a more organic interpretation of client briefing would tend to emphasise the importance of keeping design options open until the 'last responsible moment' (cf. Blyth and Worthington, 2001). This alternative view argues that design decisions are invariably made in conditions of uncertainty, and hence there is little point in making decisions any earlier than necessary. The latter argument has become even more persuasive in recent years as the broader business environment has become evermore dynamic and uncertain. The antecedents of IDEF-0 were rooted in an age when the business environment was much more stable and predictable. Nevertheless, there is a danger in continually delaying decisions, or in making decisions without the appropriate degree of sign-off from the key project stakeholders. While the overall emphasis within GDCPP was about achieving clarity of structure in the cause of better control and coordination; it was also important in terms of its focus on the use of modern IT solutions.

The key role of coordination within GDCPP is achieved by a 'process manager' who is appointed by the client. In common with many other management improvement approaches, process management must have a 'champion' if it is to be successful. Other key principles of the GCDPP include stakeholder involvement and teamwork. Kagioglou *et al.* (2000) observe that complete project teams rarely work together on more than one project, thereby inhibiting team performance. They further refer to the tendency for key contributors to be included in the process too late. Such pleas in favour of early contractor involvement can be traced back to Banwell (1964) and Emmerson (1966). Kagioglou *et al.* (2000) refer to the way the lack of continuity within the team from project to project limits their ability to learn from experience. This is yet another recurring problem with a long heritage. Firms were later to mobilise 'knowledge management' as a potential solution, although this too was to prove somewhat elusive. What is interesting is the way in which the rhetoric changes and yet the key ideas for industry improvement remain essentially the same.

One of the supposed novelties within GDCPP is the way in which it extends the boundaries of design and construction into what is referred to as the 'requirements capture phase of prebriefing client decision-making' [*sic.*] (Kagioglou *et al.*, 2000). The notion of 'capturing requirements' is immediately suggestive of an underlying machine metaphor. Kamara *et al.* (2002) also recommend 'requirements management' as more rational approach to client briefing. The influence of 1960s-style systems engineering looms large within the requirements management literature, which is by no means exempt from criticism (see Fernie *et al.*, 2003). In common with the GDCPP, sources such as Kamara *et al.* (2002) lack any explicit recognition that different interest groups within the client organisation may have very different interpretations of what is required. Elsewhere, Green and Simister (1996) suggest an approach to 'modelling client business processes' which maps competing perspectives explicitly within the client organisation while avoiding any attempt to collapse them into a single 'holistic' representation. But there was seemingly little room for such interpretive approaches to client briefing within the GDCPP.

The preceding minor quibbles should not detract from the fact that there is much merit to found within the GDCPP. It remains eminently sensible to seek a clear and consistent model of the design and construction process. The idea of stage gates as a means of implementing control would likewise seem entirely reasonable, and is an idea that has subsequently been widely adopted by design managers within the construction sector. While the authors of GDCPP acknowledge explicitly the influence of BPR, GDCPP categorically cannot be dismissed as yet another uncritical manifestation of technocratic totalitarianism. Indeed, Cooper *et al.* (2005) demonstrate a degree of sensitivity to the overblown guru-hype offered by the likes of Hammer and Champy (1993). Cooper *et al.* (2005) further recognise the difficulty in defining BPR with any precision. It would seem that the developers of the GDCPP took advantage of the popularity of 'process management' without ever quite being fully convinced by the heady rhetoric of BPR. It may well be possible to argue that the language of BPR reflects and reinforces the construction sector's pre-existing regressive tendencies, but this argument cannot necessarily be extended to innovations which are only loosely 'inspired' by BPR.

What does come across strongly within the case studies described by Cooper *et al.* (2005) is the extent to which practitioners and researchers *learned* from their collective attempts to implement GDCPP. The process of producing the various process maps stimulated a shared understanding across the project team. This was not therefore a mindless exercise in the application of a heavily prescriptive process model. The process through which the maps are produced accords much more strongly with innovation in action. That the participants found the process to be useful would seem rather more important than they fact that it might have been inspired initially by BPR. There is an important point here which will be reinforced throughout the remainder of the book: how improvement recipes are enacted in practice often differs from the idealised prescriptions which are described in the literature. It is arguably the very process of *rethinking* the advocated recipe in a specific context which is of prime importance. It would be encouraging to believe that this is what Sir John Egan had in mind when he entitled his report *Rethinking Construction* – although this is probably wishful thinking. Shifting the emphasis towards the *process* of rethinking GDCPP for each individual application implies that the Generic Design and Construction Project Protocol is not generic, and not really a protocol either. But this should not be construed as a criticism, as the benefits of such approaches often lie in the way they promote localised learning within project teams. The same argument holds equally true for a wide range of management techniques.

Perhaps the strongest commonality between GDCPP and BPR is the shared theme of 'unleashing' the power of IT. Reference has already been made to the way in which BPR boosted its credibility through association with so-called enabling technologies. The same is also undoubtedly true of GDCPP, which mobilises a beguiling list of IT solutions, including: integrated databases, electronic data interchange (EDI), artificial intelligence, neural networks, visualisation, project extranet and building information modelling (BIM). One of the most acclaimed advantages of the GDCPP is the way it facilitates the use of such new-generation IT support tools (Kagioglou *et al.*, 2000). According to Cooper *et al.* (2005) the main problem in construction is that most IT systems are purchased because of operational rather than strategic business requirements. Given the risk-adverse nature of the sector, most firms are understandably unwilling to disrupt embedded ways of working for the sake of unspecified potential. Of particular note is the argument that:

> '[t]he effective use and co-ordination of IT, people and culture interfaces should optimise the process performance which leads to eventual customer satisfaction.' (Cooper *et al.*, 2005; 43)

Unfortunately, neither people nor 'culture interfaces' can be so easily manipulated in the rationalistic cause of optimisation. The preceding quote raises a plethora of issues relating to the interaction between new technologies and pre-existing embedded organisational practices. In many respects, the authors of the GDCPP were prescient in predicting the adoption/adaptation of advanced IT solutions in the construction sector (with the notable exceptions of artificial intelligence and neural networks). It is striking that BIM has now been accepted as *de rigueur* on many major construction projects. However, it is equally striking that the implications of BIM for the embedded demarcation of professional roles in the construction sector remains an active and rich research area (Eastman *et al.*, 2008; Whyte *et al.*, 2008; Harty, 2010). The potential of BIM is undoubtedly significant, but nevertheless it seems unlikely to impact upon the accumulated problems caused by the collateral damage of the enterprise culture.

Of particular note in recent years is the way 'collaboration' and 'integration' have become linked increasingly with IT solutions. Unfortunately, such technologies have failed notably to 'integrate' the multitude of bogusly self-employed workers who lack basic employment rights in the workplace. Notions of integration, collaboration and the use of whizzy IT gadgets seem ill-placed to address the real problems of construction sector fragmentation. At least self-employed migrant workers from Eastern Europe can now send e-mails home on their Blackberries™ as they travel through Central London in their ubiquitous white vans. Needless to say, such vans no longer display CABIN stickers exhorting passers-by to 'say no to building nationalisation'.

7.5 Partnering

7.5.1 Defining characteristics

Of all the management panaceas reviewed in this book it is probably partnering which received the most widespread endorsement as a supposed solution to the ills of the construction sector (Egan, 1998; Bennett and Jayes, 1995; Bennett and Jayes, 1998; Construction Industry Board, 1997). At the time of writing, the enthusiasm for partnering has perhaps waned slightly. But for the past 15 years or so the rhetoric of partnering has pervaded throughout the infrastructure of industry improvement. The underlying principles have seemingly been unassailable since the mid-1990s. The very notion of an 'enlightened' practitioner has become almost synonymous with a commitment to partnering. The adoption of partnering supposedly improves customer-responsiveness and ensures continuous improvement. Critics of partnering have been few and far between. Indeed, following the endorsement of partnering by *Rethinking Construction* a climate developed whereby to argue against partnering has become a (commercially) dangerous pastime.

The present author unwisely stepped out of line in 2003 to present a critical perspective on partnering to an audience of civil engineering contractors. The resulting headline in the *Contract Journal* was 'Partnering ethos comes under fire'. This was quickly followed by a personal e-mail of chastisement from the chairman of the Strategic Forum, Peter Rogers, spelling out how 'appalled' he was that an academic from the University of Reading should be actively undermining the ethos of partnering. To speak out against partnering has seemingly become almost tantamount to sacrilege; the expression 'shoot the dissenters' would appear to apply equally to critics of partnering and BPR alike. It will be argued that the two ostensibly disconnected improvement recipes have more in common than is generally supposed.

McGeorge and Palmer (1997) suggest that formal partnering as a construction management concept dates from the mid-1980s. Several early partnering arrangements were established in the process engineering sector. Specific cited examples include Union Carbide with Bechtel and Du Pont with Fluor Daniel. Without a doubt one of the earliest promoted definitions of partnering was the one offered by the Construction Industry Institute (1989) in the United States:

> 'A long term commitment between two or more organisations for the purposes of achieving specific business objectives by maximising the effectiveness of each participant's resources. This requires changing traditional relationships to a shared culture without regard to organisational boundaries. The relationship is based on trust, dedication to common goals, and on an understanding of each others individual expectations and values. Expected benefits include improved efficiency and cost effectiveness, increased opportunity for innovations, and the continuous improvements of quality products and services.' (Construction Industry Institute, 1989)

It would seem therefore that partnering is concerned primarily with maximising the effectiveness of resources, thus reflecting the purpose of countless other management improvement techniques. Emphasis is also given to the importance of 'shared culture' and the need to base relationships on trust and understanding. It is perhaps the focus on mutuality of objectives coupled with trust and understanding which most notably sets a very different tone from the rhetoric of BPR. The influence of the rhetoric of Total Quality Management (TQM) is readily apparent in the reference to 'continuous improvement', which is valued in favour of radical or dramatic change. The tone of the definition offered by the Reading Construction Forum (Bennett and Jayes, 1995) reflects similar themes:

> 'Partnering is a management approach used by two or more organisations to achieve specific business objectives by maximising the effectiveness of each participant's resources. The approach is based on mutual objectives, an agreed method of problem resolution, and an active search for continuous measurable improvements.'

Mutuality of objectives again comes across as being of central importance. But here it is notable that improvements must not only be continuous, they must also be 'measurable'. The definition offered by *Rethinking Construction* provides a similar emphasis on continuous, measurable improvement:

> 'Partnering involves two or more organisations working together to improve performance through agreeing mutual objectives, devising

a way for resolving disputes and committing themselves to continuous improvement, measuring progress and sharing the gains.' (Egan, 1998)

Rethinking Construction further considers partnering to be a 'tool to tackle fragmentation' which is being used increasingly by the best firms in place of traditional contract-based procurement and project management. As was argued at length in Chapter 3, industry fragmentation was exacerbated throughout the 1980s and early 1990s as a direct result of the policy mechanisms of the enterprise culture. The end result was an industry with an institutionalised reliance on bogus self-employment. If partnering offered a tool to tackle the fragmentation of industry employment it would indeed be powerful medicine. But 'fragmentation' was left undefined as attention was quickly shifted away from the structural characteristics of the sector towards the need to adopt 'enlightened' management techniques.

The support of the Construction Task Force for partnering was surpassed in its enthusiasm by the Construction Industry Board's (1997) report on *Partnering in the Team*:

> 'Partnering is a structured methodology for organisations to set up mutually advantageous commercial relationships, either for single projects or in long term strategic relationships, which help people work together more effectively.' (Construction Industry Board, 1997; 3)

The definition above is especially notable for the lazy way in which it slips between different units of analysis. It starts off talking about organisations and ends up talking about people. That there is a direct relationship between the commercial arrangement between organisations and the way people behave is seemingly taken entirely for granted. The Construction Industry Board (1997) further suggests that partnering has three essential components:

- establishment of agreed and understood mutual objectives;
- methodology for quick and cooperative problem resolution;
- culture of continuous, measured improvement.

Of particular note is the way that Bennett and Jayes's (1995) 'agreed method' of problem resolution had now been reduced to 'quick and co-operative' problem resolution. The Construction Industry Board (1997) were particularly clear in their view that adversarial attitudes are a 'waste of time and money'. The underlying argument is that efficiency would be much better served if deviant behaviour could be eliminated. The view is further expressed that problem resolution should be based on 'win-win' solutions. It is contended repeatedly that 'all parties' benefit from partnering, although it is unclear the extent to which the philosophy extends throughout the construction supply chain.

7.5.2 *Success requires faith and commitment*

A unifying theme throughout the sources reviewed above is the belief that an appropriate 'culture' is of vital importance to the success of partnering. *Partnering in the Team* (Construction Industry Board, 1997) is especially strident in emphasising that the first step towards partnering is to ensure that the culture of the company is conducive to a 'whole-team co-operative approach'. Once again, there is a degree of ambiguity as to whether or not 'whole team' includes the construction workforce. The overriding inference is that the construction 'team' comprises organisations rather than people. The organisations involved are further seen to be unitary in nature and capable of displaying the very human characteristic of 'teamwork'. However, it is seen to be important to appoint individual 'champions' to promote the partnering concept. It is also made clear that senior management must act as exemplars of the required culture. Of particular note is the way in which success is linked continually to 'faith' and 'commitment'. This is especially notable within the rhetoric promoted by the Construction Industry Board (1997) who argue that commitment is an essential element of partnering:

> 'to succeed requires fundamental belief, faith and stamina. The commitment must start at the top and it must be shared by the senior management.' (Construction Industry Board, 1997; 8)

The strident and heavily prescriptive tone of the quotation above would surely have been frowned upon by sober-minded civil servants such as Banwell and Emmerson. The plea for 'fundamental belief, faith and stamina' echoes directly the oratory style favoured by the former Italian fascist dictator Benito Mussolini. The repeated use of the word 'must' is especially notable. The manager *must* have 'fundamental belief'; the inference is that the manager must not question. There is apparently no requirement for the recipient managers to think for themselves. 'True knowledge' is seemingly held by the small technocratic élite within the Construction Industry Board, and the rest of the construction industry is required to act on faith. Dissenters are marginalised as deviants. According to the Construction Industry Board (1997), 'cynicism and lack of commitment by the few will destroy the efforts of many'. Such a script could easily have been used to justify the Spanish Inquisition, or Mussolini's Italy, or any other regressive regime with little tolerance for dissent. Despite the warm words about 'mutual objectives' and 'relationships' the overwhelming tone of much of the partnering literature is heavily prescriptive. Partnering is seemingly a regime which is to be *imposed* on the construction industry. The 'partners' involved are ostensibly limited to commercial entities. There is on the face of it no role for independent trade unions or for any individual 'deviants' who have objectives which do not accord with those of 'top management'. Pluralist models of organisation are sacrificed in favour of crude prescription.

Writing from a self-consciously critical perspective, Green (1999) has elsewhere equated the arguments in support of partnering with corporatist propaganda. Certainly, the supporting arguments draw heavily from the lexicon of enterprise culture. The ultimate argument as mobilised by sources such as *Partnering in the Team* rests with mystical appeals to the 'customer' and the imperatives of the global market. The all-pervading nature of the enterprise culture has rendered such ideas unchallengeable. The doctrines of customer-responsiveness and continuous improvement must seemingly be accepted on faith rather than on rational argument.

Following on from the above, there would seem to be an uncanny resemblance between the factors necessary for the successful implementation of partnering and those which were described previously in the case of BPR. Both depend upon senior management making an 'unwavering' commitment to cultural change. Both also require 'champions' who are willing to ride roughshod over any criticism. In the circumstances, it is perhaps not surprising that operatives within the construction sector continue to display 'adversarial attitudes'. The continuous bombardment with mechanistic improvement recipes would probably try the patience of a saint. An obligation to sign up for BPR very quickly morphs into a necessity to demonstrate an unwavering commitment in favour of partnering. Fortunately, for the sanity of all involved, the shift from BPR to partnering is a shift in emphasis rather than a dramatic shift in direction. As observed previously, machine and team metaphors are often mobilised within the same discourse without any obvious signs of conflict. The combination is undoubtedly much easier if the underpinning model of teamwork is based on compliance. Aficionados of football may recall the all-conquering Liverpool team of the 1980s being compared with a 'well-oiled machine'. Teamwork and machine efficiency are often mutually constituted in football and construction alike.

7.5.3   *Transcending organisational boundaries*

Notwithstanding the above, partnering does, however, differ from BPR in at least one important respect – partnering is something which transcends organisational boundaries. In contrast, BPR tends to be applied within organisational boundaries; its origins lie in the prevailing concerns about 'fat' organisations which were perceived to be uncompetitive. To repeat the core argument, the imperative was to eliminate waste that does not contribute to the customer. In the case of the UK construction sector, by the mid-1990s the structural changes described in Chapter 3 had in the most part already happened. BPR arrived rather late on the re-structuring scene and was embraced by practising managers because it reflected and reinforced what was already happening. But given the contracting sector's reliance on subcontracting in the mid-1990s, the popularity of BPR was bound to be short-lived. Simply put, client organisations and contracting firms alike had already become so lean there was little in the way of downsizing which remained to be done. But client organisations still demanded better service from the construction sector, and they were prepared to use their market muscle to achieve it. Thus, efficiency through downsizing gave way to efficiency through partnering with the supply chain.

Partnering also differs from BPR in that from its very inception it was situated within project environments characterised by multiple organisations. However, partnering was soon to be extended beyond the boundaries of single projects. Several sources make the distinction between 'project specific' partnering and 'strategic' partnering, the latter phrase being used to denote a situation where the partners work together across several projects (Bennett and Jayes, 1998). Strategic partnering supposedly allows the benefits of improved understanding to be carried forward to subsequent projects. However, at the same time, the philosophy of continuous, measured improvement demands that each project exceeds the performance of the previous one. Long-term relationships are not therefore valued as an end in themselves, they are only valued if they lead directly to improvements in performance. Despite the seductive discourse of 'empowerment', 'working together' and 'relationships', the success of any partnering initiative ultimately seems to hinge on cost improvement. The emphasis on steadily improved performance might perhaps also translate too easily to a regime of management-by-stress. This of course is pure conjecture, although the literature does occasionally hint that something vaguely unpleasant might happen should the advocated targets not be met.

7.5.4   *Buying power and the rhetoric of seduction*

What cannot be denied is that the cause of partnering was championed by a number of powerful clients who had become dissatisfied with the supposed under-performance of the construction industry. The large UK supermarkets have consistently numbered amongst the most enthusiastic advocates of partnering. As regular clients of construction, they understandably wish to extend the control that they exert over the grocery supply-chain to the construction sector. It is no coincidence that Bennett and Jayes (1998) include the exemplar case studies of Sainsbury's and Asda. The Egan Report (1998) also cites the case of Tesco, who have apparently:

> '… reduced the capital cost of their stores by 40% since 1991 and by 20% in the last two years, through partnering with a smaller supply base with whom they have established long term relationships. Tesco is now aiming for a further 20% reduction in costs in the next two years and a further reduction in project time.'

If true, the benefits achieved by Tesco though partnering are indeed significant, although the Egan Report neglects to say anything about the corresponding increase in profitability achieved by Tesco's partners. Other large clients who were strong advocates of partnering include BAA and Whitbread, both of whom were represented on Egan's Construction Task Force. Other notable UK clients who claimed to be committed to partnering included (the now defunct) Rover Cars and John Lewis Partnership.

Given the collective buying power of the aforementioned clients, it is unsurprising that many leading contractors also quickly claimed to be committed similarly to partnering. To do otherwise would have been to risk attracting the label of 'adversarial', thereby denying themselves access to a significant part of the UK market. This exercise of buying power is made especially clear by the Construction Clients' Forum (1998), who at the time collectively accounted for some 80% of the construction market. The CCF document committed its members to promoting relationships based on teamwork and trust, and to working jointly with their partners to reduce costs. They also promised not to exploit their buying power unfairly, but also to look to form lasting relationships with the supply side. The overall tone is one of barely-disguised seduction. However, they then issued an unveiled threat to those dissenters who remained unconvinced:

> 'The message from the Construction Client's Forum is clear. If this Pact is concluded, clients represented on the CCF will seek to place their £40bn of business with companies that are seen to follow the approach described in this document, and will seek such commitment prior to tendering, commensurate with relevant national, European and international regulations.'

The message was indeed clear. The CCF was saying to the construction industry that in order to qualify for £40 billion worth of work their ideas on teamwork and trust must be accepted. An adherence to the language of partnering was seemingly an essential pre-requisite of doing business. This was made equally clear by the Construction Industry Board (1997), albeit on the level of individuals:

> 'If it becomes clear that anyone at the workshop is unable to adopt the spirit of partnering, that person should be replaced in the team.'

It would therefore seem that lurking behind the rhetoric of seduction was an 'iron fist'. The same implied threat lay behind the Egan report (1998). Little wonder that dissenters to partnering were so few and far between.

7.5.5 *Living up to the rhetoric*

The contrast between the rhetoric of seduction and the subsequent enforcing iron fist raises the question of whether the large clients who were advocating partnering lived up to their own rhetoric. The influence of the big supermarkets on the promotion of partnering to construction has already been noted, as has their understandable desire to exercise increased control over the construction supply chain. It is especially ironic that the industry task forces behind the two partnering reports published in conjunction with the Reading Construction Forum (Bennett and Jayes, 1995; Bennett and Jayes, 1998) were both chaired by Charles Johnston of Sainsbury. This was the same Charles Johnston who chaired the CIB working group which produced *Partnering in the Team* (Construction Industry Board, 1997). Mike Thomas of Whitbread also served on the two working groups that produced *The Seven Pillars of Partnering* (Bennett and Jayes, 1998) and *Partnering in the Team* (Construction Industry Board, 1997). For those who enjoy a good misguided conspiracy theory, the raw materials are all in place. But this is not the position which is being adopted here. Messrs Thomas and Johnston are singled out only to illustrate the way in which ideas are mobilised and legitimised across inter-organisational networks. It is categorically not being suggested that they were some sort of devious manipulators who influenced the reports to their own advantage. The point has more to do with the way power and knowledge become subtlety conflated. Charles Johnston and Mike Thomas simply rode within the same flux of ideas that many others rode within, and the present author claims no special immunity.

Nevertheless, it remains true that the big supermarkets were often cited as exemplars of the partnering ideal. In light of this it is pertinent to recall that during 1999 Sainsbury, Asda, Tesco and Safeway were all under investigation by the Office of Fair Trading (OFT). The investigation followed sustained complaints by farmers and growers that consumers were not benefiting from low farm prices. The outcome of this eight-month inquiry was that the entire £60 billion a year grocery sector was referred to the Competition Commission (previously the Mergers and Monopolies Commission). The OFT's Director General expressed a particular concern regarding the supermarkets' buying power and their exploitative influence over suppliers:

> 'I have had concerns for some time … that this power may become exploitative and the many responses from suppliers during our inquiry suggests that it is something which needs to be looked at by the Competition Commission.' (OFT, 1999)

Ironically, these were the same supermarkets which were at the time preaching 'customer-responsiveness' to the construction industry. The possibility that their supply chain management practices were directed towards earning super-normal profits, rather than serving the interests of their customers strangely had little effect on their perceived legitimacy as 'enlightened clients'.

The subsequent full report from the Competition Commission was published in 2000. It provided a sad indictment of the supply chain management practices implemented by the major supermarkets. The Competition Commission (2000) reported that they had received many allegations from suppliers about the behaviour of the 'main parties' in the course of their trading relationships. They further commented that most suppliers were unwilling to be named, presumably in fear of commercial retribution. The Competition Commission report referred specifically to a 'climate of apprehension' which seemed to prevail among many suppliers in their relationship with the big supermarkets. The Competition Commission investigators had put a list of 52 alleged practices to the main parties. They found that a majority of these practices had been carried out by many of the main parties. The practices of concern included:

> '…. requiring or requesting from some of their suppliers various non-cost-related payments or discounts, sometimes retrospectively; imposing charges and making changes to contractual arrangements without adequate notice; and unnoticeably transferring risks from the main party to the supplier.'

The point was further made that Asda, Safeway, Sainsbury, Somerfield and Tesco possessed sufficient buying power such that the practices were not only distorting competition, but were also adversely affecting the competitiveness of some of the suppliers. The Competition Commission's report concluded with the recommendation that the most effective way of addressing the adverse effects of the cited practices would be a Code of Practice. Most damningly of all, it was emphasised that a voluntary code would not be adequate. Coming from the notoriously toothless Competition Commission these were strong words indeed. The overriding message was that the big supermarkets were not to be trusted.

The impact of the publication of the Competition Commission's report on the debate about partnering in the construction sector was precisely zero. The perceived legitimacy of the big supermarkets to preach partnering to construction firms remained remarkably intact. It cannot of course be assumed that the supermarkets were consistent in their dealings with different sectors. Neither can it be assumed that contractors would have suffered any adverse effects even if such practices prevailed in their dealings with the construction sector. But if the findings of the Competition Commission are taken at face value, it does demonstrate the difficulties of achieving the required 'culture change' which was widely held to be an essential pre-requisite of successful partnering. If the self-proclaimed 'enlightened clients' could not incubate the desired culture of teamwork and trust within their own organisations, there would seem to be little chance of any wide scale culture change within the diverse and hugely fragmented construction sector.

It should also be reported that ultimately no action was taken to implement the Competition Commission's recommendation for a compulsory Code of Practice. The issue was very quickly forgotten. The New Labour government continued to nurture their close relationship with big business in the cause of wealth creation. The buying public continued to vote with their feet by purchasing an ever-increasing proportion of their weekly shopping in the big supermarkets. Few seemed to care about 'adversarial practices' in the retail sector provided the goods on the shelf offered value for money. Meanwhile the major clients of the construction sector continued to advocate partnering in the cause of lower costs. Adversarial practices in the supply chain were seemingly acceptable provided they were adversarial practices aimed at others.

7.5.6   Exemplar case studies

In light of the critique of partnering above, it is appropriate to offer a brief commentary on three of the exemplar case studies presented in *The Seven Pillars of Partnering* (Bennett and Jayes, 1998). It is particularly enlightening to focus critical attention on Bennett and Jayes' case studies of partnering as implemented by Sainsbury and Whitbread. Interest is also to be found in the case study of the now defunct Rover Cars. In many respects, Rover offers a micro-history of the British automobile industry. It had been absorbed initially into the Leyland Motor Corporation in 1967 which was partly nationalised in 1975 to form British Leyland. The British Leyland brand rapidly became synonymous with under-investment, poor management and sour industrial relations. Margaret Thatcher's predictable solution was privatisation. British Leyland was duly returned to the private sector and re-branded as the Rover Group in 1988. It passed through the hands of British Aerospace prior to being acquired by BMW. Rover benefitted from increased investment during the 1990s and enjoyed a brief revival. It was during the latter stages of the 1990s that Rover was feted as an exemplar of partnering. However, troubled times soon returned and in 2000 BMW sold both Rover and MG to the optimistically named Phoenix Consortium. Rover cars ceased to be produced in 2005 when the MG Rover Group became insolvent.

Case Study No. 1: Sainsbury's

Bennett and Jayes (1998) trace back Sainsbury's motivation for wanting to reduce the cost of construction to an intense price war with their competitors in 1994. The immediate response of Sainsbury's management was to down-size their property division from 240 to 80 staff. Although no reference is made to BPR, it would seem that the same logic was at work. The end result was that Sainsbury's property division had to out-source more of their work, thereby forcing them to work more closely with a more limited number of suppliers. At the same time, Sainsbury's decided that it had to 'change the attitudes of its own staff by adopting internal partnering'. The remaining 80 employees were therefore ominously 'instilled' with the culture of TQM and were required to be 'more flexible and out-going'. Presumably those who did not want to be 'instilled' with the new culture, numbered amongst the 160 who were down-sized. The storyline echoes that of BPR in that there is great emphasis on the way in which the remaining members of staff were 'empowered' to represent the company and to take new initiatives. There is of course a basic contradiction here. Employees and suppliers are expected to submit unquestioningly to the imposed TQM regime, whilst continuing to engage actively in the cause of continuous improvement. It would appear that the desired 'empowerment' concerns only *means* rather than *ends*. Employees are seemingly only empowered to implement Sainsbury's objectives more efficiently; they are not empowered to participate in their formulation.

Notwithstanding the above, the claims made in support of the success of partnering are impressive. The cost of Sainsbury's mainstream stores was apparently reduced by 35% and typical construction durations were reduced from 42 weeks to 15. However, these impressive achievements were not enough. Sainsbury's management thereafter established further 'tough and steadily improving cost, time and quality targets'. It would seem that the regime of continuous improvement is relentless. The conclusion that continuous improvement in this form equates to management-by-stress is difficult to avoid. Inevitably there comes a point when cost, time and quality targets cannot be improved any further. This is the point at which the client moves euphemistically 'beyond partnering'. The point at which initiatives stall also invariably involves a shift in power within the organisation. Whereas the previous dominant interest group had a vested interest in talking up the benefits of partnering, the succeeding interest group would wish to rest their credibility elsewhere.

Case Study No. 2: Whitbread

Bennett and Jayes (1998) also include Whitbread amongst their exemplar case studies. Whitbread was a hugely important client and at the time constructed around 100 new projects every year with a budget in the region of £300 million. Prior to the adoption of partnering, Whitbread's different business units (e.g. Beefeater Inns, Marriot and Travel Inns) had tended to procure construction projects through competitive tendering. The difficulty here was that this resulted in more and more contractors working for the company which resulted in each one being faced with the same learning curve. Whitbread had apparently been impressed by the Latham (1994) report and had lost patience with the 'adversarial practices' displayed by too many of their contractors. It concluded that there was much to be gained from working in partnership with a limited number of contractors. Sixteen contractors were identified as having a 'compatible culture' together with the required management and financial strength. All 16 were asked to price typical projects and the submissions were evaluated on a range of criteria, including suggestions for potential improvements and their understanding of partnering. Presumably it was not expected that the contractors' understanding of partnering should extend to the inherent paradoxes and contradictions. The list of approved contractors was subsequently whittled down to six, with whom Whitbread agreed fixed contributions for profits and overheads. The contractors quickly recognised that greater efficiency could be achieved by 'partnering down the line with a number of key specialist subcontractors' (Bennett and Jayes, 1998). By implication, it would seem that the contractors also recognised that efficiency would not be enhanced by partnering with *all* their specialist suppliers, or even with most of them. The focus is very much on fostering relations with 'key' specialist subcontractors. Presumably it was business as usual with the rest of them.

The description of Whitbread's adopted approach is especially interesting in the light of the Construction Industry Board's (1997) recommendation that partnering is only appropriate between organisations whose top management share the fundamental belief that people are honest. The case study describes how the six contractors selected as Whitbread's 'partners' meet on a bi-monthly basis to compare processes and results. However, the following comment illustrates the degree of trust:

> 'Whitbread recognises that it must attend these meetings to ensure no hint of a "cartel" emerges. It also accepts that if progress slows it will have to re-target the contractors or seek new partners.'

Behind all the rhetoric, it would appear that Whitbread ultimately does not trust its new 'partners'. They therefore fail the criterion set by the Construction Industry Board. The second sentence is also telling; once again the 'partnership' is clearly dependent upon continuous improvement, otherwise the partners are ominously 're-targeted', or otherwise simply dumped. The value of the relationships would seem to hinge entirely on performance. It would clearly be unwise for the contractors involved to depend upon Whitbread for too much of their turnover; this would not be good good risk management. The six contractors most probably continued to price work elsewhere through competitive tendering. They would have been well-advised to have done so.

Notwithstanding the paradoxes highlighted in the description above, it is once again difficult not to be impressed with the benefits secured by Whitbread. It is interesting that Bennett and Jayes (1998) cite one of the main benefits as being a reduction in the cost of Whitbread's in-house team. Although not made explicit, this presumably means that the team was 'down-sized' as had been the case with Sainsbury's. There is therefore some justification for reading partnering as an outcome of a broader

BPR-type initiative. Whitbread also apparently achieved reduced costs and much fewer defects. Apparently a saving of £335000 was achieved on Whitbread's Beefeater and Travel Inn development at The Lydiard in Swindon. This equated to no less than 19% of the capital cost. The benefits to the contractors are described in terms of the continuous flow of reliable and profitable work. The contractors also benefitted from a 40% share (together with the designers) of the cost saving achieved through 'innovation'. It is not specified how much of this was shared with the identified 'key specialist subcontractors'. Neither is it specified how much was shared with the other sub-contractors who were judged not to be key. The answer the first question is probably 'not much'; the answer the second is almost certainly 'nothing'.

Case Study No. 3: Rover Cars

During the late 1990s Rover Cars was enjoying a brief renaissance prior to being unceremoniously dumped by BMW in 2000. Bennett and Jayes's (1998) write-up of Rover's approach to partnering relates primarily to the procurement of the new Group Design and Engineering Centre (GDEC). The new building was needed to re-house the company's car designers and engineers and the initial budget was set at £12 million. Bennett and Jayes (1998) make much of Rover's link-up with Honda in the late 1980s which apparently led to the implementation of new management practices such as TQM, Just-in-Time and continuous improvement. These practices encouraged Rover's automotive buyers and suppliers to engage in partnering arrangements. The approach was driven by a new method of purchasing which was based on the Honda model of Effective Cost Management (ECM). The approach was driven by the Purchasing Department and 'empowered engineers and others to innovate and search out the best possible answers within a fixed budget' (Bennett and Jayes, 1998). Clearly the design engineers had not previously felt so empowered, even though the process of innovating and searching out solutions within given constraints would be central to most accepted definitions of engineering design. According to the description offered, it was the Purchasing Department which enabled the engineers to truly understand their own processes.

The resulting new culture was applied subsequently to the construction of the new Land Rover Discovery facility and the new production line for the Rover 800. The Rover team recognised collectively that the traditional way of procuring buildings was unacceptable, so for the GDEC project it was decided to adopt a sole supplier approach in accordance with ECM. The overriding aim was to produce buildings as if they were manufactured products. A further guiding principle within Rover was not to spend money on anything that did not contribute directly to building cars. This was a principle derived directly from the lexicon of BPR, and it resulted in a brief which laid out the specification for a very basic building. Bennett and Jayes

(1998) describe how a rigorous selection process was used to assess 35 potential contractors. Four firms were selected to make a presentation and SDC was chosen as the preferred main contractor. In choosing SDC, cost was not apparently a major issue for Rover. Emphasis instead was given to technical competence, cooperation and teamwork. Rover was also seeking an ability to live with changes and personalities with whom they could partner.

However, once SDC were appointed they were soon obliged to learn Rover's 'tough cost driven approach to partnering'. Partnering with Rover therefore seems to follow a familiar pattern. First of all comes the rhetoric of seduction, then comes the iron fist. Nevertheless, the reported benefits are once again impressive. The final building was occupied 12 months earlier than 'normal methods would have allowed' and the cost was allegedly in the region of 40% less than would have been achieved through the JCT route. Unfortunately, there is no justification for these figures and little reason to suspect they were produced independently of Rover's managers who themselves had a vested interest in 'talking-up' the benefits of partnering. Rover further apparently agreed to pay all parties an agreed profit together with their 'properly incurred direct costs'. It is easy to see how such an arrangement could indeed be attractive to contractors trying to ride out a recession. Key motivating factors included the assurance of further work together with the satisfaction of a job well done. However, the associated regime of management-by-stress is once again all too apparent:

> 'A very tough cost control and cost audit system was imposed backed up by relentless pressure from the client to look for the best possible value.'

There would therefore seem to have been little emphasis on 'mutuality of objectives'. The cost control and cost audit system were not just 'tough', they were 'very tough'. Furthermore, they were backed up by 'relentless pressure' from the client. There is no mention of any reciprocity in the direction of pressure; it was all seemingly one way. The compensation for those who are subjected to this regime of 'relentless pressure' supposedly lies in the promise of future work. This would seem to be a benefit of dubious worth in an uncertain world, and SDC would have been well-advised to avoid becoming too dependent upon Rover.

Rover's lauded commitment to its long-standing suppliers in the automotive supply chain was called into question by an article in the *Daily Telegraph* on 25 July 1998. Alongside announcing 1500 redundancies, Rover was reported to have started cutting orders with British suppliers due to the high value of sterling. One of Rover's long-standing 'partners', Plastic Mouldings (Cradley), which was dependent upon Rover for a quarter of its £17 million turnover, quickly lost about £2 million of business. In 2000 Rover was sold by BMW to the Phoenix Consortium, who were in no position to honour Rover's previous commitment to long-standing suppliers even if they wanted to. By 2005 the Rover story had come to a close. Despite the propagation of a new culture they had ultimately been unable to compete in the global marketplace. Perhaps if their designers had not been subjugated to the Purchasing Department they might have produced more attractive cars.

As a final comment, it should be pointed out that SDC still operate very successfully as a privately-owned regional contractor. They also possess a long track record of successful projects for automotive sector clients such Aston Martin, Jaguar, Land Rover and BMW. Rover Cars might have been better directed to have spent more time listening to SDC rather than subjecting them to relentless pressure. But we must also bear in mind that Bennett and Jaye's (1998) primary source for the case study were Rover's senior managers who were responsible for the implementation of partnering. It is likely that the emphasis of the case study was at least in part targeted at other factions within Rover who thought that partnering was a soft option. Hence the focus on 'very tough cost control' and 'relentless pressure imposed by the client'. The case studies therefore shed more light on the arguments which are mobilised in support of partnering, rather than the realities of what happens on the ground.

7.5.7 *Paradoxes expounded*

Up until this point, the discussion of partnering has focused essentially on the way partnering is presented in the literature. Initially, attention was given to the essential vagueness of precisely what is being advocated. Thereafter, critical attention was directed at the overly prescriptive nature of the supporting arguments. Of particular note is the way in which large clients mobilised their buying power to impose partnering on a seemingly adversarial construction sector. The possibility has been further raised that some of the major clients closely involved in the propagation of partnering fail to live up to their own rhetoric. A critique has also been presented on a selection of the exemplar case studies presented in *The Seven Pillars of Partnering* (Bennett and Jayes, 1998). The case studies were clearly written in support of partnering, and yet upon close examination they readily highlight some of the paradoxes which characterise the supporting arguments.

Bresnen (2007) also gives critical attention to the *Seven Pillars of Partnering* and has offered *Deconstructing Partnering in Project-based Organisation: Seven Paradoxes and Seven Deadly Sins* as a direct response. The emphasis on 'seven deadly sins' is perhaps somewhat tongue-in-cheek, but the underlying analysis is informed by previous empirical research into the enactment of partnering (Bresnen and Marshall, 2000a; Bresnen and Marshall, 2000b). Despite all the hype and exhortations in support of partnering, it remains the case that there has been relatively little neutral research into its implementation. Bresnen goes to some length to emphasise that his critique of partnering – although provocative – is intended to move towards 'a more realistic (and hence potentially more useful) understanding of the intricacies and dynamics of partnering in practice'. Several of the paradoxes identified resonate strongly with those highlighted in the preceding discussion, thereby adding weight to the arguments already presented. It is not necessary to repeat the full detail of Bresnen's analysis here; this would simply be to repeat many of points already made. But the chapter would not be complete without giving the flavour of Bresnen's provocative and telling contribution.

Bresnen's (2007) first paradox relates to what is described as *wishful thinking about strategic management and organisational behaviour*. He observes that the model of organisation which lies behind the various imperatives relating to the need for commitment is essentially unitary in nature. The assumption is that all parties subscribe to the same goals and objectives and the means by which they can be achieved. Hence the challenge for management is to mobilise enthusiasm and support for a pre-determined strategy. In essence, this is organisation as viewed through the lens of the machine metaphor. There is little recognition of the political nature of organisations, nor of the difficulties of achieving coherent and consistent strategic action. Such difficulties are real enough even within the boundaries of single organisations, let across in multi-organisational setting such as projects. According to Bresnen, the associated 'deadly sin' in that of *sloth*, otherwise described as a lack of awareness of the needs and perspectives of different groups. Too much of the prescriptive literature on partnering assumes lazily that those differences that do exist can be wished away on the basis of faith and commitment.

The second paradox highlighted by Bresnen relates to membership, i.e. the choice of which firms (and individuals) are judged to be appropriate partners. As we have already seen, the prescriptive literature on partnering goes to great lengths to emphasise the importance of selecting the 'right' partners with the appropriate culture. It is also made abundantly clear that those who are unable to adopt the 'spirit of partnering' should be replaced in the team (Construction Industry Board, 1997). Bresnen observes that the focus on conformity and the importance ceded to 'thinking in the right way' runs counter to the cause of innovation. As discussed in Chapter 6, there can indeed be a danger in being trapped into fixed ways of thinking. Certainly the emphasis on conformity within the discourse of partnering runs counter to all that is known about the environments which spawn creativity and innovation. Bresnen sees something inherently unhealthy in relationships which are too controlling and obsessive. The potential problems of over-involvement and over-commitment to a relationship are equated to a second deadly sin – *lust*.

The third paradox concerns the issue of equity, which is held repeatedly to be a cornerstone of the partnering philosophy (Construction Industry Board, 1997; Bennett and Jayes, 1998). The desire for equity is frequently the guiding principle in negotiating the various gain-share/pain-share arrangements which lie at the heart of many partnering agreements. Bresnan (2007) points towards the obvious temptation for the more powerful partner to negotiate an arrangement which unduly favours themselves:

> '[w]ith the best will in the world, it is difficult for a powerful client simply to give up their power to dictate terms and conditions in their favour to a smaller contractor who is dependent on them for future work.' (Bresnen, 2007; 369)

Certainly there is little evidence in the case studies reviewed above that any of the clients were tempted to give up any of their power over the contractors they had engaged as 'partners'. On the contrary, partnering became the medium through which their commercial power was exerted. The paradox coined by Bresnen (2007) is that such relationships bring an inherent danger of *encouraging exploitation or opportunism*. The associated 'deadly sin' is thus presented as *avarice*.

The fourth paradox relates to the issue of integration, and the arguments rehearsed by Bresnen (2007) are relevant to broader debates about 'integrated project teams' as promoted by the Strategic Forum (2002) (see Chapter 9). The espoused need for integration can be read in part as a reaction to the increased level of industry fragmentation. The greater the reliance on outsourcing, the greater will be the need to maintain some degree of control across organisational boundaries. It is no coincidence that the Sainsbury case study of partnering commences with a description of the downsizing of the property department and the outsourcing of functions previously performed in-house. A similar subtext is apparent within the Whitbread case study. It is thus possible to argue that partnering became popular as a means of re-asserting some of the control which had been lost as a result of outsourcing. The integration of performance controls across organisational boundaries therefore became an important imperative. The same logic can also be used to justify the popularity of supply chain management, which rests on similar notions of 'relational contracting' (Green, 2006).

Bresnen (2007) further points out that the need for integration is also seen to relate to internal boundaries within the firm. As was noted in Chapter 3, throughout the 1980s the construction market became increasingly segmented in terms of different types of client. This in turn caused firms to become increasingly differentiated internally – different business divisions specialised in the needs of different markets. The increased internal differentiation required a greater degree of integration within the firm (Lawrence and Lorsch, 1976). Integrating mechanisms commonly include the use of various IT strategies together with initiatives for cultivating a 'shared culture' across disparate operating divisions. Supporting human resource management (HRM) strategies which cut across operating divisions are also often used for the purposes of integration.

Table 7.1   Seven Pillars, Seven Paradoxes and Seven Deadly Sins

Pillar	Paradoxical effect	Deadly Sin
Strategy	Wishful thinking about strategy and behaviour	Sloth
Membership	Fostering of relationships built on exclusivity	Lust
Equity	Encouraging exploitation and opportunism	Avarice
Integration	Reinforcing a desire for control	Gluttony
Benchmarks	Setting of inappropriate targets	Envy
Processes	Over-engineering of processes	Wrath
Feedback	Failing to capture knowledge and learning	Pride

(*Source*: Bresnen, 2007)

With reference to integrating across organisational boundaries, the partnering literature is especially notable for the emphasis it gives to developing trust. However, the more critical literature highlights the fragile nature of trust within temporary project settings (Dainty et al., 2001; Bresnen and Marshall, 2002). An obvious paradox relates to the emphasis on trust being coexistent with an equally strong emphasis on continuous performance measurement. Bresnen (2007) cites the paradox in terms of a desire for cooperation and trust, on the one hand, and the *reinforcing of a desire for control*, on the other. He further associates this as 'having your cake and eating it', thereby equating to the deadly sin of gluttony.

The preceding description of the first four of Bresnen's (2007) seven paradoxes is sufficient to portray the essence of his critique. The remaining three paradoxes relate to the use of benchmarks, the tendency to 'over engineer processes' and the failure to capture knowledge and learning. The seven pillars, seven paradoxes and seven deadly sins in their entirety are summarised in Table 7.1. The unintended and undesirable consequences of 'over engineering processes' have already been addressed in respect of BPR. The fact that the same criticism can be directed at partnering is once again indicative of the impossibility of drawing sharp dividing lines between supposedly different improvement recipes. A commitment to BPR morphs seamlessly into a commitment to partnering, which in turn morphs seamlessly into a devotion to lean construction. This brings the discussion back once again to the idea of management fashions. For those who advocate the cause of industry improvement, to echo the rhetoric of the latest management fashion is perhaps more important than being clear on the substantive content of each successive idea.

7.5.8 *Trust and power*

Prior to closing the discussion on partnering, it is appropriate to address the notion of trust in a little more detail. Sabel (1992) defines trust as the *'mutual confidence that no party to an exchange will exploit the other's vulnerabilities'*. As has already been discussed, much of the debate is characterised by an assumption that firms are synonymous with individuals. Within the prescriptive literature, trust between organisations is generally seen to be a prerequisite of partnering (Green, 2006). However, others have questioned whether economic cooperation results from trust, or whether trust results from economic cooperation (Gambetta, 1988). Rarely is any attention given to the way trust is shaped by imbalances in economic power between firms in the supply chain and the broader dynamics of the marketplace (cf. Korczynski, 2000).

Cox (1999) argues that the nature of any relationship between organisations is mediated inevitably by power differentials between the contracting parties. Yet the dominant assumption within the partnering literature is that trust between individual actors is independent of the pressures imposed by the broader context. In situations where there is a power imbalance between firms, cooperation may be enforced through power rather than trust. In such circumstances, the weaker party will feel exploited and, by definition, will not trust the stronger party. In these circumstances, adversarial relationships may well be suppressed by enforced cooperation. But this has much more to do with behavioural compliance than with 'culture change'. Furthermore, enforced cooperation is likely to erode whatever trust previously existed. Such instances are further likely to generate patterns of resistance and initiate distinctive counter-cultures that may well pre-condition the response of individuals to any future advocated 'enlightened practice'. For example, the Construction Clients' Forum's (1998) attempt to impose partnering on its members' supply chains directly undermined its stated commitment to relationships based on trust (Green, 2006). Likewise, the infamous appeals of the Construction Industry Board (1997) to adopt partnering on the basis of 'fundamental belief, faith and stamina' are anathema to the concept of the 'learning organisation' – a further paradox which could have been added to Bresnen's (2007) list. There is of course a recurring theme here. Proponents of industry improvement repeatedly adopt a top-down approach to workplace change, and yet seem genuinely surprised that the advocated ideas meet with patterns of resistance. Even more bizarrely, the Egan report (1998) bypassed the entire infrastructure of construction sector engagement to promote a change programme biased disproportionately towards the interests of large repeat clients.

The legitimacy of large clients such as BAA and Tesco to impose ideas on a construction sector comprising close to two million people was accepted blithely in the cause of 'customer responsiveness'. The only seemingly acceptable role for managers and employees within construction companies was behavioural compliance. Little wonder the adversarial attitudes were hard to eradicate. Fortunately, construction sector managers have proved themselves to be consistently adept at re-inventing the advocated ideas in ways which render them meaningful in localised contexts.

Practitioners may not be able to 'do' partnering in any instrumental sense, but they are able to act out a partnering performance when required.

The final word on partnering is perhaps best left to Alderman and Ivory (2007; 392):

> '[t]o use the term partnering is to invoke a metaphoric association between relationships in the commercial and personal spheres. At its best, partnering is an earnest attempt to weave together these two worlds in the form of an appeal to work closely together and to share the benefits of so doing. At its worst, it is a discursive smokescreen behind which to conceal 'business as usual', while at the same time motivating suppliers and contractors to 'go the extra mile.'

There is of course an endless range of possibilities in between the best and worst case scenarios. In any given context the likelihood is that some actors are earnest in their attempts to enact partnering, while others deliberately deploy discursive smokescreens. Neither of these behaviours is especially stable; nor are they independent of broader contextual influences.

7.6 Summary

It is clear from the above that the explanation for the popularity of BPR and partnering does not lie in their respective substantive content. Both manage to combine a strong prescriptive argument with an essential vagueness of definition. It has further been shown to be impossible to draw precise boundaries between BPR and partnering. The two concepts both draw on broader sets of ideas relating to total quality management (TQM) and supply chain management. Rather than being understood in terms of their substantive content, it perhaps makes more sense to understand BPR and partnering as different labels which draw from a common pool of ill-defined managerial storylines. Different storylines are pulled together at different times for different purposes. In contrast to portraying BPR and partnering as discrete management techniques, they are much better interpreted as constituent parts of the enterprise culture. Furthermore, they cannot be understood in isolation from the extensive structural changes which characterised the construction sector throughout the 1980s and 1990s.

Notwithstanding the above, the rhetoric of BPR largely preceded that of partnering. The rhetoric of BPR both reflected and reinforced changes which were already happening – especially the accelerating trends of downsizing and outsourcing. The imagery of re-engineering was attractive to practising managing because it gave them a means of making sense of the changes they observed happening around them. As has been seen, the storylines of BPR played deliberately on the insecurity of middle managers.

They also resonated with the spirit of the enterprise culture. Large and bureaucratic organisations were cast aside in the cause of becoming efficient, responsive and customer-focused. Thus, the increased reliance on subcontracting, self-employment and an enlarged casualised workforce made perfect sense.

The downside of the trends above was that clients and main contractors alike suffered from a loss of control. Activities performed previously within the same organisation were now performed by others. Partnering therefore became the means by which organisations sought to maintain control without carrying the overhead costs of performing activities in-house. Partnering can perhaps be understood as the rhetorical antidote to BPR. It is striking that downsizing and outsourcing provide the backcloth to many of the exemplar case studies of partnering. This again points towards the need to understand improvement recipes such as BPR and partnering within a broader context. Both are best understood as essential components of the enterprise culture.

References

Alderman, N. and Ivory, C. (2007) Partnering in major contracts: Paradox and metaphor, *International Journal of Project Management*, **25**(4) 386–393.

Alvesson, M. and Willmott, H. (1996) *Making Sense of Management: A Critical Introduction*, Sage, London.

Banwell, Sir Harold (1964) *The Placing and Management of Contracts for Building and Civil Engineering Work*. HMSO, London.

Bennett, J. and Jayes, S. (1995) *Trusting the Team: The Best Practice Guide to Partnering in Construction*, Centre for Strategic Studies in Construction, The University of Reading.

Bennett, J. and Jayes, S. (1998) *The Seven Pillars of Partnering*, Thomas Telford, London.

Betts, M. and Wood-Harper, T. (1994) Re-engineering construction: a new management research agenda, *Construction Management and Economics*, **12**(6), 551–556.

Blyth, A. and Worthington, J. (2001) *Managing the Brief for Better Design*, Spon Press, London.

Bresnen, M. and Marshall, N. (2001) Understanding the diffusion and application of new management ideas, *Engineering, Construction and Architectural Management*, **8**(5/6), 335–345.

Bresnen, M. and Marshall, N. (2000a) Partnering in construction: a critical review of issues, problems and dilemmas, *Construction Management and Economics*, **18**(2), 229–237.

Bresnen, M. and Marshall, N. (2000b) Building partnerships: case stidies of client-contractor collaboration in the UK construction industry, *Construction Management and Economics*, **18**(7), 819–832.

Bresnen, M. and Marshall, N. (2002) The engineering or evolution of cooperation? A tale of two partnering projects, *International Journal of Project Management*, **20**(7), 497–505.

Bresnen, M. (2007) Deconstructing partnering in project-based organisation: Seven pillars, seven paradoxes and seven deadly sins, *International Journal of Project Management*, **25**(4), 365–374.

Broekstra, G. (1996) The triune-brain metaphor: the evolution of the living organization. In *Metaphor and Organization* (eds. D. Grant and C. Oswick), pp. 53–73, Sage, London.

Buchanan, D. (2000) An eager and enduring embrace: the ongoing rediscovery of teamworking as a management idea, in Proctor, S. and Mueller, F. (eds) *Teamworking*, MacMillan, Basingstoke, pp. 25–42.

Champy, J. (1996) *Reengineering Management: The Mandate for New Leadership*, HarperBusiness, New York.

Construction Industry Board (1997) *Partnering in the Team*, Thomas Telford, London.

Commonwealth of Australia (1999) *Building for Growth: An Analysis of the Australian Building and Construction Industries*, Commonwealth of Australia, Canberra.

Competition Commission (2000) Supermarkets: A report on the supply of groceries from multiple stores in the United Kingdom, CC, London.

Construct IT (1996) *Briefing and Design*. Benchmarking Best Practice Report, Construct IT, Salford.

Construct IT (1998) *Supplier Management Update*. Benchmarking Best Practice Report, Construct IT, Salford.

Construction Industry Board (1996) *Educating the Professional Team*, Thomas Telford, London.

Conti, R.F. and Warner, M. (1994) Taylorism, teams and technology in "reegineering" work organisation, *New Technology, Work and Employment*, 9, 93–102.

Construction Clients' Forum (1998) *Constructing Improvement*, Construction Clients' Forum, London.

Construction Industry Institute (1989) *Partnering: Meeting the Challenges of the Future*, CII, Texas.

Cooper, R., Aouad, G., Lee, A., Wu, S., Fleming, A. and Kagioglou. M. (2005) *Process Management in Design and Construction*, Blackwell, Oxford.

Cox, A. (1999a) Power, value and supply chain management, *Supply Chain Management*, **4**(4), 167–175.

Dainty, A.R., Briscoe, G.H. and Millitt, S.J. (2001) Subcontractor perspectives on supply chain alliances. *Construction Management and Economics*, **19**(8), 841–848.

Davenport, T.E. (1993) *Process Innovation: Reengineering Work Through Information Technology*, Harvard Business School Press, Harvard.

Davenport, T.E. and Short, J.E. (1990) The new industrial engineering: information technology and business process redesign, *Sloan Management Review*, Summer, 11–27.

De Cock, C. and Hipkin, I. (1997) TQM and BPR: Beyond the beyond myth, *Journal of Management Studies*, **34**(5), 659–675.

Eastman, C., Teicholz, P., Sacks, R. and Liston, K. (2008) *BIM Handbook: A Guide to Building Information Modeling for Owners, Managers, Designers, Engineers, and Contractors*, John Wiley & Sons, Hoboken, New Jersey.

Egan, Sir John. (1998) *Rethinking Construction.* Report of the Construction Task Force to the Deputy Prime Minister, John Prescott, on the scope for improving the quality and efficiency of UK construction. Department of the Environment, Transport and the Regions, London.

Emmerson, Sir Harold (1962) *Survey of Problems Before the Construction Industries.* HMSO, London.

EPSRC (1994) *The Innovative Manufacturing Initiative,* Engineering and Physical Sciences Research Council, Swindon.

Fernie, S., Green, S.D. and Weller, S. (2003) Dilettantes, discourse and discipline: requirements management for the construction industry, *Engineering, Construction and Architectural Management,* **10**(5), 354–367.

Fulop, L. and Linstead, S. (1999) *Management: A Critical Text,* Macmillan, Basingstoke.

Gambetta, D. (1988) Can we trust trust?, in Gambetta, D. (Ed.) *Trust,* Blackwell, Oxford, pp. 213–38.

Green, S.D. (2006) Discourse and fashion in supply chain management, in *The Management of Complex Projects: A Relationship Approach,* (eds. S. Pryke and H. Smyth), Blackwell, Oxford, pp. 236–250.

Green, S.D. and Simister, S.J. (1998) Modelling client business processes as an aid to strategic briefing, *Construction Management and Economics,* **17**(1), 63–76.

Green, S.D. (1999) Partnering: the propaganda of corporatism?, *Journal of Construction Procurement,* **5**(2), 177–186.

Grey, C. and Mitev, N. (1995) Re-engineering organisations: a critical appraisal, *Personnel Review,* **24**(1), 6–18.

Grint, K. (1994) Reengineering history: social resonances and business process reengineering, *Organization,* **1**, 179–201.

Grint, K. and Willcocks, L. (1995) Business process re-engineering in theory and practice: business paradise regained?, *New Technology, Work and Employment,* **10**(2), 99–109.

Hammer, M. and Champy, J. (1993) *Re-engineering the Corporation,* Harper Collins, London.

Hammer, M. (1990) Reengineering work: don't automate, obliterate, *Harvard Business Review,* **68**, 104–112.

Harty, C. (2008) Implementing innovation in construction: contexts, relative boundedness and actor-network theory, *Construction Management and Economics,* **26**(10), 1029–1041.

Hucczynski, A.A. (1993) *Management Gurus,* Routledge, London.

Hughes (1991) Modelling the construction process using plans of work, *Proc. International Conference on Construction Project Modelling and Productivity,* CIB W65, Dubrovnik.

Jackson, B.G. (1996) Re-engineering the sense of self: the manager and the management guru, *Journal of Management Studies,* **33**(5), 571–590.

Johansson, H.J., McHugh, P. Pendlebury, A.J. and Wheeler, W.A. (1993) *Business Process Reengineering: Breakpoint Strategies for Market Dominance,* John Wiley & Sons, Chichester.

Jones, M.R. (1995). The contradictions of business process re-engineering. In *Examining Business Process Re-engineering,* (eds Burke, G. and Peppard, J.), Kogan Page, London, pp. 43–59.

Kagioglou. M., Cooper, R., Aouad, G. and Sexton, M. (2000) Rethinking construction: the Generic Design and Construction Process Protocol, *Engineering, Construction and Architectural Management,* **7**(2), 141–153.

Kalgaard, R. (1993) ASAP interview with Michael Hammer, *Forbes,* 13 September, pp. 69–75.

Kamara, J.M., Anumba C.J. and Evbuomwan, N.F.O. (2002) *Capturing Client Requirements in Construction Projects*, Thomas Telford, London.

Korczynski, M. (2000) The political economy of trust, *Journal of Management Studies*, **37**, 1–21.

Latham, Sir Michael (1994) *Constructing the Team*. Final report of the Government/industry review of procurement and contractual arrangements in the UK construction industry. HMSO, London.

Lawrence, P.R. and Lorsh, J.W. (1967) *Organization and Environment*, Harvard Press, Cambridge, Mass.

Legge, K. (1995) *Human Resource Management: Rhetorics and Realities*, MacMillan, London.

McGeorge, D. and Palmer, A. (1997) *Construction Management: New Directions*, Blackwell, Oxford.

McGeorge, D. and Palmer, A. (2002) *Construction Management: New Directions*, 2nd edn., Blackwell, Oxford.

Micklethwait, J. and Wooldridge, A. (1997) *The Witch Doctors*, Mandarin, London.

Miers, D. (1994) Why do BPR Initiatives Fail?. *Workflow and Re-engineering International Association) Newsletter* **1**(2), 1–2.

Mohamed, S. (ed.) (1997) *Construction Process Re-engineering, Proc. of International Conference*, Gold Coast, Australia.

Mohamed, S. and Tucker, S. (1996) Options for applying BPR in the Australian construction industry, *International Journal of Project Management*, **14**(6) 379–385.

Mumford, E. and Hendricks, R. (1996) Business process re-engineering RIP, *People Management*, 2nd May, pp. 22–27.

OFT (1999) *Bridgeman Refers Supermarkets*. Press Release PN 11/99, Office of Fair Trading, London.

OST (1995) *Technology Foresight: Progress Through Partnership*, Office of Science and Technology, London.

RIBA (1967) *Plan of Work for Design Team Operation*, Royal Institute of British Architects, London.

Sabel, C. (1992) Studied trust: building new forms of co-operation in a volatile economy, in Pyke, F. and Sengenberger, W. (eds), *Industrial Districts and Local Economic Regeneration*, International Institute for Labour Studies, Geneva, pp. 215–250.

Strategic Forum (2002) Accelerating Change, Rethinking Construction, London.

Tinaikar, R., Hartman, A. and Nath, R. (1995) Rethinking business process re-engineering: a social constructivist perspective. In *Examining Business Process Re-engineering* (eds G. Burke and J. Peppard), pp. 107–116. Koran Page, London.

Taylor, F.W. (1911) *Principles of Scientific Management*, Harper & Row, New York.

Walker, A. (1989) *Project Management in Construction*, 2nd edn, BSP Professional Books, Oxford.

Whyte, J., Ewenstein, B., Hales, M. and Tidd, J. (2008) *Visualizing knowledge in project-based work. Long Range Planning*, **41**(1), 74–92.

Willmott, H. (1995) Will the turkeys vote for Christmas? The re-engineering of human resources, in Burke, G. and Peppard, J. (eds) *Examining Business Process Re-engineering*, Kogan Page, London, pp. 306–315.

Womack, J., Jones, D. and Roos, D. (1990) *The Machine that Changed the World*, Rawson, NY.

Wood, J.R.G., Vidgen, R.T., Wood-Harper, A.T. and Rose, J. (1995) Business process redesign: Radical change of reactionary thinking? In *Examining Business Process Re-engineering* (eds G. Burke and J. Peppard), pp. 245–261. Koran Page, London.

8 Lean Construction

8.1 Introduction

The preceding chapter engaged with the topics of business process re-engineering (BPR) and partnering. The purpose of this chapter is to extend the discussion to lean thinking and the way in which it is enacted in the construction sector. Lean construction started to attract interest from academics in the early 1990s. But it was only with the publication of *Rethinking Construction* (Egan, 1998) that the discourse of lean construction became central to the quest for industry improvement amongst practitioners. At the time of writing, there continues to be much debate regarding the definition of lean construction and what it means to be 'lean'. The antecedents of the concept lie in the ideas of lean production, as developed originally in the Japanese car industry (Womack *et al.*, 1990). The definition of lean production is itself contentious and is best understood as a complex cocktail of ideas including continuous improvement, flattened organisational structures, teamwork, the elimination of waste, efficient use of resources and cooperative supply chain management. However, the debate is complicated by the subsequent use of 'lean thinking' as the generic term to describe application beyond manufacturing (Womack and Jones' (1996). Further complications are added by the notion of 'lean organisation' and the metaphorical connotations associated with the word 'lean' itself. Yet the ongoing confusion of definition has not prevented 'lean construction' from becoming an established component of construction best practice (Flanagan *et al.*, 1998; Saad and Jones, 1998).

In common with BPR, there is an extensive critical literature relating to lean production which is too often ignored. It will be argued that if lean thinking is to be understood, it is important to appreciate its potential negative consequences as well as its supposed benefits. This chapter is structured in four parts. First, attention is given to the critical literature on lean production and the way in which Japan is positioned within Western discussions of management. Consideration includes the homogenising

Making Sense of Construction Improvement, First Edition. Stuart D. Green.
© 2011 Stuart D. Green. Published 2011 by Blackwell Publishing Ltd.

effects of globalisation in the automotive sector. The discussion also links lean thinking with the broader discourse of enterprise. Of particular note is the way in which the rhetoric of lean thinking reflects and reinforces the iconic status of the customer. Of further note is the overwhelmingly pre-scriptive nature of the existing literature on lean construction, with little recognition given to the socialised nature of the diffusion process. The pre-vailing production-engineering perspective all too often assumes that organisations are unitary entities where all parties strive for the common goal of 'improved performance'. The second part of the chapter develops an alternative perspective which focuses on the need to understand the processes through which ideas such as 'lean' are diffused. Attention is given to the way in which the conceptualisation and enactment of lean construc-tion differs across contexts, often taking on different manifestations from those envisaged. Consideration is extended to the function of consultants as intermediaries between the producers and users of 'new' management ideas and the role of inter-organisation networks. Given the endless exhor-tations for construction to learn from other industries, consideration is also given to responses to lean production in the automotive sector.

The third section of the chapter addresses specifically lean thinking in the construction context. Coverage includes current popular techniques such as the Construction Lean Improvement Programme (CLIP) and extends to the alleged need for an alternative 'theory of production'. Attention is given to the Last Planner technique, which in some quarters is held to be synonymous with lean construction. The aim is not to dispute that such techniques can be useful in the here-and-now, but to set them against the broader context of sectoral change. It is suggested that the concepts of 'leanness' and 'lean construction' must be understood in the context of 30 years of construction sector re-structuring. Finally, the chapter is concluded by the description of empirical research that sought to access the ascribed meanings of lean thinking by UK construction sector policy makers and practitioners.

8.2   Lean production in critical perspective

8.2.1   *The guru-hype of lean thinking*

The seminal description of lean production is provided by Womack *et al.* (1990) in their best-selling book *The Machine that Changed the World*. The advocated model draws heavily from Japanese management practices and the Toyota manufacturing system in particular. Although the book is undoubtedly well-researched and authoritative, the more critical reader can-not help but be struck by the way organisations are conceptualised as profit-maximising machines. Success is seen primarily to depend upon efficiency in meeting the needs of the customer. Womack *et al.* (1980) readily admit to giving little attention to the special features of Japanese society from which lean production emerged. *The Machine that Changed the World* was undoubtedly a significant intellectual contribution born from an extensive research effort, but the extent to which the findings can be extrapolated to other contexts remains hotly contested. Many continued to challenge the extent to which lean methods are applicable beyond the unique Japanese institutional context (e.g. Dohse *et al.*, 1985; Oliver and Wilkinson, 1992; Kenney and Florida, 1993; Morris and Wilkinson, 1995). The notion that universal management techniques can be applied irrespective of context is in harsh contradiction to the long-established principles of contingency theory (Lawrence and Lorsch, 1967). Ultimately, such concerns were marginalised by the globalisation of the automotive industry. Womack and Jones' (1996) subsequent publication on *Lean Thinking* notably ignored any discussion of contingent responses in favour of universal prescription. Most strikingly, lean thinking was not only advocated for the automotive sector, it was also advocated as being applicable to all sectors – including the provision of services. *Lean Thinking* also displays many characteristics of the popularised 'guru-hype' for which Western managers seem to have a perennial weakness (cf. Jackson, 1996). The evangelical nature of Womack and Jones' (1996) message is well illustrated by the last two sentences of the preface:

> 'In the pages ahead we'll explain in detail what to do and why. Your job, therefore, is simple: just do it!'

In other words, the reader is not required to think, or to waste time reading any other books, or indeed to waste time gaining an education. All of these are considered to be wastefully irrelevant in the quest for improved efficiency. Of particular note is the way in which the role of practising managers is reduced to implementation.

The arguments used to promote lean thinking are, in many respects, identical to those used by Hammer and Champy (1993) to promulgate business process engineering (BPR). Once again, a stark prescriptive intent is combined with a maddening degree of definitional vagueness. Lean thinking is presented as an economic imperative in the face of global competition. The only alternative to adopting leaning thinking is to go out of business. The advice to 'just do it' may well bring a comforting sense of security, but the ambiguity of what is being suggested leaves plenty of scope for different interpretations of precisely what should be done.

8.2.2   *Wizards, villains and Western hypocrisy*

While notably ignored by Womack and Jones (1996) and *Rethinking Construction* (Egan, 1998) the 1990s saw an explosion of interest in the merits and de-merits of so-called Japanese management methods. In essence, the debate hinges on whether Japanese methods are based on nice things like loyalty, empowerment and consensus or whether they are based on nasty things like management-by-stress and exploitation. It seems to suit Western authors to sometimes portray the Japanese as wizards of modern management, and at other times to portray them as exploitative villains. The truth of course is more problematic, and certainly accords with neither of these two extremes. There is certainly much hypocrisy among Western commentators who attribute the success of the Japanese motor manufacturers to Japan's protected home market. It is conveniently forgotten that the hegemony of the US and European manufacturing industry was founded on similar protectionist principles. For example, the most successful sectors of the US economy have been hugely supported through prolonged state intervention. It is difficult to deny that the United States' commercial lead in advanced technology has benefited hugely from a substantial public subsidy via the Pentagon's defence budget.

Much of the early critical literature draws from Marxist theories that depict stark battle lines between 'capital' and 'labour'. Hampson *et al.* (1994) certainly provide a very different account of the Japanese 'culture' of industrial consent and cooperation to that provided by *The Machine that Changed the World*. They argue that an essential requirement for the success of lean production was the weakness of Japanese organised labour in the aftermath of World War Two:

> 'the Japanese labour movement was "systematically disembowelled" in the immediate post-war period by an alliance of Japanese capital, the Japanese state and the United States occupying forces – the latter alarmed by the Communist influence in the Japanese labour movement.' (Hampson *et al.*, 1994)

These of course are not the kinds of issues which are normally raised during discussions of lean production. According to this particular critical storyline, it was the weakness of organised labour that provided the context for so-called enlightened practices such as lifelong employment. The argument is that throughout Japan's economic recovery workers were tied effectively to single companies through firm-specific training and so-called 'enterprise unions'. Dissident Japanese workers have apparently long resented the loss of individual freedom associated with in-company unions. Kamata (1982) is often cited to illustrate how Toyota's single-minded drive for success in the 1970s was accompanied by significant personal deprivation on the part of the workforce. It seems that workers were often required to live in guarded company camps hundreds of miles from their families and suffered high levels of stress at the workplace as they struggled to meet company work targets. But the real concern was that relentless Japanese competition throughout the 1970s and 1980s forced Western corporations to adopt similar methods. The end result was a widespread reduction in industrial democracy and corresponding intensification of work. Yet it seems somewhat farfetched to lay the blame for such trends solely on the shoulders of the Toyota manufacturing system. Karl Marx of course developed his arguments of oppression and exploitation entirely independently of Japanese management practices. And the industrial combines of the Soviet Union notably failed to win many awards for their enlightened human resource management (HRM) practices.

The arguments promoted by the likes of Hampson *et al.* (1994) are heavily shaped by the metaphor of organisation as an instrument of domination. As commented in the preceding chapter, condemning every new initiative as a means of technocratic totalitarianism ultimately becomes sterile and threatens to replace one set of supposed dogma with another. Arguments that lean methods necessarily produce 'bad' outcomes share the same instrumental logic with the contention that lean methods necessarily produce 'good' outcomes. Both perspectives display the same tendency towards technological determinism by denying managers any active role in shaping how such ideas are implemented. Ultimately, both perspectives are equally impoverished. In the final analysis, critiques of Japanese management methods become conflated with critiques of globalisation. The broader story is therefore perhaps less about the Japanisation of Western management practices, and rather more about the globalisation of Japanese management practices.

8.2.3 Globalisation and Japanese transplants

What is clear is that the reality of global competition impinged upon the UK car industry with dramatic effect throughout the 1980s and 1990s. It is equally clear that the enthusiastic adoption of Japanese management methods was not enough to save Rover Cars. Rather more successful have been the Toyota and Nissan 'transplants' which continue to mass produce cars in the United Kingdom (albeit under foreign ownership). The reasons for the success of these so-called transplants are again debated within a hotly contested literature. The authors of the Egan report were especially impressed by the implementation of lean thinking within the Nissan plant in Sunderland. Yet Nissan posted a succession of massive global losses throughout the latter part of the 1990s. The Sunderland plant may well have been the most efficient in Europe, but Nissan globally were at the time in serious trouble. In 1999 French carmaker Renault took a 37% stake in Nissan, and installed Carlos Ghosn (nicknamed 'Le Cost Killer') as president and CEO. The subsequent turnaround in Nissan's fortunes has indeed been impressive; but this was not achieved through the implementation of esoteric notions of 'lean thinking'. It was achieved by the unprecedented closure of several Nissan plants in Japan coupled with a series of global cost-cutting purges. While the UK construction industry was looking to learn lessons from the Japanese automotive industry, Nissan were relying for their salvation on a French-Brazilian 'cost killer'.

Published case studies of Japanese manufacturing transplants also fall consistently short of the 'lean utopia' envisaged by Womack and Jones (1996). For example, Fucini and Fucini (1990) point to the gradual disillusionment of the American workforce at Mazda's plant in Michigan. Despite the relatively high wages available, workers complained frequently about poor safety standards, stress of work, loss of individual freedom and discriminatory employment practices. Similar criticisms have been levelled at the Nissan plant in Sunderland. Garrahan and Stewart (1992) and Turnbull (1988) argue that Nissan's supposed regime of flexibility, quality and teamwork all too often translates in practice to one of control, exploitation and surveillance. It is, however, difficult to sustain the latter argument in the case of the construction sector. Control and surveillance seem quaintly attractive when compared to the anarchy of employment relationships which characterise many major construction projects (Green, 2006). The exploitation thesis also fails to explain the growth of self-employment in the construction sector. Many operatives embraced self-employment with enthusiasm, although many others became self-employed because there was no viable alternative.

Beale (1994) further describes how the Nissan system of continuous improvement is directly dependent upon the existence of a single union agreement that is in effect a 'no-strike' deal. The acceptance of such an agreement was a condition of Nissan's initial location in Sunderland. Nissan also received significant government subsidies to locate in the North East of England, strangely at odds with Thatcher's stated commitment to free-market economics. The relatively high levels of local unemployment continue to provide Nissan with significant negotiating power over the workforce. There is arguably always an implicit threat that production might be switched elsewhere if the workforce refuses to conform. Whilst the workforce may well be grateful for the relatively high-paid jobs that Nissan provides, critics would argue that there is a price to pay in terms of worker autonomy. Beale's (1994) perspective would of course have been far too 'Old Labour' to have been of any interest to the authors of *Rethinking Construction*. His arguments may have been well-received in the 1960s. But in 1994 he was just another dinosaur holding out stubbornly against the enterprise culture.

The stark reality is that the Construction Task Force was simply not interested in hearing arguments against lean thinking. They had been tasked by the Deputy Prime Minister John Prescott to come with solutions, and the solutions had to be compatible with the spirit of the enterprise culture. Thus, the critical literature was systematically ignored in favour of a literature which was equally one-sided in the other direction.

8.2.4   *Lean thinking and the enterprise culture*

The authors of *Rethinking Construction* could perhaps be forgiven for ignoring the hyper-critical literature on Japanese management practices. But the report could usefully have been more sensitive to mainstream criticisms such as those proposed by Rehder (1994):

'Japan's industrial work hours are among the longest in the world and the quality of life is poor and not improving. Public and recently government sentiment in Japan is growing increasingly critical of the "lean system", citing its drain on human and natural resources, its stressful and wasteful short model cycle and its street-congesting and polluting just-in-time system.'

Unfortunately none of the concerns above was even acknowledged. The preceding quote is of particular interest in that it raises concerns about the sustainability of lean thinking. Many would now accept that *Rethinking Construction*'s emphasis on narrowly-defined efficiency in isolation from any consideration of sustainability was its biggest failure. Certainly there was no mention of the disparate needs of multiple stakeholders within *Rethinking Construction*. Notions of stakeholder theory had been cast aside by the remorseless rhetoric of customer responsiveness. This was the enterprise culture at its zenith and the cult of the customer was in full sway. Satisfying the needs of future generations was classified implicitly as waste; and waste of course is to be eliminated.

The assumption which underpinned *Rethinking Construction* was that the UK construction sector had been slow to adopt the sophisticated management techniques that had revolutionalised the automotive sector. Within the broader context of UK manufacturing, it is clear that poor performance cannot be blamed entirely on out-dated management practices. Any meaningful diagnosis must surely take broader institutional factors into account; decades of under-investment and poor industrial relations cannot be solved by an instrumental set of management techniques. Any serious consideration of the relative decline of British manufacturing must take into account the long-term effects of the decline of empire and associated loss of protected markets. Of particular importance in explaining the prolonged lack of investment was the bankrupt state of the British economy post-World War Two. But such broader considerations do not lend themselves to simple solutions. Thus, they seldom appear in accounts which are orientated towards performance improvement. Managerial audiences are rarely interested in understanding problems; they want solutions which bolster their own sense of self-identity.

A much better way of understanding the influence of lean thinking is to focus on the rhetoric in which it is presented, rather than its supposed substantive content. The language of lean thinking undoubtedly resonated with that of the 'enterprise culture' which came to dominate in both the United Kingdom and the United States during the 1980s (see Chapter 2). The prevailing political climate made strategies based on 'cutting out the fat' much more socially acceptable than they would have been during previous decades. In the United Kingdom, the doctrine of enterprise was primarily based on 'wealth creation' in the hope that the resultant benefits would trickle down to other levels of society While it is easy to bemoan the intensification of work in the cause of efficiency, few would deny that many

Western workers had become complacent in the belief that the world owed them a living. Whichever standpoint one adopts, it is surely naïve to advocate that the implementation of lean methods in Western manufacturing can be understood in isolation from these broader social and political changes.

8.3   Understanding diffusion

8.3.1   Perspectives on organisation

An essential starting point for reviewing the models of organisation that are implicit in the published models of lean construction is provided by the unitary, radical and pluralist frames of reference (cf. Fox, 1974; Burrell and Morgan, 1979; Morgan, 2006). The unitary perspective assumes that all parties strive to achieve common objectives for the organisation. It is this perspective that dominates the production-engineering literature and is especially evident amongst the advocates of lean construction (e.g. Koskela, 1992; Ballard and Howell, 1998; Ballard, 2000). Emphasis is given to efficiency, control and leadership with little recognition of conflict or power. Organisations are seen to comprise homogeneous entities with no variation between the interests of individuals (Marchington and Vincent, 2004). Burgoyne and Jackson (1997) suggest that an appropriate metaphor to capture the essence of the unitary perspective is that of a 'parade of individuals marching purposefully forward in step in one direction to the same tune'. From this perspective, the implementation of 'lean construction' in the cause of waste elimination is seen to be in everyone's best interests. Whilst such assumptions are implicit within many improvement recipes, they sit ill-at-ease with the fragmentation and occupational diversity of the construction sector. Of further note amongst those who adopt a unitarist perspective is the assumption that the implementation of lean construction falls within the remit of 'management', i.e. there is an assumption that management are able to implement lean construction irrespective of the actions of others.

In harsh contrast to the unitary perspective described above, Morgan (2006) describes the 'radical' frame of reference as influenced by 'old-fashioned' structural Marxism. From this perspective, society is viewed as 'comprising antagonistic class interests, characterised by deep-rooted social and political cleavages held together as much by coercion as by consent'. Such deep-rooted conflicts are seen to be played out within organisations, which are in themselves viewed as 'instruments of oppression'. The appropriate metaphor here is that of a 'battlefield' where rival forces such as management and trade unions strive to achieve ends that are ultimately incompatible (Burgoyne and Jackson, 1997). As noted above, there is a significant literature that analyses lean production from this perspective, primarily within the context of the automotive sector (cf. Dohse *et al.*, 1985; Garrahan and Stewart, 1992; Beale, 1994). Green (1999a, 1999b) is one of the few authors to mobilise these critical arguments in the domain of lean construction. However, such provocations have had minimal impact on the accepted discourse. Howell and Ballard (1999) provide one of the few attempts from mainstream authors to engage with critical arguments:

> 'We argue that Green misses the key foundations of lean which are drawn from a long history of production management thinking which first attempts to manage the physics of production in the service of higher performance.'

The rejoinder above is primarily of note in its acknowledgement of lean construction's intellectual heritage and the associated unitary frame of reference. The overriding concern with the 'physics of production' illustrates clearly the technocist orientation of the mainstream lean construction literature. It is taken for granted that 'higher performance' is an aspiration shared by all parties. The lack of engagement with critical perspectives is perhaps not surprising. Those who insist on seeing the world in terms of exploitation and domination will inevitably marginalise themselves from the mainstream conversation. Critical arguments based on a materialist understanding of power have in recent years been challenged by postmodernist interpretations that advocate the need for a discursive understanding of power (cf. Alvesson and Deetz, 1996; Fournier and Grey, 2000). The rejection of 'grand narratives' in favour of an emphasis on multiple voices and localised contexts resonates better with a pluralistic frame of reference.

The interpretation of organisations in accordance with a pluralistic frame of reference resonates strongly with the political metaphor introduced in Chapter 6. Within the domain of political science, the term pluralism is usually associated with liberal democracies where potential authoritarian tendencies are constrained by the free-interplay of interest groups with a stake in government (Morgan, 2006). The pluralist view of organisations emphasises the diversity of individual and group interests. In the words of Morgan (2006): 'the organisation is regarded as a loose coalition which has just a passing interest in the formal goals of the organisation'. Morgan's words resonate strongly with Cherns and Bryant's (1984) model of temporary multi-organisations (TMOs) in the construction sector; both are underpinned by the same pluralist perspective. Conflict is accepted as an inevitable characteristic of organisations, although this is not seen to be dysfunctional. Power is the medium through which conflicts of interest are mediated, and interest groups draw power from a plurality of sources (Clegg, 1979; Morgan, 2006). Such perspectives are by no means new, but they have been largely ignored in the debate about lean construction.

8.3.2 *Arenas of enactment*

Burgoyne and Jackson (1997) build on the pluralist perspective to suggest that organisations can usefully be understood as an 'arena'. The arena concept creates an image of an organisation as a space where differences come together, are contested and to some extent, reconciled. The outcomes comprise a partial reconfiguration of the pre-existing factions and alliances. Events in the arena are to some extent visible; all parties can observe what takes place. Furthermore, observers may choose to become active participants if they perceive that their interests are at stake. Perhaps most pertinently, events constitute an element of performance (cf. Clark and Salaman, 1996). Individuals act out 'roles' and utilise 'scripts'. However, Burgoyne and Jackson (1997) go beyond the dramaturgical metaphor to suggest that the appropriate metaphor for understanding the pluralistic perspective is that of the 'carnival':

> 'Within the carnival, dazzling array of seemingly unrelated activities are being simultaneously undertaken by individuals and groups with diverse agendas seeking to satisfy diverse needs and desires. This frenzied activity, however, invariably takes place with more synergy than conflict, and with a dynamic complexity that is beyond the intelligence of any single agent to understand.'

Any appointed (or self-appointed) 'champion' of lean construction, is therefore seeking to act out a role in the above arena. Published 'best practice' guidelines and prescriptive textbooks such as Womack and Jones (1996) provide the scripts against which they improvise. The lexicon of 'lean thinking' therefore provides the language of the performance: 'value must be generated', 'flow must be managed' and 'waste must be eliminated'. Others within the arena will challenge the meaning of such 'jargon' and may marshal other resources as a means of resistance.

Language is therefore mobilised as a source of power whereby individuals compete for influence and resources. The task of 'management' is to shape the debate and convince competing parties to follow their chosen course of action. Language in its narrative form therefore dictates the agenda. It frames the way that people understand and act. But 'management' itself will be a pluralistic arena characterised by competing interest groups. There is rarely any certainty of outcome. Even if lean construction were a coherent recipe for industry improvement, it is doubtful whether this coherence would survive its progressive mediation through the multitude of arenas that characterise each construction project. The likelihood is that the language of lean will mix with the language of other scripts in unique and transient combinations. But any such interactions about the meaning of lean construction will only comprise a small part of the 'dazzling array of seemingly unrelated activities' that characterise the arena.

Notwithstanding the above, it is important to recognise that the acceptability of different scripts will depend upon their persuasiveness as
sense-making devices (Weick, 1995). Practitioners will attach more legitimacy to those narratives that help them make sense of their experienced
reality. The discourse of any improvement initiative is more likely to be
accepted if it resonates with the observed changes already underway.
Managers are increasingly overwhelmed by externally-driven change
over which they have little control. Narratives such as lean construction
may provide some degree of comfort that they are in control of events
whilst boosting their sense of self-identity. More importantly, such narratives may serve to sustain and accelerate structural changes that are
already underway. Such issues are notably ignored by the technocist literature on lean construction. Questions concerning the complex interaction between action and structure over time echo the structure-agency
debate that has long characterised the broader domain of social theory
(cf. Giddens, 1984). Issues worthy of consideration include the relationship between language and action, the manner in which human agency
relates to structural aspects of society and the way that action is structured
in everyday contexts.

8.3.3   Consultants, intermediaries and inter-organisation networks

The pluralist perspective developed above provides a standpoint on the
diffusion of new management ideas that is notably missing from the current literature on lean construction. A further essential point of reference is
provided by Bresnen and Marshall's (2001) consideration of the problems
of transferring and applying new management ideas in the construction
sector. Their observations on the contested nature of mainstream management knowledge are especially pertinent in the case of lean, which continues to defy universal definition and is repeatedly criticised for its lack of
coherence (cf. Kinnie *et al.*, 1996; Legge, 2000). Bresnen and Marshall (2001)
further cite the highly socialised and politicised nature of the knowledge
diffusion process. Such a diagnosis is readily compatible with Burgoyne
and Jackson's (1997) notion of a 'carnival' of activities being played out
across a succession of arenas.

The role of management gurus in the promotion of management fashions was touched upon in the preceding chapter. There is an extensive
literature on management fashions and the way in which they are mar-
keted to practising managers (e.g. Huczynski, 1993; Abrahamson, 1996;
Clark and Keiser, 1997; Salaman, 1998). In many respects 'lean production' follows the pattern of previous management fashions such as total
quality management (TQM) and business process re-engineering (BPR)
(cf. Fincham, 1995; Legge, 2002). All were presented initially as major
innovations that were indispensable for modern managers. Selected
aspects of the associated terminologies were absorbed subsequently into

the discourse of practising managers and in some cases triggered change programmes with direct material consequences (Benders and van Bijsterveld, 2000). The process through which such fashions are generated and diffused therefore impacts directly upon the 'reality' of modern organisations. Of particular importance is the range of actors involved in the fashion-setting process. Active participants extend beyond management gurus to include business school academics, practising managers, the business media and government bodies. Scarbrough (2003) gives particular emphasis to the intermediary roles performed by consultants and professional groups in terms of mediating between management gurus and end consumers. This mediating function is seen to be enacted through their involvement in inter-organisational networks and their ability to legitimise new ideas in the eyes of practising managers.

Within the context of the construction sector, the networks that are mobilised include a diverse range of interest groups, quangos, government outreach bodies and membership clubs committed purportedly to promoting 'change'. The evolving infrastructure of industry improvement has been described at various stages throughout the preceding chapters, but particular emphasis has been given to the various networks that were spawned in the wake of *Rethinking Construction* (see Chapter 5). The networks of industry improvement are invariably in a constant state of flux with little inherent stability. Legge (2002) argues that consultants use such networks to build the client base for their products. In the case of TQM, consultants are seen to have developed three 'good stories' that enable them to sell TQM in response to a range of different concerns. The first is termed the operational management 'story' and emphasises conformance to the requirements of customers. The second presents quality in terms of 'value for money' and is used to counter concerns about low price. The third story draws from the 'excellence literature' (e.g. Peters and Waterman, 1982) and emphasises employee involvement and empowerment as a means of achieving a 'quality culture'. Management groups within client organisations are seen to mobilise different stories to suit their own political agendas. In this respect, consultants and users have a shared vested interest in the 'interpretative viability' of management fashions (cf. Benders and van Veen, 2001). The key point is that actions undertaken under a fashion's label vary significantly across contexts. Indeed, the inherent ambiguity of a management fashion is essential for its effective diffusion. The possibility of generating alternative storylines makes the label much more marketable in a wider variety of contexts. If promoters were only able to mobilise one 'good story' it would be perceived as relevant in far fewer cases. Even if there were an uncontested core of technological innovations that are central to lean production, such innovations would still be subject to complex processes of social shaping (cf. Bijker *et al.*, 1987; Williams and Edge, 1996).

8.3.4   Responses to lean production in the automotive sector

As has already been observed, construction practitioners are exhorted repeatedly to learn lessons from other sectors, especially the automotive sector. The lessons to be learned, however, are not quite as straightforward as the advocates of the Egan agenda would have us believe. Scarbrough and Terry (1998) make a useful contribution in providing specific insights into the diffusion of lean practices in the UK automotive sector. Drawing from a series of previous studies, they outline two competing models to explain a firm's response to lean production. The first is derived from the prescriptive literature and focuses on the diffusion of the 'Japanese paradigm' of lean production as a supposed unique socio-technical innovation that offers significant advantages over other methods. This is very much the line of argument advocated in the prescriptive literature (cf. Wickens, 1987; Womack *et al.*, 1990), and also echoes previous arguments in favour of BPR. The contention is that competitive forces in the marketplace leave firms with little choice other than to adopt the most efficient techniques and practices. The key elements of lean production are seen to be universally applicable irrespective of context, with little importance accorded to the history of individual firms or sectors. Scarbrough and Terry (1998) emphasise that the assigned role of management is that of acceptance; there is little scope for managerial discretion. It is this same storyline of technological determinism that dominated the plethora of consultant-led courses on lean construction which followed in the wake of *Rethinking Construction*. The fact that courses of this nature continue to attract audiences suggests that the message is popular with at least some industry representatives. However, there is little evidence in support of the model of universal adoption which was largely discredited even before lean thinking became popular with construction sector policy makers (cf. Abo, 1994; Morris and Wilkinson 1995; Delbridge 1998).

The second model outlined by Scarbrough and Terry (1998) follows Storey and Sisson's (1989) diagnosis and is described as the 'bolt-on' model of change. From this perspective, lean production is seen to be an addition to the technical fixes already available. Emphasis is placed on the constraints of the cultural and institutional context. Storey and Sisson (1989) cite the limitations imposed by de-regulated labour markets and the institutionalised allegiance to short-term cost reduction policies. They further note a deeply ingrained aversion to risk amongst British managers, who are condemned to adopt short-term cost reduction policies as a result of external factors beyond their control. Managers can therefore give lip-service to the language of lean production, whilst persevering with established practices and routines. At best, individual 'lean techniques' are piloted alongside other accepted techniques. The bolt-on model rejects any radical re-orientation in operating practices as infeasible given contextual constraints within which managers operate. In essence, context and local history are of primary importance in conditioning a firm's response to lean production. This view resonates strongly with Bresnen and Marshall's (2001) argument that the institutional-cultural environment surrounding the construction process must be taken into account when seeking to import approaches developed in other sectors. It is this argument which has shaped the structure of this book. Improvement techniques such as lean construction are not directed at a blank piece of paper, but at specific institutional-cultural environments with pre-existing dynamics of change.

Scarbrough and Terry (1998) subsequently move beyond the above dichotomy to offer an alternative 'adaptation model' informed by empirical research within Rover and Peugeot-Talbot during the mid-1990s. The adaptation model credits management with a more proactive role and highlights the possibility that lean production may act as a catalyst for workplace change. However, the resultant changes are by no means predetermined by the pre-existing discourse of lean production. In this respect, Scarbrough and Terry's (1998) findings concur with Benders and van Bijsterveld's (2000) study of the diffusion of lean production in Germany. Both studies give prominence to the tendency for innovations to be re-invented within localised contexts. The outcomes of the adaptation process will therefore differ across contexts with little uniformity, and often in different forms to those envisaged. Hence attempts to generalise about the outcomes of 'lean production' become highly problematic. Historical and plant-level factors continue to be important and change is construed as highly path dependent. The dynamics of the wider institutional context are also seen to be critical in shaping technological innovation (cf. Pettigrew, 1997). This observation is equally relevant to the ways in which 'new' management ideas are enacted in construction firms. The history of individual firms and their existing embedded practices will influence how ideas such as lean thinking are interpreted and enacted. The dynamics of the broader institutional context are of further importance in shaping and constraining responses to new ideas. But the heterogeneous nature of the construction sector means that ideas such as lean thinking are inevitably enacted in an almost endless variety of different ways.

8.4   Lean thinking in the construction context

8.4.1   Lean improvement techniques

The preceding discussion provides sufficient background to engage with the literature relating to lean thinking in the construction context in a more informed way. It is clear that the implementation of lean is influenced crucially by the terrain at which it is directed. The experience of lean production in the automotive sector suggests that the most likely outcome is that managers give lip-service to the language of lean production, whilst perse-vering with established practices and routines. However, the adaptation model opens up the possibility of more creative approaches whereby the ideas of lean production are used as a catalyst for change. In this respect, the fact that lean production comprises a transient and ill-defined cocktail of different ideas can be seen as a positive advantage. It has already been argued that 'interpretative flexibility' is an essential characteristic of any management fashion. In other words, the persuasiveness of lean thinking is dependent upon its ability to take on different meanings in different con-texts. Thus, the absence of any universally agreed definition of 'lean con-struction' ceases to be a problem and becomes an essential criterion of success.

In common with TQM, there are sufficient 'good stories' with in the lit-erature on lean thinking to cater for a wide range of different tastes. Lean production is variously understood as a set of techniques, a discourse, a 'socio-technical paradigm' and even a cultural commodity. This coexist-ence of different interpretations is no less evident within the specific domain of the construction sector. The Egan (1998) report was responsible for popu-larising the lean label amongst construction professionals, and it is clear that the Construction Task Force saw lean thinking primarily as a set of techniques:

> 'Lean thinking presents a powerful and coherent synthesis of the most effective techniques for eliminating waste and delivering significant sustained improvements in efficiency and quality.' (Egan 1998)

Of particular note is the strong reliance on an underlying machine meta-phor. Also of note is the way in which 'lean thinking' could easily have been replaced by 'business process re-engineering' with little loss of meaning. The underlying allegiance is to instrumental techniques directed towards eliminating waste and improving efficiency. The reference to quality would appear to have been tagged on at the end as an afterthought. If the advocated techniques do not result in the promised efficiency improvements, it is always possible to blame those responsible for imple-mentation. They obviously were not applied with sufficient enthusiasm. Hence the faith in instrumental management techniques remains intact irrespective of the outcome. What is clear is that practising managers – and policy makers – are continuously on the look out for new techniques to demonstrate that they are up-to-date with the latest thinking. And there is always a plethora of management consultants striving to meet the latent demand. One example amongst many is the Construction Lean Improvement Programme (CLIP) offered as a consultancy service by the privatised BRE Ltd. CLIP's focus on process improvement through the elimination of waste points towards its antecedents in BPR. The advocates of CLIP also lay claim to benefits within the areas of communication, planning and logistics. Collaborative planning is seen to be especially important in terms of bringing 'the whole team together at an early stage: client, designers, main contractor, subcontractors and specialists, in order to plan the optimum sequence of works'. There is an unstated assumption that the identified 'optimum sequence of works' serves equally the interests of all parties. In other words, it is taken for granted that the context of implementation is unitary.

The notion of early contractor involvement (ECI) is hardly a new idea, having been advocated previously by Banwell (1964) However, what is notable is that construction planning is no longer seen to be the sole responsibility of the main contractor; planning is now seen to be a collaborative process which requires inputs from subcontractors and specialists. The benefits of better planning are of course presented as a 'win-win', but it is difficult not to read the popularity of CLIP as further evidence of the changing role of the main contractor. Having retreated initially from the physical activities of construction to focus on planning and coordination, it now seems that main contractors are increasingly determined to delegate their responsibilities even further. Whether subcontractors and specialists receive any extra payment for sharing this additional expertise is not made clear. Experiences of partnering suggest that subcontractors are squeezed ever more tightly to contribute more input for less return. Embedded industry recipes are not so easily overcome. What remains in short supply are independent evaluations of the way CLIP is enacted in practice. Such evaluations are unlikely to be conducted by BRE given they have a commercial interest in promoting CLIP as an effective technique directed towards eliminating waste and improving efficiency.

8.4.2 From factory physics to last planner

For those who wish to extend their understanding beyond instrumental improvement techniques, there are plenty of other takes on lean construction which warrant exploration. The Lean Construction Institute (LCI) in the United States has a sustained track record of promoting lean construction as a system of production control (e.g. Ballard and Howell, 1998; Choo *et al.*, 1999; Ballard, 2000). Tommelein's (1998) work on pull-driven scheduling and simulation is also often cited as an exemplar of lean construction. Alternatively, Koskela (1992) has attracted many admirers within the International Group for Lean Construction (IGLC) for his advocacy of the need for an alternative conceptual model of the production process (e.g. Koskela, 1992). A consistent point of commonality is their shared reliance on an underlying unitary conceptualisation of organisation. Such authors tend to see lean construction as being 'inspired' by lean production, rather than being concerned with the slavish application of the Toyota manufacturing system. Koskela (2000), for example, is clear that his alternative 'theory of production' is distinct from the supposed Japanese model.

He is notably critical of the lean cocktail proposed by Womack and Jones (1996), whose five core 'principles' are unceremoniously dismissed as slogans. Koskela (2000) further criticises Womack and Jones' terminology for being imprecise and unsystematic; few would disagree with him in this respect. Koskela's response is to retreat to the level of abstract theorising, with little interest in how such theories play out in practice. According to Clegg *et al.* (2006) the theories on offer are highly rationalistic with little recognition that projects rarely unfold evenly and smoothly in accordance with pre-determined plans. Koskela may wish to distance himself from the guru-hype of lean thinking, but the machine metaphor nevertheless remains dominant.

To theorise on how the 'physics of production' might be applicable to construction may be commendable as an ivory-tower exercise. But the overriding challenge must be to connect with the day-to-day challenges faced by practising construction managers (Green and Schweber, 2008). Theories of course come in many shapes and sizes. Yet there seems to be little interest in theories which emphasise the way in which improvement recipes such as lean thinking are diffused in practice. Little recognition is accorded to the way ideas are re-shaped by the cultural-institutional context of the construction sector, or by the unique path dependencies of different firms. The heterogeneous nature of the construction sector renders generalisations few and far between. It must also be borne in mind that any attempt to learn 'new methods of working' also depends upon the 'unlearning' of previously embedded practices. Unpicking existing ways of working inevitably entails risk, with little in the way of guaranteed return. Hence the 'bolt-on model' tends to be more feasible than radical change. Firms are not condemned to replicate existing ways of working forever, but change tends to occur as a process of evolution rather than top-down design (Green *et al.*, 2008).

The project-based nature of the construction environment also means that processes of innovation often extend across organisational boundaries; the fact that the innovation process is rarely within the control of a single organisation adds further to the uncertainty of outcome. But for present purposes, the real point of interest is that innovation is far more situated across multi-organisation networks than it was 40 years ago. The large contractors of the 1970s were much less reliant on engaging others than the hollowed-out firms of today. The paradox is that the monolithic contractors of the past operated in a much more stable environment, and thus had little reason to innovate.

Ballard's (2000) *last planner* technique is perhaps more immediately relevant to the challenges faced by industry practitioners. The basic idea of the last planner is to ensure that tasks only commence when preceding tasks are 100% complete and all necessary resources are available. Such tasks are labelled 'quality assignments' and the project manager is expected to ensure there are always a buffer of such assignments available – even if not necessarily on the critical path. As Winch (2002) points out, the difficulty is that work eventually grinds to a halt should there be an insufficient number of quality assignments. The focus on self-organising gangs as discrete resource units once again resonates strongly with an increased reliance on subcontracting. Contractors' managers who have retreated from the supervision of directly-employed operatives may find the last planner persuasive in helping them make sense of their revised role. In this sense, the last planner goes with the grain rather than against it. The broader context is that main contractors' site managers are no longer able to re-assign individuals from one gang to another in response to contingent circumstances. Gangs now arrive on site as inseparable units; even if the gang members only met when they were picked up at dawn in the proverbial white van. The implication is that construction planning is now routinely dictated by the pattern of subcontracting, rather than the immediate imperatives of the physical activities of construction.

Of further note is that the last planner is essentially tactical in its orientation and, as such, can easily be combined with existing practices. Practitioners can therefore lay claim to 'lean construction' with little in the way of radical change. Even if the last planner were a radical new innovation, it would be unrealistic to expect it to retain any coherence once implemented in real contexts. In the wake of *Rethinking Construction,* many pre-existing techniques aimed at eliminating waste were also opportunistically relabelled 'lean construction', thereby rendering before-and-after analysis untenable. Empirically, it becomes impossible to discriminate between new innovations and old innovations which have been re-labelled. There are, incidentally, a number of points of commonality between the last planner and CLIP. Both depend upon a model of collaborative planning and both take it for granted that all parties benefit equally. But in both cases it is more meaningful is to focus attention on the localised processes of innovating rather than the supposed substantive content. And localised processes of innovating will be shaped by what has gone before. Attempts to implement BPR may evolve progressively into attempts to implement lean construction. At the same time, advocates of partnering will be keen to influence at least the language that is used to enact the innovation. Thus, the material manifestation of what is being implemented may evolve over time; but the biggest changes are likely to relate to the fashionable status of the labels around which localised innovations occur.

8.4.3 *From leanness to anorexia*

Moving beyond the narrow confines of the literature on lean construction, the term 'lean organisation' has a much wider currency and carries with it a number of metaphorical connotations. This wider debate extends beyond idealised models of production to consider the defining characteristics of 'lean organisation'. Legge (2000) suggests that the popularity of 'lean' can in part be credited to the nuances of the word itself. 'Lean' is the opposite of 'fat', but different from 'thin'. Whilst both 'fat' and 'thin' have negative connotations, 'lean' equates with healthiness and fitness. In this respect, 'lean' is a metaphor for a desirable bodily condition. To say that an organisation is 'too fat' would be to invite a degree of 'downsizing'. On the other hand, to say that an organisation is 'too thin' (or even anorexic) would be to imply that it lacks some sort of nutrition and, by implication, the capacity to perform the required functions. Leanness of course also resonates with the discourse of the enterprise culture. The dominant model of competitiveness in the construction sector has already been seen to rest on leanness and agility in the market place (see Chapter 3). What is clear is that the advantages of being lean were recognised two decades prior to the publication of *Rethinking Construction*.

The argument in favour of leanness in essence rests on the dangers of being fat and unresponsive in a competitive marketplace. Fat organisations carry too many overheads, caused typically by owning their plant and employing their own operatives. In consequence, fat organisations struggle to be enterprising. Lean organisations, in contrast, are much quicker on their feet. Lean organisations are designed specifically to survive fluctuations in demand. But leanness runs the risk of being an unstable state, with disadvantageous implications for the sector at large. Lean organisations do not invest in training, or implement enlightened human resource management (HRM) practices. Lean organisations rely on subcontractors, which in turn rely on LOSCs which too often depend upon agency labour. Contracting firms which adhere to the lean model cease to take responsibility for the operatives who perform the work. The lowest common denominator is the vulnerable self-employed operative with limited rights to welfare provision, sick pay and pension contributions. The UK construction sector embarked on the road to leanness in the late 1970s. *Rethinking Construction* served to endorse and reinforce the trends that were already deeply embedded within the dominant industry recipe, if only by omission. Construction sector leaders could at least point towards the *Respect for People* initiative as evidence that 'something was being done' about the damaging side effects of lean construction (see Ness, 2010).

8.5   The meaning of leanness

8.5.1   *Stages of leanness*

One of the most comprehensive attempts to define 'leanness' is that offered by Kinnie *et al.* (1996), who define leanness in terms of three overlapping phases. Stage 1 is professed to be a transitional phase where the focus is on helping the organisation become leaner. Stage 2 concentrates on the achieved state of leanness as an end point, whereas Stage 3 focuses on leanness as a process for managing the lean organisation so that it remains lean and responsive. They further suggest that the most negative connotations are associated with the first stage, and the most optimistic with the last. Despite not being directed specifically at the construction sector, their diagnosis is nevertheless highly pertinent, and goes some way towards providing a contextually sensitive interpretation of leanness.

The adoption of Kinnie *et al.*'s (1996) three-phase model of leanness provides the basis for bringing together many of the arguments presented previously. Under Stage 1, key topics include restructuring, downsizing, de-layering and changes in the contractual status of employees. In this respect, the onset of leanness in the UK construction sector can be traced back at least 30 years. Since the mid-1970s, the industry has seen extensive re-structuring involving the outsourcing of labour and increased reliance on subcontracting (see the detailed description in Chapter 3). Coupled with this was the significant shift in the employment status of the industry's workforce. Self-employment grew from 30% of the total workforce in 1977 and peaked at 60% in 1995 (ILO, 2001). The major contractors that characterised the 1970s have since developed into exemplars of the lean 'hollowed-out' firm. As such, they have largely removed themselves from the physical work of construction, preferring to concentrate on management and coordination functions. This trend has significantly undermined the industry's capacity for training and innovation (cf. Gann, 2001). Recent years have seen a series of intermittent attempts by government to tighten up the taxation regimes governing self-employment (Harvey, 2003). These partial clampdowns initiated a limited shift back to direct employment during the late 1990s, but in truth the long-term accumulative effects of downsizing, de-layering and outsourcing were hardly dented. At the time of writing the level of self-employment is slowly rising once again (Harvey and Behling, 2008). The downward pressure of the unregulated competitive processes described in Chapter 3 became even more ferocious in times of recession. Nevertheless, it would therefore seem that many construction firms completed Stage 1 long before 'leanness' came into vogue.

Stage 2 – leanness as an outcome – was arguably achieved for the majority of firms in the mid-1990s. According to Kinnie *et al.* (1996) the critical characteristic of a lean organisation that has gone though a phase of delayering, re-structuring and downsizing is its assumed structural flexibility. The notion of firms being 'staffed for troughs, not peaks' will be familiar to anyone with experience of the construction sector. Forms of flexibility such as reliance on subcontracting, overtime, temporary and seasonal working have long since been part of the construction sector's 'industry recipe' (cf. Spender, 1989). These initiatives are invariably centrally directed and imposed. Business process re-engineering (BPR) can also have a role here. Task flexibility is further enhanced through team-working and work intensification as a means of managing with fewer employees. In preference to carrying a fixed overhead of directly-employed workforce, subcontractors are assembled on a project-by-project basis in accordance with local market conditions. However, once this lean outcome is achieved, Kinnie *et al.* (1996) argue that it immediately creates problems in that the assumed flexibilities are not sufficient to allow firms to respond to new market demands. This observation echoes similar criticisms by Naim and Barlow (2003), who argue that the lean model frequently offers efficiency at the expense of effectiveness; i.e. lean organisations may lack the 'ability to give customers exactly what they want, when they want it'.

Stage 3 of Kinnie *et al.*'s (1996) model is concerned with the ongoing process of managing the lean organisation so that it remains lean and responsive. The model focuses attention on the weaknesses of 'leanness' as an outcome in the context of a turbulent external environment subject to strong competitive pressures. Leanness as a process focuses attention on the attributes required by organisations if they are to respond to environmentally induced change. Tactics here include continuous improvement, TQM and team-working, which are described as the means of instilling a capacity for change at the lower levels of the organisation. Such techniques are seen to counter-act the lack of responsiveness of lean organisations. Cultural change programmes and supply chain management (SCM) are also placed within this stage. In the case of the construction sector, ongoing attempts to implement partnering and collaborative working could also be included as specific tactics for counter-acting the loss of control associated with an over-reliance on leanness. The inherent fragmentation of the construction process would also account for continuing pleas for 'integration' by bodies such as the Strategic Forum (2002) (see Chapter 9). The same might also be said of techniques such as CLIP and last planner. It is perhaps not surprising that SCM is so much in vogue among main contractors' managers given that it is almost the only task they have left to do. Certainly the task of managing the main contractor's direct employees cannot be especially onerous.

Kinnie *et al.* (1996) are ambiguous in terms of whether the original recipe of lean production promoted by Womack *et al.* (1990) fits into Stage 2 or 3, although they do concede that there is much overlap between these stages. In many respects Stages 2 and 3 are best understood as largely concurrent and mutually adjusting. In this respect, Rees *et al.*'s (1996) interpretation of lean production is useful in that it distinguishes between the 'hardware' of production systems such as just-in-time (JIT) and the 'software' of 'high commitment' HRM practices. Womack *et al.* (1990) notably gave little emphasis to the importance of integrating HR practices into the lean production model (cf. MacDuffie, 1995). One possible interpretation of how lean production fits into Kinnie *et al.*'s model is that the 'hardware' falls within Stage 2 whereas the 'software' falls within Stage 3. Given the continuing low status and influence of HRM in the construction sector (Cully *et al.*, 1999), this raises the question of whether there is any significant number of construction firms who have reached Stage 3. However, this comment must be judged against Kinnie *at al.*'s (1996) observation that few organisations of any sort appear to get to this stage.

Finally, it is pertinent to consider the relationship between the previously outlined models of diffusion and the three-stage model of leanness described above. Stage 1 is clearly driven by economic imperatives with little discretion on the part of middle managers. However, management responses during Stages 2 and 3 will be highly path dependent and different organisations will tend to adapt the lean production paradigm to suit their own circumstances. Possibilities include both the 'bolt-on' and 'creative adaptation' models. The notion of organisational arenas also remains useful in understanding how different ideas compete for attention. However, – if Kinnie *et al.* (1996) are right – it is only the achievement of leanness as an outcome (with the associated backcloth of downsizing and de-layering) that makes 'lean production' meaningful. The overall picture is one of significant complexity; but the important point is that lean construction cannot be understood as a 'best practice' which is implemented in the same way across a diversity of different contexts.

The framework developed here differs starkly from those provided in the lean construction literature. However, it should be noted that Ballard and Howell (1998) start their argument in favour of 'shielding production' with the observation that project managers in construction 'manage contracts' rather than task execution. There is therefore an implicit assumption that the organisation in question has already become 'lean' in accordance with Stage 1 of Kinnie *et al.*'s (1996) model. It should further be noted that the interpretation above resonates with Pettigrew's (1997) notion of 'processual analysis' that seeks to understand organisations as a sequence of individual and collective events, actions and activities unfolding over time in context. Of particular importance is the recognition of a complex interplay between human action and the context within which it occurs. Managerial actions are not only shaped by context, but are also *shaping*. To some extent at least, the current context of the UK construction sector has been shaped by government and management actions over the last 30 years. From this perspective, 'leanness' is a systemic outcome and 'lean construction' (in all its variants) is the means by which managers seek to eradicate the damaging side effects.

8.5.2 *Perceptions of lean amongst industry policy makers*

There is relatively little research which sets out to explore lean construction as advocated above. Nevertheless, it is possible to access the storylines propagated by those policy makers who advocate lean thinking for the UK construction sector. Green and May (2005) report on 25 interviews which were conducted with construction sector policy makers in the period May 2001 to April 2003. The interviewees were drawn from a range of quangos, institutions and sector 'clubs', including Egan's original

Construction Task Force. In addition to those directly involved in The Egan report, senior representatives were interviewed from the following organisations:

- Strategic Forum for Construction
- Construction Industry Training Board (CITB)
- Royal Institute of British Architects (RIBA)
- Construction Research and Innovation Strategy Panel (CRISP)
- Design Build Foundation (DBF)
- Construction Industry Council (CIC)
- Construction Round Table (CRT)
- Construction Best Practice Programme
- Reading Construction Forum (RCF)

Whilst it is appropriate to list the organisations above, it should also be emphasised that the views expressed were personal opinions rather than sanctioned policy statements. It is interesting to note that several of the above named organisations have ceased to exist since the interviews were conducted. The fractured and ever-changing landscape of the industry's lobby groups stands as a testament to the difficulties of harnessing disparate interest groups to the 'common good'. The interviewee sample also included senior representatives of construction firms and client organisations who were at the time recognised as exemplars of lean construction. Whilst these companies were frequently cited in numerous 'sanitised' case studies promoting the benefits of lean, it is not appropriate that they should be named here. It should be noted that several interviewees had multiple allegiances, i.e. they were active within at least one of the policy arenas above whilst also promoting lean thinking within their own organisation. A further two interviewees were sourced from within consultancy firms actively involved in the promotion of lean construction as an improvement recipe.

It is perhaps not surprising to record that the perceptions of 'lean' amongst industry policy makers shared all the vagueness of definition, contradictions and ambiguities found in the literature. The variety of lean perspectives and 'theories' described by the interviewees was striking. They ranged from narrowly-defined operational techniques implemented at project level to strategically orientated, sometimes industry-wide models. Some of the descriptors used were:

- *'... it is a philosophy, a way of thinking.'*
- *'.... an infinite stream.'*
- *'... a set of principles.'*
- *'... a set of techniques.'*
- *'... a process where there is no going back, the task and the pull of the next phase drives the pace ...'*

Just-in-time (JIT), partnering and supply chain management (SCM) were all variously positioned as over-arching themes as well as tools that are used in the process of 'becoming lean'. What was notably absent was any interpretation of leanness in the more generic sense as proposed by Kinnie *et al.* (1996).

Beyond the dedicated lean champions, few of those interviewed admitted to having read the source texts on lean production (e.g. Womack *et al.*, 1990; Womack and Jones, 1996). Even fewer were familiar with, or even aware of, the work of the Lean Construction Institute (LCI) or the International Group for Lean Construction (IGLC). However, several had met Dan Jones in discussion seminars preceding the publication of the Egan report, and all were aware of the origins of lean production in the Toyota manufacturing system. One interviewer remarked upon the absence of explicit 'lean' references in *Rethinking Construction* although he went on to emphasise that it was 'of course, all about lean'.

Without exception the interviewees were thoughtful and reflective. There was scepticism amongst many regarding the extent to which the Japanese model of lean production could be applied directly to the construction process. Certainly there were few advocates of the 'universal applicability' model of diffusion. This was summed up nicely in the comment: 'simple idea, but very difficult to put into practice given the nature of construction'. Others were somewhat measured in their support for lean thinking even in principle and preferred to see it more as a rallying call for industry development rather than as a specific recipe for improvement. It was interesting to note that there was an almost unanimous agreement on the need to improve performance, which was construed invariably to be identical to waste elimination. Several were fond of using the Japanese word '*muda*' to signify waste (usually accompanied by a raising of the eyebrows). *Muda* was seen to take a variety of forms and its elimination was generally seen as a healthy way of reducing costs.

Whilst many were sceptical about the application of lean production to the construction process, several suggested that the real opportunities for application of the 'purer' models lay with the off-site manufacture of components and pre-fabricated units. The lean model of off-site manufacturing extended to the just-in-time delivery of components to site. Two of the clients interviewed were very clear that they had been 'doing lean' even before the term was invented. Both of these organisations were at the time cited repeatedly as exemplars of lean construction. One claimed to have achieved especially impressive productivity improvements for repetitive building through prefabrication, modularisation and multi-skilling. The interviewee claimed that their buildings are 90–95% complete in the factory:

'all we have to do on site is fit the "lego" parts together, and this takes a matter of hours rather than days.'

However, the motivation for investing in prefabrication did not lie within the pages of *Rethinking Construction*. The interviewee representing one of the internationally-known clients had initially been motivated by the company's experiences of industrial unrest in London, dating back to the early 1970s. Simply put, the shift to modern methods of construction (MMC) had been motivated by a determination not to be 'held to ransom by trade unions'. Thus, the company had deliberately shifted pre-assembly to non-unionised locations in the North of England as a long-term risk-management strategy. Prefabrication, modularisation and pre-assembly can therefore be seen as material manifestations of the enterprise culture.

A significant proportion of the interviewees associated lean thinking with partnering, which was sometimes described on a project level and sometimes at a more strategic level in terms of framework agreements. Several interviewees also referred to supply chain management, which was often used interchangeably with partnering. It is interesting to note that partnering was either seen to be a means of achieving leaner working, or alternatively as an outcome of leaner working. Several attributed Latham (1994) as the main promoter of partnering. Indeed, several expressed the view that they felt much more comfortable with the Latham report than they did with the subsequent Egan report. It was also of note that a number of interviewees felt the need to emphasise that they were talking about 'real' partnering, thereby acknowledging that some contractors talked about partnering 'whilst screwing discounts out of sub-contractors'.

The expression 'integration' was used a great deal by several interviewees (cf. Strategic Forum, 2002). Some conceptualised integration in terms of increased trust, teamwork and cooperation amongst contracting parties; thereby leading to a reduction in 'adversarial waste'. Others described it in a more instrumental way, focusing on the need for better planning and task organisation. In either case, integration was seen as an antidote to the fragmentation of construction, both in terms of the supply base and the construction process. Some interviewees linked fragmentation to flexibility, describing them as 'different sides of the same coin'. Flexibility to expand and contract in accordance with fluctuations in demand was seen to be an important characteristic of contracting firms. Such comments echo the notion of 'structural flexibility' that Kinnie *et al.* (1996) contend to be the central condition of 'leanness as an outcome'. They also demonstrate that the structural changes described in Chapter 3 continue to cast a long shadow over the construction improvement landscape.

Authors such as MacDuffie (1995) emphasise the importance of supporting human resource management (HRM) practices in ensuring the successful outcome of lean production Yet HR issues were rarely mentioned by the interviews reported by Green and May (2005), either spontaneously or explicitly. When prompted specifically by the interviewer, the most common response related to the need for training. However, one interviewee rejected the idea that training was needed as a precondition of the effective implementation of lean:

> '... low-level tasks will increasingly be accomplished by new technology, standardisation and off-site fabrication. Fewer people will be needed and those who are will not require much training since the work will comprise mainly prefabricated unit assembly.'

At the same time, any suggestion of perceived de-professionalisation or de-skilling of design and construction was firmly rejected. The potential for any consequent reduction in motivation and commitment was treated as irrelevant. The long-term implications of the trend towards pre-fabrication should be considered in the context that approximately 45% of the construction sector's turnover is concerned with the maintenance and refurbishment of the existing building stock. Such work relies upon traditional craft skills that will be increasingly eradicated from the new-build sector should the projected dominance of modern methods of construction (MMC) ever become a reality. As an aside, it is interesting how policy sources have increasingly emphasised MMC, rather than 'prefabrication' and 'off-site manufacture', as a means of overcoming memories of the quality problems experienced during the 1960s (Lovell and Smith, 2009). As before, the shadow of past events still looms large over the construction improvement debate.

More generally, interviewees provided scant evidence of confidence in 'high-commitment' HRM practices. One interviewee expressed the view that:

> '... it's a tough industry out there. It's about survival. We really don't have the luxury of investing in namby-pamby HR stuff. We have to get things built.'

Whilst the opinion expressed above was at the extreme end of the spectrum, the interviewees were generally much more comfortable talking about efficiency than they were on the topic of supporting enlightened HRM practices. Paradoxically, several talked at length about the need for better 'soft skills' relating to human issues. Perhaps most pertinent is the fact that only two out of the 25 interviewees talked about HR issues without direct prompting. None of the interviewees made any direct reference to the level of self-employment in the sector; this was simply not judged to be relevant.

8.5.3   Concurrent and competing models of lean

The diversity of the views within the interviews described above make any summary prone to over-generalisation. Nevertheless, Green and May's (2005) analysis reveals three dominant models of lean construction.

Box 8.1   Lean model 1: Waste elimination

Lean Model 1: Waste elimination
- Waste elimination is paramount.
- Technical/operational focus.
- Espoused aim is to ensure smooth uninterrupted flow of activities.
- Assumed that cost savings made at the operational level will aggregate to the corporate level.
- Further assumed that all parties will benefit equally from 'improved performance'.
- Discourse dominated by machine metaphor.
- Underlying unitary perspective on organisations.

However, it should be emphasised that these stylised accounts were frequently coexistent, i.e. individual interviewees were by no means consistent in their interpretation of lean even over the course of a single interview. The three models derived from the interviews should therefore be understood as characterisations rather than accurate depictions of the views expressed. Whilst overly simplified and neat, they could be portrayed as depicting a progression in conceptualisation from a narrow, operational project-level point of view to a strategic perspective on the industry at large. Credit for the initial derivation of these models should go to Susan May, who conducted the majority of the interviews with senior practitioners from which they were derived. The descriptions which follow are once again adapted from Green and May (2005).

Lean Model 1 (see Box 8.1) to a large extent reflects and reinforces previous long-established improvement storylines. For the advocates of this model, the discourse of lean construction has seemingly been assimilated into a pre-existing industry recipe. There is a progressive and seamless transition from scientific management to operational research to value engineering to BPR to lean thinking. The favoured slogans are adjusted accordingly: *'eliminate needless movements'*, *'optimise workflow'*, *'cut out unnecessary cost'*, *'obliterate non value-adding activities'*, *'eliminate muda'*. The words change, but the underlying metaphors remain the same. The worldview is to sort out the inefficiencies first, and thereafter we can worry about other things (if we ever get round to it). The storyline is limited notably to the 'hardware' of lean production. Advocates of this approach are often embarrassed by reference to supporting HR practices. It is indeed a *'hard, cold world out there'*. In terms of diffusion models, Lean Model 1 is very much representative of the 'bolt-on' model.

If Model 1 is characterised as 'assimilation by the old guard', then Model 2 (see Box 8.2) can be characterised as 'assimilation by the new guard'. The

---

Box 8.2   Lean model 2: Partnering

Lean Model 2: Partnering

- Emphasis on relations between firms partnering and supply chain management.
- Project/corporate view.
- Aim is to eliminate adversarial relationship/change culture.
- Leanness is seen to be the outcome of better relationships.
- Less conflict, more trust equals improved collaboration.
- Emphasis on knowledge sharing, learning.
- Dominant metaphors: teamwork, cybernetic.
- Underlying pluralistic perspective on project organisations, unitary perspective on firms.

---

underlying storyline arguably owes more to Latham than to Egan. Most interviewees talked about both 'project partnering' and 'strategic partnering'. Many also talked about 'framework agreements' as a means of formulating long-term relationships. The advocates of this model also had a great deal of faith in the instrumental benefits of 'partnering workshops' as a means of resolving conflict. Effective facilitators were seen as key in this respect, especially in terms of being trained in the 'appropriate soft skills'. A limited number of major clients were seen to be taking the lead; 'if you want to work with us, then you have to work the way we want to'. Other interviewees expressed concerns that clients were becoming too autocratic and that the 'partnerships' on offer were too often somewhat one sided. Contractors are seemingly expected to trust the client, but are themselves subject to 'key performance indicators'. Trust and a commitment to 'measuring performance' are in many respects strange bedfellows (see Chapter 7). Yet the advocates of Model 2 acknowledged the notion of the 'project coalition' to a greater extent than the advocates of Model 1. Several interviewees clearly recognised that different parties in the 'project team' had different objectives that would need to be reconciled. However, the unit of analysis rarely strayed below the level of the firm. The unitary perspective on firms remained very much in place. In terms of diffusion models, Lean Model 2 is once again reflective of the 'bolt-on' model. Furthermore, its advocates were again broadly silent on the need for supporting HR practices and the required shift in the established 'industry recipe'.

Lean Model 3 (see Box 8.3) combines elements of the previous two models, but focuses much more strongly on the institutional mechanisms that shape and condition the context within which participants interact. The emphasis was on the framework within which construction is delivered, and was seen by its advocates to be a radical break from the previous *modus operandi* of project delivery. From this perspective, lean cannot be achieved by considering construction, design and building operation in isolated compartments.

Box 8.3   Lean model 3: Structuring the context

Lean Model 3: Structuring the context
- Lean is about structural change in the way projects are delivered.
- Long-term contractual relationships are an essential pre-requisite.
- Implementation of lean requires 'complete re-think of design and construction'.
- Technology clusters, integrated teams, integrated processes.
- Great emphasis on simplification of design, standardisation, pre-fabrication, application of IT.
- Dominant metaphors: psychic prison, organic, cybernetic.
- Underlying pluralistic perspective on organisations.

Lean is seen to require a re-arrangement of the contractual boundaries between parties, and an incentivisation system that rewards both firms and individual team members. There was a much stronger storyline on innovation as the engine of competitiveness. The industry was therefore encouraged to break out of its long-established industry recipe and operate on a different basis. In contrast to the two preceding models there was at least some recognition of the need to overcome institutionally-embedded practices. Crucially, there was not seen to be a single model of lean that is appropriate in all cases; different solutions were seen to be necessary for different circumstances. The commitment was less about 'continuous improvement' and more about 'continuous evolution'. There was much talk of 'system integration', especially in terms of the processes of design and construction but also in terms of the product. There was also a far greater emphasis on technology than in the previous models. Product and process innovation were seen to be essential components of improved performance.

There was a recognition that training was essential at all levels if the preceding vision is to become a reality. In contrast to the other two perspectives, there was a much stronger emphasis on the role of individuals in ensuring success, and the commitment to 'responsible innovation' rested on the need for diversity in terms of skills and outlook. In terms of the previously described diffusion models, Lean Model 3 is suggestive of 'creative adaptation'. Several interviewees claimed to be involved constantly in horizon scanning on the look out for new ideas from which they could learn. One client had adapted *last planner* to suit their specific purposes and the interviewee was positive about the learning that had been achieved from its implementation.

Lean Model 3 is clearly more sophisticated than the other two models. It is also much more optimistic in terms of its vision for the future. Its advocates claimed to be committed to research and development, but it is notable that the vision was dependent upon some degree of shelter from short-term competitive forces. Others emphasised that the revised way of working is also dependent upon a repeat client with a significant capital expenditure programme. The clients in question were prepared to engage proactively with the construction industry, apparently with a view to ensuring that 'it could achieve its own objectives'. Whilst Lean Model 3 was certainly in a minority in comparison to the other models, it was by no means limited to a single organisation. However, it must be conceded that the advocates of Lean Model 3 were highly ambivalent towards the label 'lean construction'. One interviewee actually suggested that the label was avoided because of the 'lean is mean' connotation: 'we don't really call what we do lean construction. I think we've moved beyond lean now'. One suspects that many of the interviewees would tell very similar stories irrespective of whether they were asked about lean construction or not. There is also a nice irony in moving 'beyond lean' before anybody has quite defined what it is. Perhaps the point at which the meaning of such concepts becomes accepted is the point at which they cease to be useful.

8.6 Summary

The starkest conclusion to be drawn from this chapter is that lean construction defies universal definition. The meaning of 'lean' is characterised by an empirical elusiveness. Nevertheless, the essential vagueness of lean construction has not prevented its acceptance as a recommended component of 'best practice'. The continuing popularity of lean construction techniques is indicative of an interactive collusion between the promoters and receivers of the lean discourse. However, the 'receiving' managers should not be cast as passive, gullible recipients. Management practitioners have a continuous need for persuasive scripts against which they can act out the role of improvement champions. The legitimacy of lean construction is underpinned by sources such as Womack and Jones (1996) and *Rethinking Construction* (Egan, 1998). The lean brand has been further institutionalised by its endorsement by a succession of government outreach bodies and sector membership clubs dedicated to 'industry improvement'. The reported interviews with construction sector policy makers confirm the essential ambiguity of lean construction, even amongst those who promote its adoption. In essence, lean construction has been shown to draw from the same cocktail of ideas as business process re-engineering (BPR) and partnering. All three storylines are mutually conflated, and they must therefore be understood as part of the same discourse. In many respects, the inherent ambiguity of all three improvement recipes is essential for their effective diffusion. The coexistence of different storylines makes them relevant to a much broader range of contexts. As with any other management fashion, the 'interpretive flexibility' of lean construction is an essential condition for its longevity.

The essential vagueness of lean thinking renders generalisations regarding its outcome problematic. In this respect, the critical literature is frequently as impoverished as the practitioner-based prescriptive literature. Arguments that lean thinking necessarily produces 'bad outcomes' rest essentially on the same instrumental logic as arguments that lean thinking necessarily produces good outcomes. In practice, lean construction continues to be conceptualised and enacted differently in different contexts, often taking on different manifestations from those envisaged. The multiple storylines of lean construction also compete for attention with other 'improvement' recipes as managers strive to find pragmatic compromises while demonstrating themselves to be up-to-date with the latest thinking. The most likely outcome is that organisations give lip-service to the language of lean, whilst persisting with established practices and routines. The wholesale implementation of any 'new production paradigm' would entail not only the cost of learning the new approach, but also the risks associated with unlearning the old one.

More optimistically, there is a possibility that lean construction may act as a catalyst for change in the workplace. The 'creative adaptation' option credits middle managers with a much greater capacity for innovation. It also assumes that they have sufficient domain control to break with the institutionally embedded practices that characterise the construction sector. This may well depend upon the creation of a business context that is sheltered from the short-term competitive pressures that shape the sector at large. It is perhaps no coincidence that the 'structuring the context' model was advocated predominantly by interviewees from within large client organisations. As such, they enjoy an unusual degree of market power to influence the behaviour of the construction supply chain. BAA plc is one of the few clients who have been able to operate in such a way, working in close cooperation with a limited number of framework partners. BAA for many years benefited from the relative certainly which resulted from its quasi-monopolistic position in a heavily regulated industry. Here again we see the same paradox alluded to in Chapter 2. Government policies of privatisation all too often led to heavily regulated quasi-monopolies. Enterprise it was not. It is notable that in recent years, BAA has come under increasing pressure from the Competition Commission. In October 2009 BAA agreed to sell Gatwick, the UK's second busiest airport, to Global Infrastructure Partners (GIP) for a fee of £1.51 billion. BAA as a client organisation would never quite be the same again. There is of course an irony in that the client which made most progress in leading change in the construction sector was able to do so as a result of its relative shelter from the rigours of market competition. Had the leading contracting firms been similarly sheltered from the marketplace, they also might have been better placed to invest in long-term construction improvement.

Notwithstanding the above, considerable emphasis has been given to the way the meaning of 'lean construction' is reinvented repeatedly within localised arenas. Practitioners can be seen to engage in a continuous search for meaning. As part of this process they negotiate rules for action which in turn shape the context within which they operate. Ideas of lean construction are thereby shaped by the context within which practitioners operate, but they also act to shape the opportunities for future action. This is not to say that the construction industry is doomed to follow the model of leanness and agility which became dominant during the 1980s. Each generation of managers has the opportunities to re-negotiate the way the industry operates. But it is important not to keep repeating the mistakes of the past. Of central importance is the relationship between language and action, the way that human agency relates to structural aspects of society and the way that action is structured in everyday contexts.

What is clear is that lean construction cannot be understood in isolation from long-term structural changes at the sector level. Managers seek continuously legitimising discourses to help them make sense of experienced reality. The legitimacy of lean discourse is rooted in 30-year trends of corporate re-structuring, de-layering and outsourcing. Such trends are characteristic of the transitional stage of 'becoming lean' in search of structural flexibility, as described by Kinnie *et al.* (1996). In this respect, the construction sector embarked upon the quest for leaner ways in the late 1970s, long before the terminology of lean thinking became fashionable. Lean construction therefore cannot be separated from the widespread reliance on non-standard forms of employment. Of particular concern in the UK construction sector is the inherent weaknesses of 'leanness as an outcome'. The unstable endgame of lean construction equates too easily to corporate anorexia.

References

Abo T. (ed.) (1994) *Hybrid Factory: The Japanese Production System in the United States*, Oxford University Press, Oxford.

Abrahamson, E. (1996) Management fashion, *Academy of Management Review*, **21**(1), 254–85.

Alvesson, M. and Deetz, S. (1996) Critical theory and postmodernism approaches to organizational studies, in *Handbook of Organization Studies*, (eds S.R. Clegg, C. Hardy and W.R. Nord), Sage, London, pp. 191–217.

Ballard, G. and Howell, G. (1998) Shielding production: an essential step in production control, ASCE, *Journal of Construction, Engineering and Management*, **124**(1), 11–17.

Ballard, H.G. (2000) *The Last Planner System of Production Control*, PhD Thesis, University of Birmingham, UK.

Banwell, Sir Harold (1964) The Placing and Management of Contracts for Building and Civil Engineering Work. HMSO, London.

Beale, D. (1994) *Driven by Nissan: A Critical Guide to New Management Techniques*, Lawrence & Wishart, London.

Benders, J. and van Bijsterveld, M. (2000) Leaning on lean: the reception of a management fashion in Germany, *New Technology, Work and Employment*, **15**(1), 50–64.

Benders, J. and van Veen, K. (2001) What's in a fashion? Interpretative viability and management fashions. *Organization*, **8**, 33–54.

Bijker, W., Hughes, T. and Pinch, T. (1987) *The Social Construction of Technological Systems: New Directions in the Sociology and History of Technology*, MIT Press, Cambridge, MA.

Bresnen, M. and Marshall, N. (2001) Understanding the diffusion and application of new management ideas in construction, *Engineering, Construction and Architectural Management*, **8**(5/6), 335–345.

Burgoyne, J. and Jackson, B. (1997) The arena thesis: management development as a pluralistic meeting point, in *Management Learning: Integrating Perspectives in Theory and Practice*, (eds J. Burgoyne and M. Reynolds), Sage, London, pp. 54–70.

Burrell, G. and Morgan, G. (1979) *Sociological Paradigms and Organisational Paradigms*, Gower, Aldershot.

Cherns, A.B. and Bryant, D.T. (1984) Studying the client's role in construction, *Construction Management and Economics*, **2**, 177–84.

Choo, H.J., Tommelein, I.D., Ballard, G. and Zabelle, T.R. (1999) WorkPlan: constraint-based database for work package scheduling, *Journal of Construction Engineering and Management*, **125**(3), 151–160.

Clark, T. and Salaman, G. (1996) The use of metaphor in the client-consultant relationship: A study of management consultancies, in *Organization Development: Metaphorical Explorations*, (eds. C. Oswick and D. Grant), Pitman, London, pp. 154–174.

Clark, T. and Salaman, G (1998) Telling tales: management guru's narratives and the construction of managerial identity, *Journal of Management Studies*, **35**(2), 137–161.

Clegg, S. (1979) *The Theory of Power and Organization*, Routledge & Kegan Paul, London.

Clegg, S., Pitsis, T.S., Marosszeky, M. and Rura-Polley, T. (2006) Making the future perfect: constructing the Olympic dream, in *Making Projects Critical*, (eds. D. Hodgson and S. Cicmil), Palgrave Macmillan, pp. 265–293.

Cully, M., Woodland, S., O'Reilly, A. and Dix, G. (1999) *Britain at Work*, Routledge, London.

Delbridge, R. (1998) *Life on the Line in Contemporary Manufacturing: the Workplace Experience of Lean Production and the Japanese Model*, Oxford University Press, Oxford.

Dohse, K., Jurgens, U. and Malsch, T. (1985) From Fordism to Toyotism? The social organization of the labour process in the Japanese automobile industry, *Politics and Society*, **14**(2), 115–146.

Egan, Sir John (1998) *Rethinking Construction*. Report of the Construction Task Force to the Deputy Prime Minister, John Prescott, on the scope for improving the quality and efficiency of UK construction. Department of the Environment, Transport and the Regions, London.

Fincham, R. (1995) Business process re-engineering and the commodification of management knowledge, *Journal of Marketing Management*, **11**(7), 707–720.

Flanagan, R., Marsh, L. and Ingram, I. (1998) *Bridge to the Future: Profitable Construction for Tomorrow's Industry and its Customers*, Thomas Telford, London.

Fournier, V. and Grey, C. (2000) At the critical moment: conditions and prospects for critical management studies, *Human Relations*, **53**(1), 7–32.

Fox, A. (1974) *Beyond Contract: Work, Power and Trust Relations*, Faber & Faber, London.

Fucini, J. and Fucini, S. (1990) *Working for the Japanese*, The Free Press, New York.

Gann, D. (2001) Putting academic ideas into practice: technological progress and the absorptive capacity of construction organisations, *Construction Management and Economics*, **19**, 321–330.

Garrahan, P. and Stewart, P. (1992) *The Nissan Enigma: Flexibility at Work in a Local Economy*, Mansell, London.

Giddens, A. (1984) *The Constitution of Society: Outline of the Theory of Structuration*, University of California, Berkeley, CA.

Green, S.D. (1999a) The missing arguments of lean construction, *Construction Management and Economics*, **17**(2), 133–137.

Green, S.D. (1999b) The dark side of lean construction: exploitation and ideology, in *Proc. of the 7th Conference of the International Group for Lean Construction (IGLC-7)*, (ed. I.D. Tommelein), University of California, Berkeley, pp. 21–32.

Green, S.D. and May, S.C. (2005) Lean construction: arenas of enactment, models of diffusion and the meaning of 'leanness', *Building Research & Information*, **33**(6), 498–511.

Green, S.D. (2006) The management of projects in the construction industry: context, discourse and self-identity, in *Making Projects Critical*, (eds. D. Hodgson and S. Cicmil), Palgrave Macmillan, pp. 232–251.

Green, S.D. and Schweber, L. (2008) Theorising in the context of professional practice: the case for middle range theories, *Building Research & Information*, **36**(6), 649–654.

Green, S.D., Larsen, G.D. and Kao, C.C. (2008) Competitive strategy revisited: contested concepts and dynamic capabilities, *Construction Management and Economics*, **26**(1) 63–78.

Hammer, M. and Champy, J. (1993) *Re-engineering the Corporation*, Harper Collins, London.

Hampson, I., Ewer, P. and Smith, M. (1994) Post-Fordism and workplace change: towards a critical research agenda, *Journal of Industrial Relations*, June, 231–257.

Harvey, M. (2003) Privatization, fragmentation and inflexible flexibilization in the UK construction industry, in *Building Chaos: A International Comparison of Deregulation in the Construction Industry* (Eds. G. Bosch and P. Philips), Routledge, London, pp. 188–209.

Harvey M. and Behling F. (2008) *The Evasion Economy: False self-employment in the UK Construction Industry*, UCATT, London.

Howell, G. and Ballard, G. (1999) Bringing light to the dark side of lean construction: a response to Stuart Green, in *Proc. of the 7th Conference of the*

International Group for Lean Construction (IGLC-7), (ed. I. D. Tommelein), University of California, Berkeley, pp. 33–37.

Huczynski, A.A. (1993) Management Gurus: What Makes Them and How to Become One, Routledge, London.

ILO (2001) *The Construction Industry in the Twenty-first Century: Its Image, Employment Prospects and Skills Requirements*, International Labour Office, Geneva.

Jackson, B.G. (1996) Re-engineering the sense of self: the manager and the management guru, *Journal of Management Studies*, **33**(5), 571–590.

Kamata, S. (1982) *Japan in the Passing Lane: An Insider's Account of Life in a Japanese Auto Factory*, Pantheon Books, New York.

Keiser, A. (1997) Rhetoric and myth in management fashion, *Organization*, **4**(1), 49–74.

Kenney, M. and Florida, R. (1993) *Beyond Mass Production: The Japanese System and its Transfer to the U.S.*, Oxford University Press, New York.

Kinnie, N., Hutchinson, S., Purcell, J. (1996) The People Implications of Leaner Ways of Working. Report by the University of Bath, *Issues in People Management, No. 15*, Institute of Personnel Development, London, pp. 6–63.

Koskela, L. (1992) *Application of the New Production Philosophy to Construction*, Technical Report No. 72, Center for Integrated Facility Engineering, Stanford University, CA.

Koskela, L. (2000) *An Exploration Towards a Production Theory and its Application to Construction*, VTT Publications No. 408, Technical Research Centre of Finland.

Latham, M. (1994) *Constructing the Team*. Final report of the Government/ industry review of procurement and contractual arrangements in the UK construction industry, HMSO, London.

Lawrence, P.R. and Lorsch, J.W. (1967) *Organization and Environment*, Harvard Press, Cambridge, Mass.

Legge, K. (2000) Personnel management in the lean organization, in *Personnel Management: A Comprehensive Guide to Theory and Practice* (eds K. Sisson and S. Bach), 3rd edn, Blackwell, Oxford, pp. 43–69.

Legge, K. (2002) On knowledge, business consultants and the selling of total quality management, in *Critical Consulting: New Perspectives on the Management Advice Industry* (eds T. Clark and R. Fincham) Blackwell, Oxford, pp. 74–90.

Lovell, H. and Smith, S.J. (20010) *Agencement* in housing markets: The case of the UK construction industry, *Geoforum*, **41**(3), 457–468.

MacDuffie, J.P. (1995) Human resource bundles and manufacturing performance: organisational logic and flexible production systems in the world auto industry, *Industrial and Labor Relations Review*, **48**(2), 197–221.

Marchington, M. and Vincent, S. (2004) Analysing the influence of institutional, organizational and interpersonal forces in shaping inter-organizational relations, *Journal of Management Studies*, **41**(6), 1029–1056.

Morgan, G. (2006) *Images of Organization*, Thousand Oaks, CA.

Morris, J. and Wilkinson, B. (1995) The transfer of Japanese management to alien institutional environments. *Journal of Management Studies*, **32**(6), 719–30.

Naim, M. and Barlow, J. (2003) An innovative supply chain strategy for customized housing, *Construction Management and Economics*, **21**, 593–602.

Ness, K. (2010) The discourse of 'Respect for People' in UK construction, *Construction Management and Economics*, **28**(5), 481–493.

Oliver, N. and Wilkinson, B. (1992) *The Japanization of British Industry*, 2nd edn. Blackwell, Oxford.

Peters, T.J. and Waterman, R.H. (1982) *In Search of Excellence: Lessons from America's Best Run Companies*, Harper & Row, New York.

Pettigrew, A.M. (1997) What is processual analysis? *Scandinavian Journal of Management*, **13**(4), 1–31.

Rees, C., Scarbrough, H. and Terry, M. (1996) The People Implications of Leaner Ways of Working. Report by IRRU, Warwick Business School, University of Warwick, *Issues in People Management, No. 15*, Institute of Personnel Development, London, pp. 64–115.

Rehder, R.R. (1994) Saturn, Uddevalla and the Japanese lean system: paradoxical prototypes for the twenty-first century, *International Journal of Human Resource Management*, **5**(1), 1–31.

Saad, M. and Jones, M. (1998) *Unlocking Specialist Potential*, Reading Construction Forum.

Scarbrough, H. (2003) The role of intermediary groups in shaping management fashion, *International Studies of Management and Organisation*, **32**(4), 87–103.

Scarbrough, H. and Terry, M. (1998) Forget Japan: the very British response to lean production, *Employee Relations*, **20**(3), 224–236.

Spender, J.-C. (1989) *Industry Recipes*, Blackwell, Oxford.

Storey, J. and Sisson, K. (1989) The limits to transformation: human resource management in the British Context, *Industrial Relations Journal*, **21**, Spring, 60–65.

Strategic Forum (2002) *Accelerating Change*, Rethinking Construction, London.

Tommelein, I.D. (1998) Pull-driven scheduling for pipe-spool installation: simulation of lean construction technique, *Journal of Construction Engineering and Management*, **124**(4), 279–288.

Turnbull, P. (1988) The limits to Japanization – just-in-time, labour relations and the UK automotive industry, *New Technology, Work and Employment*, **3**(1), 7–20.

Weick, K.E. (1995) *Sensemaking in Organizations*, Sage, Thousand Oaks, CA.

Wickens, P. (1987) *The Road to Nissan*, Palgrave Macmillan, London.

Winch, G. (2002) *Managing Construction Projects*, Blackwell, Oxford.

Williams, R. and Edge, D. (1996) The social shaping of technology, *Research Policy*, **25**, 865–899.

Womack, J.P., Jones, D.T. and Roos, D. (1990) *The Machine that Changed the World*, Rawson Associates, New York.

Womack, J.P. and Jones, D.T. (1996) *Lean Thinking*, Simon and Schuster, New York.

9　From Enterprise to Social Partnership

9.1　Introduction

The years following the election of Tony Blair as Prime Minister in 1997 saw a plethora of improvement initiatives aimed at the construction sector. The activities in the immediate wake of *Rethinking Construction* were described in Chapter 7. Of particular note was the Movement for Innovation (M4I) and the Construction Best Practice Programme. But the unfolding improvement agenda must be understood within the context of New Labour's policy direction. *Rethinking Construction* was undoubtedly born from the enterprise culture, but its emphasis on performance management reflected the managerialism of New Labour. The narrow-focus on technical efficiency was also soon to be outflanked by New Labour's policy agenda which was to have a significant influence on the improvement debate within the construction sector.

In the early years of the Blair administration much was made of a 'new' combination of enterprise and social democracy known as the 'Third Way'. Although ridiculed by many for lacking intellectual coherence, the rhetoric of the Third Way captured the mood of optimism which prevailed as Britannia was suddenly cool once again. Where once it had been trade union leaders eating sandwiches at 10 Downing Street, it was now pop stars drinking champagne. But behind the glitz and glamour, New Labour demonstrated an obsession with presentation previously unheard of. The agenda in the early years was one of relentless reform across a range of policy agendas. In the foreword to the 1999 White Paper *Modernising Government*, Tony Blair laid out the government's mission:

> '[t]he Government has a mission to modernise renewing our country for the new millennium. We are modernising our schools, our hospitals, our economy and our criminal justice system. We are modernising our democratic framework, with new arrangements for Scotland, Wales, Northern Ireland, the English regions, Parliament and local authorities.'

Making Sense of Construction Improvement, First Edition. Stuart D. Green.
© 2011 Stuart D. Green. Published 2011 by Blackwell Publishing Ltd.

The commitment to 'modernisation' is clear. Modernisation notably did not mean reversing the privatisation policies implemented previously under the Conservative administration. But modernisation did mean an unprecedented level of investment in schools and hospitals, delivered in 'partnership' with the private sector. The National Audit Office (2001) report *Modernising Construction* directed the mantra of modernisation at the construction sector. There was much continuity with what had gone before, but there were also important points of difference. The first significant report to be published after Blair's re-election as Prime Minister in 2001 was *Accelerating Change* (Strategic Forum, 2002). The title was once again borrowed directly from the New Labour lexicon. Those who were hoping for a return to the Labour policies of the 1970s were to be sadly disappointed, but the movers and shakers within the construction sector were mightily relieved. The advent of New Labour did not precipitate the decline of the enterprise culture, but it did initiate the evolution of the enterprise culture into a concept which embraced social democracy with a corresponding emphasis on 'social partnership' between the public and private sectors. Consideration of externalities was back in fashion. Ultimately, the sense of progress and modernisation fell victim to cynicism and government sleaze, as Tony Blair finally gave way to Gordon Brown in 2007. The construction improvement agenda under Gordon Brown will be addressed in the final chapter.

9.2   Building Britain with New Labour

9.2.1   From hope to cynicism

It is difficult now to remember the sense of optimism which followed the election of the New Labour government in 1997. New Labour enjoyed a huge parliamentary majority and Tony Blair immediately adopted a presidential style of government. Government power became evermore centralised on the Prime Minister's personal office (Jenkins, 2006). The one exception related to the degree of authority conceded to Gordon Brown at the Treasury in respect of the financial implications of policy. The government also became increasingly obsessed with the need for ministers to be fully 'on message'. In the latter regard, unprecedented influence was wielded by the Prime Minister's political advisors. This was especially true of the press officer, Alastair Campbell, former political editor of the *Daily Mirror* and avid supporter of Burnley Football Club. The age of spin had arrived and presentation was seen to be everything. While the language was that of modernisation, the way that style triumphed consistently over substance owed more to postmodernism. Deputy Prime Minister John Prescott was one of few senior members of government seemingly immune to communication via media sound bite.

The climate which surrounded the implementation of the Egan Agenda within construction (see Chapter 5) was directly reflective of the prevailing climate within government. Policy was decided by a tightly-knit cabal, and thereafter the emphasis switched to the means of implementation. Ministers and MPs were expected to fall into line. Within the corridors of Whitehall, the prime issue of importance was that the machinery of government should be 'on message'. In the case of the construction sector improvement agenda, the culture of compliance strayed beyond government to include the plethora of government-supported quangos orientated towards implementing the Egan Agenda. The need to align with the Egan 'principles' applied equally to client representative groups, although there was frequently a significant credibility gap between rhetoric and reality. Many contracting firms were similarly more compliant in words than in deeds, but most had discovered the benefits of 'lean thinking' long before it became fashionable (see Chapter 3).

In the first couple of years following the 1997 election, the Blair government secured some notable wins, not least of which was the 1998 Good Friday agreement in Northern Ireland. Foreign policy initiatives were a further characteristic of Blair's first term of office, including military adventures in Kosovo and Sierra Leone. It was during this period of relative optimism that *Modernising Construction* was published. New Labour was duly re-elected for a second term in June 2001, with only a slightly reduced majority. The Conservative Party had still not recovered from the infighting which caused John Major's downfall, and was a long way from being electable under the leadership of William Hague. New Labour's re-election was undoubtedly aided by the stable economic growth which prevailed throughout their first period in office. Gordon Brown as Chancellor of the Exchequer very quickly established his authority and reputation for sound economic management. Margaret Thatcher may have been the Iron Lady, but it was Gordon Brown who was the Iron Chancellor.

The epochal event of the time occurred on 11 September 2001 with the al-Qaeda attack on the World Trade Center in New York. With US president George Bush in the driving seat, Britain participated in the invasion of Iraq on the basis that Saddam Hussian was alleged to possess weapons of mass destruction (WMD). The invasion of Iraq in 2003 was of questionable legality and was hugely divisive within British society. British troops were also engaged, comparatively surreptitiously, in Afghanistan where they were destined to remain active for much longer. The controversy of the invasion of Iraq undoubtedly stained Blair's legacy as Prime Minister, which for many would forever be associated with the 'dodgy dossier' on WMD. The invasion resulted in the resignation of Foreign Secretary Robin Cook, and the somewhat less prompt resignation of the Secretary of State for International Development, Claire Short. For many the gloss of New Labour wore off all too quickly.

Despite the continuing controversy about Iraq, Blair succeeded in being elected for a third term in 2005, albeit with a significantly reduced majority. The Conservative Party were searching around desperately for a leader to return them to power. William Hague had been replaced by Ian Duncan Smith following the Conservatives' election defeat in 2001. Duncan Smith did not bring around the reversal in fortune that the Tories hoped for and was subsequently replaced by Michael Howard in November 2003. Howard had served previously in government under both Thatcher and Major and was too strongly connected with memories of a previous era to generate a sufficient swing to oust New Labour.

9.2.2 PFI, schools and hospitals

However, the legacy of New Labour's first term was not all about foreign policy adventures. While Tony Blair courted George Bush in Washington DC, Gordon Brown was left to focus on domestic economic policy. At the time there seemed little danger of Brown going the way of previous high-spending Labour Chancellors. He seemed deliberately to nurture a public image of Scottish Presbyterian prudence. In common with Blair, Brown relied heavily on a small clique of personal advisors often to the detriment of career Treasury officials. Brown also slipped remarkably easily into the rhetoric of the enterprise culture, with much emphasis on competitiveness and wealth creation. Prior to the 1997 election, Brown had committed himself to staying within the Conservatives' spending targets for three years beyond the election. This was a promise which he duly kept. Early fund raising initiatives included the 'windfall' utilities tax. This had been in the election manifesto and was directed at the excess profits realised by the privatised utilities. The tax was a strictly one-off arrangement which raised in the region of £5 billion. The proceeds were invested directly in the progressive 'New Deal' employment initiative (Marr, 2007). Brown's other extravagant gesture was to delegate the fixing of interest rates to the Bank of England. The effective outcome was a continuation of the low-inflation economic policy pursued by the preceding government.

All of this was fine, but left Brown with the problem of how to fund ambitious health and education programmes without raising public borrowing. The solution was to shift borrowing 'off-balance sheet' through the expanded use of the previously vilified private finance initiative (PFI). Prior to taking power, Blair and Brown had steadfastly warned against the Tories' privatisation policy. But once elected, New Labour embraced PFI with even more enthusiasm than the preceding Conservative administration. The spin-mongers tried hard to re-label it as public-private partnership (PPP), but with only limited success. The government's enthusiasm for PFI/PPP was demonstrated in 1998 when the Department of Social Security transferred the ownership and management of most of its estate to a private sector consortium. This was part of an early PPP deal known as PRIME and was worth over £1 billion. By 2003 Brown was boasting extensively about his achievements in health and education, amounting to 23 hospitals and 239 schools (HM Treasury, 2003). The new infrastructure was real and comprised a lasting legacy for the facilities within which patients were treated and children were taught. But the money to pay for it was essentially mortgaged against the future. PFI prisons were a further area of growth within which a limited number of contractors in partnership with security firms had established quasi-monopolistic positions. Prisons of course were much simpler to operate than schools or hospitals, with much more straightforward output specifications.

Roe and Craig (2004) estimate that the accumulated off-balance sheet debt by 2003 amounted to £100 billion. If this was prudence, it was of a strange variety. Numerous commentators continued to express concerns about the windfall profits secured by PFI contractors, although in time such concerns were alleviated partly by a maturing market. A flourishing secondary market in PFI concessions began to develop, thereby exposing the lie that the PFI procurement arrangement necessarily ensures an 'optimal balance' between capital cost and operating cost. The operating concessions were of course not always sold on, but even the possibility that they might be sold on would inevitably cause capital cost to be prioritised over through-life operating costs. In any case, responsibilities were often spilt between different operating companies and integration could not be taken for granted. PFI further acted to create a significant distance between those who would use the building and those who were responsible for its design, thereby militating against good briefing practice. Although PFI would subsequently be presented as an 'integrated procurement approach', the reality was that the institutional barriers between design, construction and building operation were replicated rather than alleviated (Leiringer *et al.*, 1999).

Following on from the above, it is important to recognise that PFI is a game whose rules have been in almost constant evolution since the scheme's initial inception. The one guiding principle has been the desire to keep public sector borrowing 'off balance sheet', coupled with a seemingly axiomatic belief that the private sector can deliver more innovative and efficient solutions. In this respect, privatisation policy under New Labour remained broadly consistent with that of the preceding Tory government.

9.2.3   *Enterprise meets social democracy*

But it would be misrepresentative to suggest that the policy discourse of the New Labour government was identical to the preceding Conservative administration. One important point of difference relates to the way New Labour intertwined the discourse of enterprise with a modernising discourse of social democracy. According to Hall (2003), the New Labour government had:

> '... adapted the fundamental neo-liberal paradigm to suit its conditions of governance – that of a social democratic government trying to govern in a neo-liberal direction while maintaining its traditional working-class and public-sector middle class support, with all the compromises and confusions that entails'. (Hall, 2003; 14)

The interpretation above could be seen to apply to many of the New Labour government's policy objectives in respect of construction. It can also be taken as an explanation of the continued focus on enterprise while at the same time emphasising notions of corporate social responsibility (CSR). The iconic status of the customer remained firmly in place, as did the broader commitment to outsourcing and the privatisation of service delivery. However, the New Labour policy agenda differed starkly in the way it strove for greater social justice. It further differed in the way it saw a major role for the state in its achievement. This revised policy orientation progressively became apparent through the commitment to drive sustainability through government procurement (e.g. GCCP, 2000; DEFRA, 2006). Fears about global warming meant that a stronger focus on environment sustainability was inevitable. This was an issue of scientific fact which could not be ignored. But it was the government's social democratic credentials which shaped the promotion of social sustainability as part of the procurement agenda.

The combination of enterprise with social democracy was encapsulated within the ideas of the so-called 'Third Way', as advocated by the eminent sociologist Anthony Giddens (1998). In many respects, New Labour was following in the wake of a number of overseas politicians, including Bill Clinton (United States), Lionel Jospin (France) and Gerhard Schroeder (Germany). Nevertheless, in the early years of the Blair administration the Third Way was promoted as the big idea, even if Deputy Prime Minister John Prescott remained stubbornly resistant. What was notable was the way in which the advocates of the Third Way accepted the trend of previous policy initiatives, and then sought to mobilise them towards a greater degree of social justice. It is also necessary to acknowledge that the Third Way remains much criticised in terms of its essential vagueness and lack of intellectual coherence (e.g. Hall, 1998; Driver and Martell, 1999; Faux, 1999). Powell (2000) has further cast doubt on the extent to which the Third Way can be held to be either distinctive or new. In these respects, it shares many of the defining characteristics of a management fashion, and is perhaps best understood as a political fashion. Connolly *et al.* (2008) argue that the Third Way must be understood in the context of the times, and positioned against what had gone before. The enterprise culture had previously been positioned against the failings of the 'Social Contract' and the 'winter of discontent' (see Chapter 2). Policy up until 1997 was seemingly driven by the ideological belief that the private sector is more efficient than the public sector. But the core difficulty was that the private sector will only invest if there is a prospect of making a profit, and this remained problematic in the context of traditional public services such as health and education. There was also the problem that by 1997 everything that could conceivably be privatised had already been outsourced to the private sector. Hence if the private sector was to be engaged further, it became necessary to make further steps towards guaranteeing both demand and funding, thereby making investment more attractive to private sector firms (Ietto-Gillies, 2006).

9.2.4   Efficiency and risk in service provision

The opening up of even more public services for privatisation created a raft of new opportunities for private contractors, many of which originated in construction. The expansion of traditional construction firms into the provision of public services further eroded the feasibility of any attempt to define precisely what is meant by the 'construction sector'. A significant proportion of the income of the UK's leading contracting firms became secured increasingly from activities outside the definition of construction. Many engineering consultancies diversified similarly into service provision, even to the extent whereby firms once associated primarily with civil engineering were now negotiating contracts for the provision of catering services; from designing concrete waffle floor slabs to sandwiches in two short steps. The Association of Consulting Engineers (ACE) was notably forced to rename itself the Association of Consultancy and Engineering in an effort to remain relevant to its members' increasingly diverse business activities. At least they were able to retain the cool acronym.

Some privatisations were very successful, and reportedly succeeded in delivering better public services for reduced cost. Outsourcing companies such as Capita and Serco expanded rapidly in response to the plethora of opportunities. Such firms were particularly notable for the way in which they promoted themselves as working in 'partnership' with the public sector. Construction companies also knew a thing or two about outsourcing, and firms such as Balfour Beatty, Carillion and Skanska similarly expanded very successfully into service provision. Strangely enough, this was at the time that the construction sector was still being roundly criticised for the poor calibre of its managers. The firms themselves had never quite believed Egan's jibes about their supposedly out-of-date management practices, and the author can personally vouch for the quality and dedication of the vast majority of managers from within firms such as these. But then it must also be conceded that there are also many inspirational managers within public sector organisations. In both cases, the debate is rather less about the capabilities of the managers themselves, and rather more about the institutional context within which they operate. Nevertheless, there is a deep-rooted paradox in stigmatising the construction sector's management as 'dinosaurs' on the one hand, while entrusting it with an increasingly large proportion of the nation's public services, on the other.

Notwithstanding the many successful privatisations, it is necessary to acknowledge that there were also some important failures. WS Atkins notably walked away from a contract to run Southwark's education service on the basis it could not generate sufficient income to make it worthwhile (Jenkins, 2006). A further high profile case concerned the failure of the Metronet, which went into administration four years into a £17 billion contract to modernise the London Underground. It was later confirmed that London Underground would remain responsible on a permanent basis for the renewal and maintenance of the track for which the Metronet 'infracos' had previously been responsible. Jarvis had been one of the earliest firms to specialise in PFI markets and also became heavily involved in railway maintenance. On both counts, the firm suffered a succession of adverse headlines until finally going into administration in March 2010. Firms which default on PFI contracts inevitably leave the taxpayer with the bill for picking up the pieces. It is therefore dubious to claim that PFI shifts the risk of failure to the private sector. If PFI firms really are operating in a marketplace, some will succeed and others will inevitably fail. This is how markets work; and it is the possibility of failure which acts against management complacency and inefficiency. But if a PFI firm fails, the risk reverts to the taxpayer. In the case of Metronet, the cost of failure was reputed to be in the region of £410 million.

The issue of risk has been central to the debate on the merits of PFI from the outset. In essence, the debate hinges around the extent to which commercial risk can meaningfully be transferred to the private sector when the taxpayer is ultimately required to pick up the bill in the event of failure. Early entry PFI contractors during the Major administration were severely criticised for earning super-normal profits. In mitigation, the major contractors cited the significant risks involved, together with the huge costs associated with the bidding process. As the PFI market matured, the risks gradually declined and the possibility of windfall profits declined accordingly. For a government committed to public-private partnerships, it undoubtedly made sense to reduce the level of risk associated with major PFI projects. But on too many occasions, the transfer of risk remains illusory. Demand risk is routinely borne by the taxpayer in that a certain through-put of patients or students is guaranteed from the outset. There have been cases of some schools closing as soon as they opened due to declining numbers of pupils, thereby necessitating the need to terminate the contract at significant cost to the taxpayer. In the absence of any guarantee relating to ongoing demand, the risk of declining pupil numbers would have to be priced by the private sector, and the cost would be astronomical. Hence enterprise gives ground to public-private partnership.

9.2.5   *Partnerships in delivery*

Notwithstanding the above, one of the central rhetorical themes of the Third Way was the notion of 'partnership'. The metaphor of partnership was aimed primarily at eroding the barriers between public and private sectors and promoting the cause of collaborative working. The attempt to re-brand PFI in terms of public-private partnerships (PPP) has already been noted. But the idea of partnership can also be taken to imply the need for collaborative working between management and workforce. Within the context of employment relations, Collins (2001; 21) argues that the metaphor of partnership is 'invoked in order to express the highly cooperative ideal expressed by the flexible employment relation' (Smith and Morton, 2006). The same source also quotes Collins (2002) who suggests that the purpose of partnerships is:

> 'to enhance competitiveness, through improvements in quality and efficiency. This purpose requires the exchange of information: management needs to explain its product and marketing plans to the workforce, and the workers need to use their human capital to suggest how production and products can be improved. (Collins, 2002; 456–9).

The quote above succeeds in capturing the essence of the Egan improvement agenda as advocated in the construction sector. The headline emphasis on quality and efficiency is immediately suggestive of *Rethinking Construction*, likewise the attempt to mobilise the expertise of the workforce to improve productivity. The expertise of the workforce might have been mobilised a little easier had the contracting sector not been so keen to avoid direct employment. But the enterprise culture was by now hardwired into the dominant industry recipe. There is nevertheless a delicious irony in that several large hollowed-out construction firms involved in highway maintenance increasing found themselves acquiring large directly-employed workforces as a result of TUPE-transfers from local authorities. The same was also true of the large outsourcing companies which had assumed responsibility for contracted-out public services. However, it should be emphasised that this applied only to very small number of large firms, and the TUPE-transferred operatives were undoubtedly perceived as a risk which was priced appropriately. The marginal strengthening of the TUPE regulations had been New Labour's sole concession to trade union opposition to its privatisation policies.

As the New Labour government moved into its second term of office, the rhetoric of partnership became even more prominent. In 2003 the Department for Children, Schools and Families (DCSF) announced the ambitious *Building Schools for the Future* (BSF) programme. The programme aimed originally to renew all 3500 English secondary schools over the 15-year period 2005–20. The initiative was launched with great fanfare with a budget of £2.2 billion in its first year (2005–06). Of central importance were the supposed 'long-term partnering efficiencies' to be achieved between the public and private sectors through the establishment of local joint ventures called Local Education Partnerships (LEPs). The LEPs were to have exclusive rights for ten years to deliver new and refurbished school facilities and related services. Unsurprisingly, the majority of LEPs were secured by large national contractors, thus serving to force local contractors to operate as subcontractors. This was reflective of the way framework agreements in general acted to squeeze out local firms on the basis of size, despite, in many cases, admirable track records of client satisfaction coupled with localised engagement.

The objectives of BSF were to be exposed subsequently as being too ambitious (NAO, 2009) The NAO further criticised the high cost of establishing the first 15 LEPs, which were seen to have been inflated by local authorities' extensive reliance on consultants. However, it was acknowledged that *Partnerships for Schools* had since 'taken measures to control capital costs so that BSF capital costs are similar to most other school building programmes and cheaper than Academies built before their integration into BSF'. This was not a view which was destined to survive beyond the New Labour government. The procurement of schools had undoubtedly become a politicised agenda. The construction sector is tasked ultimately with delivering such politicised agendas, and hence the rhetoric of industry improvement became inherently politicised. Certainly construction improvement was no longer focused solely on efficiency improvement; firms were also required to demonstrate their credentials in terms of social sustainability. Many simply employed consultants to develop and maintain the latter storyline, while they themselves concentrated on 'managing the supply chain'.

NAO (2009) further pointed towards the difficulties of establishing effective working arrangements and relationships between local authorities and private sector partners. The paradoxes of partnering and collaborative working seemingly cannot simply be wished away in the cause of value for money. Perhaps most telling of all was the observation that there is a general lack of skills in procurement and programme management across the public sector, and that this constrains capacity within BSF. It seems that irrespective of the extent of privatisation the public sector cannot achieve value for money without its own expertise in procurement and programme management. Local authorities may in the future choose to take back in house some of the expertise that they have lost. Perhaps some future government may even return to the idea of a public sector 'property services agency'. The reliance on consultants within BSF was indicative of a much broader problem; Craig and Brooks (2006) have estimated that the New Labour government spent £70 billion on consultants between 1997 and 2006. This was indicative of a trend established several decades previously under Margaret Thatcher. The big picture is one of relentless privatisation coupled with the imposition of centralised performance monitoring to ensure efficiency and value for money. The same approach has been applied by stealth to the construction sector. Thus, the culture of enterprise goes hand-in-hand with the culture of audit. The paradox exists on the level of national policy and shapes the improvement agenda in construction. In essence, it is the same paradox which characterises individual partnering initiatives.

The health sector has been similarly characterised by the rhetoric of 'partnership'. The Department of Health followed a parallel path to DCSF and established a national joint venture with Partnerships UK 'to develop and encourage a new market for investment in primary care and community based facilities and services'. The joint venture company was initially called Partnerships for Health (PfH), before giving way to Community Health Partnerships (CHP) which is a wholly-owned company of the Department of Health. CHP exists for the purpose of delivering the NHS Local Improvement Finance Trust (LIFT) initiative. This involved primary care trusts (PCTs) in setting up LIFT companies in partnership with the private sector. The so-called 'LIFTco' then becomes the vehicle for improving and developing frontline primary and community care facilities. Each LIFTco is a limited company with the local PCTs, CHP and the private sector partner as shareholders. It is the LIFTco which takes responsibility for owning and maintaining the required buildings and for leasing the premises to PCTs, GPs, local authority social services, dentists, pharmacists, etc. The local PCTs are tasked with protecting the public interest. If it sounds complicated, it is because it was complicated. Hence the need for all those expensive consultants to explain what all the acronyms meant and to advise on 'change management'.

The first task of a LIFT was to develop a 'strategic service development plan'. The local health economy would then seek private sector partners to set up a LIFTco, which in turn would contract to deliver the required services over a 20 or 25-year period. Hence the required facilities and supporting services are delivered in partnership between the public and private sector. The boundaries between the two sectors thereby become evermore blurred, and the risks of private sector involvement become evermore opaque. Notions of public sector clients and private sector contractors become conflated around ideas of 'LIFTcos', 'opcos' and 'infracos'. The ethos of public service became interspersed with the profit motive. The Third Way was not only unclear in theory, but was also very confused on the ground. Everyone seemingly works in partnership; everyone 'partners' with the supply chain. Collaborative working abounds. But the contractors still work as contractors, and the subcontractors still operate as subcontractors. The concept of 'partnership' did not of course extend to the construction workforce, and the competitive drivers on employment practices described in Chapter 3 remained entirely unaffected. Leanness and agility in the marketplace continued to be the passwords to success on the construction site.

Similar issues characterised the services which were contracted to outsourcing specialists. Catering, cleaning and security were all services routinely outsourced to private sector contractors, and all three sectors were – and still are – characterised by casualised employment, low pay and poor conditions. The report of the TUC *Commission on Vulnerable Employment* (TUC, 2008) is especially damning. It describes how the widespread use of labour-supply agencies results in the unequal treatment of agency workers and directly employed staff. It further describes how many 'workers' and bogusly self-employed operatives are routinely denied the welfare rights enjoyed by employees. Agency workers are held to be especially vulnerable to abuse, even more so when migrant workers are forced into vulnerable employment by immigration regulations. The TUC *Commission on Vulnerable Employment* provides a damning incitement of employment practices in twenty-first century Britain. In the region of two million workers are held to be trapped in a continual round of low-paid and insecure work where mistreatment is the norm. Perhaps the construction sector is not so unique after all.

9.2.6   *A blot on the landscape*

In the same way that the focus on leanness and agility described in Chapter 3 can be read as the material manifestation of the enterprise culture, the preceding description can be read as the material manifestation of so-called Third Way partnerships'. PFI was relabelled public-private partnership (PPP) and multiple variants continued to evolve. The New Labour government notably failed to show any enthusiasm for reversing any of the preceding Tory government's privatisations. The trend towards ever-increasing levels of privatisation continued unabashed, extending even to the day-to-day administration of public services. This was despite Blair and Brown's virulent criticism of privatisation policy while in opposition. Neither did New Labour repeal any the industrial relations legislation of which they had previously been so critical.

The railways continued to be problematic, and the after-effects of the botched privatisation presented significant operational challenges. Railtrack plc remained heavily regulated until it was eventually forced into bankruptcy following the general seizure of the railway system in the wake of the Hatfield rail crash in October 2000. It was replaced by Network Rail, a unique 'not for dividend company' which is immune from bankruptcy and effectively state-owned. According to the railway historian Christian Wolmar (2007), by 2005 the railway was receiving five times as much public subsidy than it had as a nationalised industry. In 2006, Network Rail's chairman, John Armitt, announced that track maintenance was being taken back in-house from the private sector contractors who had been criticised for poor quality work and spiralling costs. This in effect amounted to the re-nationalisation of track maintenance in response to a perceived lack of control caused by the previous reliance on private-sector subcontractors. At the time of writing, the trains continue to be operated by private sector train operating companies (TOCs) on the basis of what increasingly approximates towards risk-free contracts. The privatised railways thereby reverted to a 'partnership' between the public and private sectors. It was a partnership which required significant state subsidy, and it was a partnership which offered little risk to private sector operators. The shift back into direct employment of the track maintenance workforce did indeed go against the grain, and is thus described as a 'blot on the landscape'.

However, the return to direct employment within the railways had minimum long-term impact on the construction sector. Despite the much-vaulted 'shift back' to direct employment in the late 1990s, by 2002 the level of self-employment was rising inexorably once again. Network Rail had acted decisively against fragmentation through the effective re-creation of a railway track maintenance DLO. But this was categorically not on the agenda for the construction sector at large. Integration remained something that was talked about as an abstract concept; it was not something that was in any danger of being implemented. Policy makers continued to prefer vague exhortations in favour of 'collaborative working'.

Laing O'Rourke was cited frequently at the time as going against the grain in terms of its alleged commitment to direct employment. But this was categorically not a return to direct employment in the style of the 1970s. Many workers initially resisted Laing O'Rourke's offers of direct employment on the basis of the terms and conditions which were on offer. Few of the contracts on offer extended longer than two years, thereby avoiding the statutory requirement for redundancy payments. Basic rates of pay were also drastically reduced, with a corresponding drop in overtime pay. On the positive side, there was a 'discretionary bonus' which rested entirely upon the discretion of management. For the cynics, the new contracts were labelled as 'contricks'. But their implementation was fully supported by UCATT, which did its best to encourage its members to comply. However, many operatives continued to resist the new contracts, and there were several wild-cat strikes in response to the pressure to sign up. Laing O'Rourke's shift to direct employment should perhaps be understood in terms of Atkinson's (1984) model of *financial* flexibility, whereby firms seek to adjust employment costs in response to fluctuations in demand. Hence Laing O'Rourke sought to move away from standardised pay structures towards more individualised systems, with a greater linkage to performance. Such an approach can hardly be described as inferior to the model of *numerical* flexibility which dominated elsewhere, but neither does it deserve the plaudits which it received at the time. It should be noted that Laing O'Rourke also seemingly continued to use the services of an illegal blacklisting agency to ensure that no 'trouble makers' were employed. But this was equally true of other leading construction firms, including Balfour Beatty, Costain and Sir Robert McAlpine (Evans *et al.*, 2009). The shadow of the Economic League continued to fall over construction sector employment practices, despite its abolition in 1994. As an aside, it is important to emphasise that the author claims no moral high ground, having himself used the blacklisting services of the Economic league while working for a national contractor in the 1980s.

9.3   Modernising Construction

9.3.1   *On message in an expanding market*

The preceding discussion of New Labour's policy initiatives provide the context within which the post-Egan construction sector improvement agenda can be evaluated. *Modernising Construction* was published in January 2001 by the National Audit Office (NAO) in the final month's of Tony Blair's first term of office. It was concerned primarily with the industry's performance in the delivery of public sector projects, and in the performance of government agencies in construction procurement. It also offered a useful commentary on the performance of the various initiatives that lay within the remit of the Movement for Innovation (M4I) and the equivalent *Achieving Excellence* programme in the public sector. The report provides an invaluable insight into the dominant themes of the improvement debate in the immediate aftermath of the Egan Agenda.

Despite the role of the National Audit Office as a quasi-independent watchdog of government efficiency, it was never likely that the authors would conclude that *Rethinking Construction* was deeply flawed. The overriding necessity was to be 'on message'; this was how New Labour operated, and this is how they expected their civil servants to operate. The only surprise was the extent to which the construction sector had seemingly accepted the Egan agenda, despite its repeated criticisms of the sector's management capability. But in truth construction firms were more interested in pursuing work in an expanding market. CEOs were less interested in whether PFI was a good deal for the UK taxpayer, and rather more interested in how they could secure a position on the gravy train. These were good times for construction firms, and there were few who were prepared to rock the boat. The *cognoscenti* could exhort 'best practice' all they liked, provided that the opportunities to win work and make money continued to be forthcoming. And the ongoing emphasis on privatisation and the contracting-out of public services continued to provide new opportunities for innovative firms to expand their business portfolios. Such opportunities reinforced the previously established tendency for contracting firms to organise in terms of business sector. Schools and healthcare were certainly areas which required contracting firms to embed themselves within the respective policy contexts, continuously monitoring and responding to the ever-changing landscapes. Such opportunities required new capabilities beyond being 'good builders', but once the work was secured the competitive model remained broadly unaffected. The essential challenge was to subcontract the risks of construction down the supply chain at minimum cost. But effective supply chain management was about much more than efficient contract trading. It was also about positioning in the marketplace. As noted previously, the financial barriers to entry posed by PFI work increasingly resigned smaller, regional contractors to the role of subcontractors in public sector work. The same also applied to the national framework agreements established by large private sector clients. Subcontractors felt progressively more squeezed by ever-increasing demands for continuous improvement, and in turn squeezed the sub-subcontractors increasingly. At the very end of the supply chain were the gangs of bogusly self-employed tradesmen from Eastern Europe, who worked hard and posted good money home. Until the boom years came to an end and they all went home again, taking their skills and enterprise with them.

9.3.2   Partnering 'takes great strides'

Modernising Construction was first and foremost about the modernisation of the procurement and delivery of construction projects for the benefit of public sector clients. The report shared its theme of modernisation with the 1999 White Paper *Modernising Government.* Relationships between the construction industry and public sector clients were seen to have been long since characterised by conflict and distrust. This had now been repeated so often it had become a truism. Yet the original solution of creating public sector direct labour organisations (DLOs) was clearly no longer on the agenda, other than the special case of railway track maintenance. *Modernising Construction* repeated Latham's (1994) previous claim that the resulting inefficiencies comprised nearly 30% of the cost of construction, the evidence for which remains unclear. Such sound bites seem to enter the ether of the construction improvement debate without the need for any verification. The aspiration was to overcome decades of conflict and distrust through widespread 'culture change'. This was a bizarre circumstance. On the one hand, the government was promoting PFI on the basis of the efficiency gains that could be achieved through the private sector. And yet, on the other, the privately-owned construction sector was being lambasted for its inefficiency and adversarial attitudes.

However, *Modernising Construction* recognised that improvement was also dependent upon the adoption of appropriate behaviour by public sector clients. The recognised interdependence between public sector clients and providers militates against the possibility of the construction sector implementing change in isolation. It also militates against the possibility of the industry's major private clients instigating change through coercion. Neither of these points is made explicit within *Modernising Construction*, but the implication is clear for the discerning observer. Under the previous Conservative government, private was 'good' and public was 'bad'. But such simple dichotomies were no longer fashionable under the New Labour regime. Emphasis instead was given to public-private partnerships, policed through an endless bureaucracy of key performance indicators.

It was significant that the foreword to *Modernising Construction* was written by Sir Michael Latham. This would seem to have been a deliberate attempt to emphasise the continuity between Latham (1994) and Egan (1998) reports, and gloss over the differences. It commenced with a reminder of his previous recommendation that the client should be at the core of the construction process. He further highlighted the central message from *Constructing the Team* that the best route to achieving client satisfaction was through teamwork and cooperation. Creative design was conceded to be important, but the machine metaphor was in evidence in terms of the emphasis given to 'stripping out' waste and inefficiency.

Latham went on to express the view that partnering had made 'great strides in recent years'. He considered that fastest growth had occurred within housing associations and some other (unspecified) parts of the public sector. He further described the response from private sector clients as mixed, accusing many as not yet understanding that 'lowest price' did not always equate to value for money. This was a marked departure from *Rethinking Construction*, which had seemingly held private sector clients to be beyond criticism. Lowest price procurement was seen to often result in commercial responses from contractors who thereafter tried to claw back a return through adversarial practices such as variations, claims and the 'Dutch auctioning' of subcontractors and suppliers. Partnering was somehow held to turn such processes around by offering a 'win-win scenario' for all parties.

Latham progressed to repeat the now established mantra that the whole team should sign up to achieve reasonable margins on an open book basis. Mutually of objectives was seen to be key, and great weight was placed on how this could be achieved through the initiation of a project charter. He further emphasised the importance of problem-solving through the whole team working collaboratively. Latham was a consistent supporter of the core ideas of teamwork and collaboration, and he undoubtedly saw partnering as the means through which these could achieved. But his endorsement of the principles of *Rethinking Construction* was also highly selective. He strongly endorsed its emphasis on client satisfaction through partnering, but he was noticeably silent in respect of Egan's various other recommendations.

9.3.3   *Improving construction performance*

Modernising Construction set out to highlight existing good practice that was being adopted by government clients and industry. The belief was that if these good practices could be more widely adopted they would realise sustainable improvements in construction performance, thereby achieving better value for taxpayers. Emphasis throughout was given to the need for a greater concentration on meeting the needs of the end user at lower through-life costs. But insufficient emphasis was given to the resultant benefits that could expected to accrue to contractors and consultants, thereby sowing the seeds of further distrust. *Modernising Construction* was further notable for adopting the cause of supply chain integration:

> '[t]he entire supply chain including clients, professional advisors, contractors, subcontractors and suppliers of materials must be integrated to manage risk and apply value management and engineering techniques to improve buildability and drive out waste in the process.'

It was argued repeatedly that the process above would reduce though-life and operational costs. Integration of the supply chain would apparently also lead to greater certainty in terms of project time and cost, fewer accidents and more sustainable construction. The quest to drive out waste from the process was immediately reminiscent of the emphasis within *Rethinking Construction*. The NAO also displayed the same touching faith in the effectiveness of ill-defined management techniques such as 'value management and engineering'. The evidence to support the claims made was remarkably thin given the remit of the NAO. The overriding inference was that the evidence had already been provided by the Latham (1994) and Egan reports (1998). There seemed to a prevailing feeling that if such claims were repeated often enough, and with sufficient enthusiasm, they would eventually become true of their own accord. The culture of spin had seemingly extended to the NAO, perhaps under the tutelage of Alastair Campbell.

The NAO's summary of what needed to be done is reproduced in Figure 9.1. The analysis was based apparently on the recommendations of the Latham (1994) and Egan (1998) reports, taken together with the Levene (1996) report *Efficiency Scrutiny into Construction Procurement by Government*. It was the latter report which had concluded that government bodies were partly to blame for the poor performance of the industry.

Figure 9.1 is useful in that it offers a coherent narrative which summarises the findings of the three reports in terms of their lowest common denominators. But it also disguises the differences between the reports in terms of their stark differences in emphasis. Especially noticeable is the way in which the more fanciful of Egan's recommendations are ignored. For example, there is no mention of lean thinking, or of the need for the advocated industry-wide regime of performance measurement. Perhaps most interesting of all is the slippage between different units of analysis. *Rethinking Construction* was very much an agenda for changing the *modus operandi* of the construction sector at large, but the summary offered in *Modernising Construction* was directed primarily at public sector clients and the ways in which they could improve project procurement and delivery. The shared solution lay in the eradication of adversarial relationships in favour partnering. If only life were so simple (see Chapter 7).

9.3.4   *Improving the performance of departments and contractors*

Modernising Construction was especially firm in its assertion that traditional forms of contracting based on lowest-price tendering do not provide value for money in the longer term. It was emphasised that private sector clients were seeking increasingly to establish long-term collaborative relationships with construction firms, or were seeking to implement partnering on individual projects. The endorsement of partnering by the NAO was

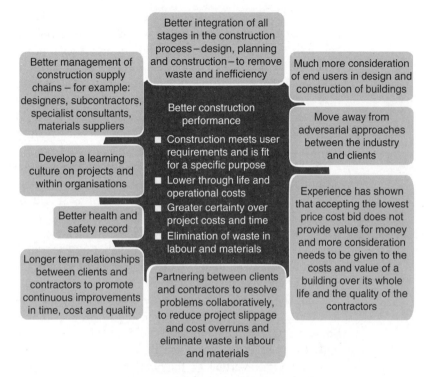

Figure 9.1   Better construction performance – what is needed? National Audit Office analysis of Latham, Levene and Egan reports (*Source: Modernising Construction,* NAO, 2001)

significant in that it was further legitimised as a central component of construction best practice. But the approach was still advocated essentially on the basis of *potential*, rather than convincing evidence. Perhaps in response to concerns about the lack of supporting evidence, the NAO commissioned two academics to produce an anodyne review of the partnering literature (Fisher and Green, 2001). It was understood at the outset that the review would have to be at least neutral in terms of the espoused benefits of partnering. And in truth the overwhelming majority of the academic literature at the time was indeed orientated favourably towards the benefits of partnering. As has been demonstrated throughout this book, academics are often implicated directly in the social construction of best practice – and the present author is no exception.

Notwithstanding the above, NAO sought to collate its own evidence in support of partnering. *Modernising Construction* included details of different forms of partnering used by various government departments and agencies. Evidence was also gleaned from five Movement for Innovation (M4I) demonstration projects: Anglian Water, Kingston Hospital, Notley Green Primary school, Dudley Southern by-pass and the *Building Down*

Barriers initiative presided over by Defence Estates. However, it should be noted that none of the case studies was seemingly subject to any independent audit. In each case, the supporting information appeared to have been provided by those responsible for promoting the partnering initiative in question. What was most noticeably absent was any attempt to position the case studies against the extensive restructuring which had taken place – and was still underway – of the public sector client base. The case studies were therefore offered in isolation of any broader evaluation of government policy *vis-à-vis* outsourcing, downsizing and privatisation. All five examples could be read more meaningfully in terms of the third stage of 'leanness' as described in Chapter 8. This broader long-term analysis was entirely missing in *Modernising Construction*, as it had been previously missing in *Rethinking Construction*. The previous 25 years of change was seen as irrelevant. The problem was reduced to 'adversarial relationships must be overcome by partnering and collaborative working'.

In common with many other sources, *Modernising Construction* was careful to emphasise that partnering was not an easy option, and that it was not the same as having a cosy relationship with contractors. Many within the public sector remained sensitive to the possibilities of corruption. But at the time, cosy relationships with contractors were of course increasingly the rage, especially when they were dressed-up as public-private partnerships. Nevertheless, the NAO was clear that partners should still be appointed competitively, and that clear improvement targets should be set. It was recommended that both parties should be committed to continuous improvement and open book accounting. It was further seen to be important that clients should have access to contractors' records so they could verify the contractors' costs and efficiency improvements. Trust is purported to be an essential pre-requisite of partnering, but behind the rhetoric of collaborative working contractors were still seemingly not to be trusted.

9.3.5   *Performance measurement*

Modernising Construction further pledged its support for the notion of performance management as advocated by *Rethinking Construction*. Reference was made to the key performance (KPIs) produced by the DETR 'in conjunction with the construction sector'. In respect of the public sector, reference was made to KPIs produced by the Government Construction Clients' Panel (GCCP) and the Office of Government Office (OGC). The propagation of endless performance targets was highly characteristic of the New Labour government. KPIs in various forms were extended across all policy domains and were by no means limited to construction. But there is a certain irony in advocating private sector enterprise on the one hand, while insisting that performance should be monitored through the use of KPIs, on the other.

The NAO further claimed to have assessed the impact of construction improvement initiatives on performance within four organisations: NHS Estates, Defence Estates, the Highways Agency and the Environment Agency. Tellingly, they went on to say that it was too early to quantify the benefits being achieved. In other words, some three years after the publication of *Rethinking Construction* there was still no hard evidence in support of the approaches which had been advocated. Strangely, the absence of any such evidence had not prevented the Construction Best Practice Programme from disseminating its best practice 'factsheets'. Likewise, the lack of evidence did not prevent all four organisations from predicting confidently significant savings in construction costs and improvements in quality. Once again, it is likely that the NAO assessors spoke with the individuals involved intimately in the four improvement initiatives. Thus, the interviewees would understandably take an optimistic view of the benefits likely to be achieved. Given that they had invested a significant amount of their own energy (and credibility) in the respective initiatives, they would be unlikely to offer a balanced view on the chances of a positive outcome.

The NAO assessors further spoke with 11 construction firms and 17 specialist contractors and consultants and asked them to describe the initiatives which they were taking to improve their services to clients. The sound bites listed as responses were derived directly from the accepted lexicon of industry improvement:

- establishing longer-term relationships and partnering arrangements;
- providing more value to customers with greater consideration of their needs;
- better supply chain management;
- learning from promoting good practice;
- identifying better ways of reducing whole life costs.

Irrespective of what was being achieved in practice, contractors had at least learnt to 'talk the talk' of the Egan Agenda. As discussed previously, this was arguably the most important part of convincing clients that they were up-to-date with the latest thinking. But the interviewees did also apparently concede that there were a numbers of factors which could limit improvements in construction performance. Concerns were expressed about the declining number of applications to construction courses at universities from 1994–98. Contractors and consultants alike were concerned that the industry was already becoming more reliant on a less skilled work force. Strangely, they did not seem to relate this to their own retreat from provision of training. Disquiet was also expressed that staff within government departments were insufficiently trained to act as intelligent construction clients, thereby echoing one of the themes of the Latham (1994) report. Further doubts were expressed about the extent to which maximum use was being made of information technology as a means of improving the cost effectiveness of construction. It seems there were some who were still not convinced of the benefits of IT-enabled business process improvement.

9.3.6 Recommendations for action

The failure of the NAO to find any concrete evidence that the current improvement agenda was producing the required benefits did not prevent it from recommending more of the same. The recommendations were directed at four key groups: (i) the Department of The Environment, Transport and the Regions (DETR), (ii) the Office of Government Commerce (OGC), (iii) government departments who are clients of the construction sector (the so-called 'line' departments), and (iv) the construction industry itself.

The DETR was encouraged to provide more coordinated direction to the plethora of initiatives which were at the time in existence. There was considered to be some duplication in effort and the existence of so many initiatives was seen to be a source of confusion to many in the industry. According to *Modernising Construction*, the Construction Industry Board (CIB) had acquired the responsibility for the provision of strategic direction, although this was to be short lived. The DETR was further exhorted to use its influence with M4I to ensure that the demonstration projects were truly innovative. Apparently the M4I team had originally accepted all suggestions for demonstration projects that had been put forward by the industry. Hence the fears of those who had at the time questioned the rigour of the demonstration projects can be seen to have been justified. The M4I Board were asked to tighten the criteria for inclusion in the future. The final recommendation to the DETR related to the perceived need for ever more sophisticated performance measures. What had been achieved to date was seen to be little more than an important first step. The NAO's advice here is symptomatic of the classic response when performance targets fail to achieve the desired effect: 'we need more sophisticated performance indicators'.

The recommendations directed towards the OGC were noticeably modest; they were simply encouraged to disseminate good practice more widely. The so-called 'line departments' were encouraged to measure improvements in construction performance actively. The Egan Report, *Rethinking Construction*, was cited as having 'estimated' that a reduction of 10% per year in construction costs was achievable. At best, this was a misrepresentation of the status of the Construction Task Force's suggested targets for improvement. At worst it was outright lie. It should be recalled that Egan's original targets were based on 'an impressionistic and partial assessment' as a result of solid data being hard to come by. Undeterred, *Modernising Construction* went on to suggest that the desired improvement would only be achieved if the good practice initiatives promoted by DETR and OGC were implemented actively and widely. Given the vagaries of the advocated recipes, and lack of supporting evidence, this continued to be an expression of faith rather than as assessment based on hard evidence. Line departments were further exhorted to ensure that all staff involved in procuring and managing construction attended appropriate training courses. Whether the advocated recipes on offer were of any use was an issue which was notably ducked. As tends to be the case with all management fashions, if they initially fail to achieve the required impact, the inevitable response is to advocate that they should be implemented with more enthusiasm.

The final recommendation was directed at the construction industry, which was taken to include consultant architects, engineers, quantity surveyors and project managers. The continued use of the term 'construction industry' to cover such a heterogeneous sector is indicative once again of an assumption that generic 'best practice' is applicable universally. The overriding plea was that the 'industry' should make greater use of innovation in the cause of improving public sector procurement. The industry was seen to have much to gain from the range of improvement initiatives which were at the time under way. Treasury of course would have been somewhat disappointed had the NAO concluded otherwise. Particular note was made of the benefits of remuneration on the basis of target price with associated opportunities to share in efficiency gains. The benefits of longer-term relationships were emphasised yet again, together with partnering with a commitment to continuous improvement. The NAO clearly thought that the benefits of partnering would be accepted provided that they kept being emphasised. The possibility that contractors had mixed experiences of partnering was not acknowledged. Neither was it acknowledged that subcontractors' experiences of partnering had very little to do with the mutuality of objectives espoused repeatedly in the literature.

9.4   Accelerating change

9.4.1   Egan rides again

Despite being tasked tentatively with providing 'strategic direction' to the host of improvement initiatives in *Modernising Construction*, the Construction Industry Board (CIB) was dissolved formally within six months, primarily as a result of a lack of support among the industry's clients. Plans had been underway for some time to replace the CIB, while at the same time bringing together the various strands of the Rethinking Construction movement, including M4I, CABE, the Housing Forum, the OGC and the remnants of CRISP (Adamson and Pollington, 2006). The Strategic Forum formally came into being on 1 July 2001 in the immediate aftermath of the June 2001 general election. Sir John Egan was duly invited to become the first chairman of the new strategic body, whose initial membership is listed in Table 9.1.

Under the tutelage of the new Minister for Construction, Brian Wilson, the Strategic Forum set about producing a new report. The direction of

Table 9.1   Strategic Forum membership

British Property Federation (BPF)
Commission for Architecture and the Built Environment (CABE)
Confederation of Construction Clients (CCC)
Construction Industry Council (CIC)
Construction Industry Training Board (CITB)
Construction Industry Products Association (CPA)
Construction Research and Innovation Strategy Panel (CRISP)
Constructors Liaison Group (CLG)
Department of Trade and Industry (DTI)
Design Build Foundation/Reading Construction Forum (DBF/RCF)
Health and Safety Executive (HSE)
Housing Forum (HF)
Local Government Task Force (LGTF)
Major Contractors Group (MCG)
Movement for Innovation (M4I)
Office of Government Commerce (OGC)
Rethinking Construction (RC)
Trades Union Congress (TUC)

(*Source*: Strategic Forum, 2002)

change had of course already been decided upon, the task in hand was how to accelerate the pace of change by seeking to 'inculcate the principles of Rethinking Construction throughout the industry'. Hence the report was entitled *Accelerating Change* (Strategic Forum, 2002).

From the outset it was understood that the Strategic Forum should seek to build on and reaffirm the principles of *Rethinking Construction*. The main thrust of the report was the extent to which the targets for improvement outlined in *Rethinking Construction* were being achieved in practice. *Accelerating Change* therefore re-confirmed the culture of performance measurement, justified on the basis of 'if you do not measure how can you demonstrate improvement?' The seven original targets were expanded to 12 with the addition of client satisfaction measures, thereby strengthening further the primacy of the client over other stakeholders in the construction process. In the preamble to *Accelerating Change*, Sir John Egan stated confidently:

'[t]he demonstration projects clearly show that the targets we set were realistic, and that when achieved the results bring benefit to all.' (Strategic Forum, 2002; 7)

The targets to which he was referring, it should be recalled, were those which the Construction Task Force had described previously as being based on an 'impressionistic and partial assessment" (Egan, 1998; 18). But the original caveats had now seemingly been forgotten as the 'illustrative' targets were legitimised and adopted verbatim as policy. Egan's faith in the demonstration projects, however, was not shared by others. Of particular note was the lack of any objective process of verifying the claimed innovations (Gray and Davies, 2007). But this was seemingly unimportant. It had already been decided that the implementation of the *Rethinking Construction* 'principles' (in all their glorious vagueness) would lead to improved performance. The task in hand was to demonstrate what had already been decided. The same approach was adopted by the New Labour government in 2003 in respect of Saddam Hussain's WMD. It had already been decided that they existed because this suited the advocated policy of 'regime change'. Hence the challenge was reduced to providing supporting evidence; likewise, the Strategic Forum's approach to *Accelerating Change*.

The vision of *Accelerating Change* was set out as follows:

> '[o]ur vision is for the UK construction industry to realise maximum value for all clients, end users and stakeholders and exceed their expectations through the consistent delivery of world class products and services.'

Despite the stated intention to build on and reaffirm the principles of *Rethinking Construction*, the vision above demonstrates how the debate had moved on since 1998. The task of maximising value had taken over surreptitiously from the previous focus on eliminating waste. The introduction of end users and stakeholders also represented a significant shift away from the ruthless emphasis on the client which characterised *Rethinking Construction*. Despite these immediate differences in focus, the authors of *Accelerating Change* were at pains to emphasise its continuity with *Rethinking Construction* rather than the points of divergence. There was clearly a sensitivity to the charge of 'initiative overload', as alluded to previously by *Modernising Construction* (NAO, 2001).

9.4.2   Progress since Rethinking Construction

Accelerating Change commences its review of progress with a reminder of *Rethinking Construction's* objectives:

> '[t]o achieve radical improvements in the design, quality, sustainability and customer satisfaction of UK construction. And for the industry to be able to recruit and retain a skilled workforce at all levels by improving its employment practices and health and safety performance.' (Strategic Forum, 2002; 14)

Unfortunately, the quote above once again represents a retrospective rewriting of the original Egan report, which had very little to say about either design or sustainability. Neither did *Rethinking Construction* give any great deal of emphasis to employment practices in the construction sector, other than a passing reference to the need to provide staff with 'employment conditions to enable them to give of their best'. *Rethinking Construction* therefore fell some way short of advocating fair and equitable employment practices. In some sense, the re-writing of the objectives of *Rethinking Construction* could be read as an acknowledgement of its deficiencies. But in truth, the 'principles' of *Rethinking Construction* were undergoing a process of continuous re-negotiation.

The authors of *Accelerating Change* described confidently how the programme of demonstration projects lay at the heart of the *Rethinking Construction* initiative. It was reported that there were 400 demonstration projects in the programme with a total value of £6 billion. They were claimed to provide examples of off-site fabrication, standardisation, use of new technology, partnering and supply chain integration and other (unspecified) areas of process improvement. The need for the industry to adopt a culture of performance measurement was re-affirmed on the basis of the argument that without measurement there is no way of demonstrating improvement. It was further described how an industry-wide group had developed a set of simple headline key performance indicators (KPIs) based on the seven *Rethinking Construction* targets, but with the addition of five new measures of client satisfaction. Designated champions on the Demonstration Projects were apparently required to measure their own performance against the specified KPIs and to report annually. Taken together, the demonstration projects were seen to outperform significantly the industry at large (see Table 9.2).

The demonstration projects were apparently subject to 'independent research' which reviewed the impact of the projects among participants:

- More than two thirds reported improved partnering, procurement or supply chain management skills.
- More than half reported that their organisations had made changes in eight specific arenas of their business
- More than two-thirds of participating individuals felt that they had been at the cutting edge of construction innovation and learned new skills.

Unfortunately, no details of the supposed 'independent research' were provided, so it is difficult to confirm the extent to which it had been conducted rigorously. It would certainly seem that the demonstration projects had captured the enthusiasm of the survey respondents, and this in itself was a significant achievement. The lessons learned were later disseminated through the websites of M4I and the Housing Forum in the form of case studies, progress reports and themed reports. Nevertheless, despite the plethora of information which appeared on the internet, the aspiration to turn the Construction Best Practice Programme into a 'knowledge centre' which provided the whole industry with information on the lessons learned never came to meaningful fruition. Knowledge, of course, comprises rather more than an accumulation of

Table 9.2   M4I demonstration project performance compared to all construction for 2001

Headline key performance indicator	Measure	All construction	M4I	M4I Enhancement
Client satisfaction [–]–[–] Product	Scoring 8/10 or better	73%	85%	+16%
Client Satisfaction – Service	Scoring 8/10 or better	65%	80%	+23%
Defects	Scoring 8/10 or better	58%	86%	+48%
Safety	Mean accident rate/100k employed	990	495	+100%
Cost Predictability – Design	On target or better	63%	81%	+29%
Cost Predictability – Construction	On target or better	50%	71%	+42%
Time Predictability – Design	On target or better	46%	81%	+76%
Time Predictability – Construction	On target or better	61%	70%	+15%
Profitability	Median profit of turnover	5.6%	7.6%	+2% percentage points
Productivity	Median value added/ employee (£000)	28	34	+21%
Cost	Change compared to 1 year ago	+2%	–2%	+4%
Time	Change compared to 1 year ago	+4%	–8%	+12%

(*Source*: *Accelerating Change*, Strategic Forum, 2002)

assembled information on the supposed benefits of instrumental management techniques. Some would argue it has much more to do with developing the ability to discriminate between the competing claims of different perspectives. But knowledge of this kind was not seemingly valued by the Strategic Forum; the only knowledge considered worthwhile was that which related to the implementation of the principles of *Rethinking Construction*.

In reflecting back on the lasting contribution of the demonstration projects, to the author's knowledge there are very few independently-evaluated case studies of the lessons learned which remain accessible in the public domain. The tendency to publish the findings on the internet is understandable; but the transient nature of the internet means that relatively little of the information which was posted remains available for public critique. In truth, most of the case studies comprised little more than an entirely unreflective narrative from the appointed (or self-appointed) 'innovation champion'. Such case studies served their immediate purpose in terms of dissemination, but it is difficult to argue that they made any lasting contribution to knowledge.

Nevertheless, any doubts about the rigor with which the demonstration projects had been evaluated did not prevent *Accelerating Change* from endorsing enthusiastically what had been achieved:

> '... results show that the demonstration projects are consistently exceeding the targets in Rethinking Construction.'

However, the vagueness of precisely what was being implemented in the demonstration projects remained elusive. Indeed, beyond the managerialist rhetoric of radical change, the defining principles of *Rethinking Construction* were vague from the outset. And this essential vagueness was only exacerbated by the way in which they were being constantly re-written as the *Rethinking Construction* bandwagon rolled on. The authors of *Accelerating Change* had clearly identified some notable omissions within the original *Rethinking Construction* report, and in consequence issues such as design quality, sustainability and enlightened employment practices suddenly found their way into the lexicon of industry improvement.

Despite the widespread pressure to conform to the Egan Agenda, *Rethinking Construction* had received extensive criticism from the architectural profession for giving insufficient emphasis to issue of design quality. This criticism was subsequently alleviated by M4I's encouragement to the Construction Industry Council (CIC) to develop a set of 'design quality indicators', which were launched in July 2002. Likewise, criticisms that the Egan report had failed to take sustainability seriously resulted in the establishment of a 'Sustainability Working Group'. This resulted in the provision of a set of project-based Environmental Performance Indicators (EPIs) in 2001. A further example is provided by the response to the criticism that *Rethinking Construction* failed to give sufficient emphasis to the espoused commitment to people. M4I launched a working group on 'respect for people' which had published its report *A Commitment to People – Our Biggest Asset* in November 2000. Here again the emphasis was on toolkits and KPIs. Indeed, this seemed to be the generic response to every criticism. In response to any criticism that design quality, sustainability or employment conditions were not being addressed, the Eganites could simply point towards the proliferation of toolkits, DQIs, KPIs and EPIs. However, at least the debate had progressed slightly since the publication of *Rethinking Construction* and the parameters of best practice were being expanded continuously. No wonder the Construction Best Practice Programme struggled to keep up.

Notwithstanding some of the above mentioned deficiencies, the broader *Rethinking Construction* initiative continued to receive financial support from the Department of Trade and Industry (DTI). A further two years' funding was announced to extend the initiative's duration through to April 2004. *Accelerating Change* also claimed that the initiative was being:

> '... backed solidly through the direct engagement of hundreds of companies and industry organisations, as well as other government departments.' (Strategic Forum, 2002; 17)

Enlightened clients were also apparently seeking to work with people who were committed to the *Rethinking Construction* agenda. This claim of course owed as much to the adopted definition of 'enlightened' as to those who were committed to *Rethinking Construction*. If a particular client was not adopting the principles of *Rethinking Construction* then it could not, by definition, be classified as enlightened. And of course the only other category that was available was that of 'dinosaur'.

In summarising the progress made since the Egan report, *Accelerating Change* states that the four key objectives of *Rethinking Construction* remained as follows:

1. Providing and selling the business case for change.
2. Engage clients in driving change.
3. Involve all aspects of the industry.
4. Create a self-sustaining framework for change.

The need to sell the 'business case' for change was subsequently to become one of the main mantras of industry reform. However, if the case for change really was so compelling, those firms which adopted 'best practice' most effectively would be those which went on to be most successful. And those which were resistant to change would be those which went out of business. Industry was seemingly a bit slow on the uptake and needed to be persuaded. An alternative explanation is that the principles and targets of *Rethinking Construction* were of limited application, and the evidence from the demonstration projects was not as convincing as had been claimed. Clients seemingly needed separate encouragement for them to engage in the change process. The Strategic Forum clearly believed in the universal applicability of 'best practice', which was apparently applicable for 'all aspects of the industry'. However, there was also an ominous warning that the industry should take responsibility for developing and maintaining continuous improvement. This represented a clear indication that the government was not prepared to fund the *Rethinking Construction* initiative indefinitely.

Accelerating Change's emphasis on continuous improvement also contrasted starkly with Egan's original call for a radical change. The rhetoric of business process re-engineering had seemingly slipped out of fashion by 2002. The members of the original Construction Task Force had been even more effusive in their support of lean thinking. Yet strangely, it was not even mentioned in *Accelerating Change*. Perhaps someone has discovered that lean thinking was not quite the panacea that had been suggested. Unfortunately, nobody told the Construction Best Practice Programme.

9.4.3 Strategic direction and targets

Accelerating Change further detailed how the Strategic Forum had identified 'three main drivers to accelerate change and secure a culture of continuous improvement':

- The need for client leadership.
- The need for integrated teams and supply chains.
- The need to address 'people issues', especially health and safety.

These were seen to embrace customer focus, supply chain integration and respect for people. The three issues were further seen to be linked strategically, although it was unclear how they related to the drivers for change which had been previously specified in *Rethinking Construction*. Clients were seen to be the starting point of the change process and they were exhorted to procure on the basis of best value rather than lowest price. Nevertheless, the focus on the elimination of waste remained a favourite theme. Delivery of the espoused vision was seen to require collaboration between an extensive list of parties, including the whole of the entire supply chain, government, the finance and insurance sector and the research institutions. The quest to align all parties to the common cause of industry improvement is again reminiscent of *Progress through Partnership* (OST, 1995).

It was further claimed to be important to secure the 'right sort of people' with the 'right blend of skills and competences'. Presumably those individuals who remained sceptical of the Egan Agenda were the 'wrong sort of people'. The predetermined model of the 'right sort of people' was hardly compatible with the cause of increasing diversity in the workforce. Neither is it compatible with an environment within which innovation is likely to flourish.

During New Labour's second term of office the use of performance targets became ever more popular as instruments of public policy. There was also greater enthusiasm for inventing new targets than there was for investigating why previous targets had not been achieved. The Strategic Forum followed this broader trend by setting out yet another set of strategic targets to be achieved by the end of 2004:

(1) 20% of construction projects by value should be undertaken by integrated teams and supply chains;
(2) 20% of client activity by value should embrace the principles of the Clients' Charter.

It was further exhorted that both of the above should rise to 50% by 2007. The Strategic Forum also expressed its determination to reverse the long-term decline in the industry's ability to attract and retain a quality workforce. Of additional note was the promise to put in place a means of measuring progress towards the advocated targets. Seemingly, some four years after the publication of *Rethinking Construction*, little progress had been made in terms of implementing the required regime of performance measurement. It was notable that this had now been prioritised over the need for the construction industry to produce its own structure of objective performance measures, as had been advocated originally. The industry, it seems, had taken the easy way out, and had simply accepted the targets suggested initially by the Construction Task Force. Given the imposed requirement for the industry to 'agree' their suggested measures with clients, this was probably a pragmatic response. However, it also suggested that the response of the construction sector to *Rethinking Construction* was more akin to passive behavioural compliance than it was to the desired cultural change. In truth, construction sector chief executives thought the advocated targets were nonsense. They were just too polite to say so.

9.4.4 *Accelerating client leadership*

Despite the repeated emphasis on the need to comply with the principles of *Rethinking Construction*, precisely what constituted the 'right sort of people' was something of a movable feast. Clearly, it was necessary to sign up to the Egan Agenda, but it remained unclear which particular principles one was required to endorse. *Rethinking Construction* had previously given scant attention to sustainability, and yet it was now claimed to be 'self-evident' that clients should enter the construction process with a 'clear understanding of the environmental and social responsibilities of clients' (Strategic Forum, 2002; 20). Clients were further exhorted to have a clear understanding of what value means to them. Any lack of clarity at the outset was seen to imply changes throughout the delivery process, resulting in waste, duplication and general dissatisfaction. The underlying conceptualisation of the briefing process was heavily shaped by the machine metaphor. Particular care was taken to advise inexperienced clients on the need to follow a defined process, commencing with the verification of need. Inexperienced clients were advised to access independent advice, with the predictable proviso that the advice on offer should meet the principles of *Rethinking Construction*. In other words, advice should only be sought from the 'right sort of people'.

Clients were further encouraged to create an environment which would deliver excellence in health and safety. Clients, it was claimed, would increasingly be judged by their customers and financial analysts on their 'ethical stance in relation to safety'. The view was that this would be applied in the same way as was already happening for environmental performance and sustainability. The Strategic Forum's recurring emphasis on safety can only be applauded, but there was little attempt to analyse the root causes of the industry's poor safety record. Emphasis instead was placed on the need for training so that all parties were aware of the 'demonstrable business, efficiency and safety benefits of integrating teams and processes'. Needless to say it was not made clear where the demonstrable evidence to support such claims was going to be found. The responsibility for this was conveniently shifted to the Construction Best Practice Programme (CBPP).

Lean thinking may have been forgotten, but partnering and 'integrated procurement' were still very much in vogue:

> 'Too many organisations continue to believe that partnering and integrated procurement are experimental techniques and that the majority of their mainstream projects can still be effectively procured through traditional arrangements.' (Strategic Forum, 2002; 22)

In light of the discussion of partnering in Chapter 7, it would seem that organisations were justifiably cautious of adopting partnering solely on the basis of 'faith and commitment'. What is clear is that partnering cannot be understood as a definable 'technique', experimental or otherwise.

Accelerating Change went on to give particular emphasis to public sector clients, which at the time accounted for 40% of construction orders. The public sector was seen to have a vested interest in getting best whole life value from construction. Suggestions for action even included the linking of government funding of projects to the principles of *Rethinking Construction*. Public sector clients were further encouraged to take a lead in the procurement of sustainable construction, despite the fact that this received very little emphasis within *Rethinking Construction*. The Strategic Forum's focus on whole life value represented a return to the script advocated previously by *Progress through Partnership* (see Chapter 4). This debate might have been progressed more quickly had it not been stifled prematurely by *Rethinking Construction*'s sharp focus on short-term waste elimination.

Advice was also directed toward private sector clients, who were exhorted to understand how construction projects should contribute to their business needs. The Construction Clients' Confederation was at the time working on a 'Starter Charter' aimed at inexperienced clients. But, given the widely varying experience of clients, it was recognised that any guidance could not be universally relevant. The reluctance to recommend 'one-size fits all' guidance for clients did not extend to the KPIs aimed at the construction sector, despite its considerable heterogeneity. It would seem that clients were not willing to be dictated to in the same way that they liked to dictate to the construction sector.

Client action was further recommended to support the development of long-term integrated supply chains with the aim of increasing productivity, reducing time, increasing cash-flow efficiency and minimising risk. Despite such recommendations for action, there were no KPIs aimed at major private

clients to ascertain the extent to which they were implementing the principles of *Rethinking Construction*. Clients seemingly deserved a much softer
form of encouragement. Given the demise of the Construction Industry
Board (CIB) due to lack of client support, it was perhaps inevitable that the
Strategic Forum would tread carefully in respect of its advice for clients.

9.4.5 *Accelerating supply side integration and integrated teams*

Notwithstanding the above, the major contribution of *Accelerating Change*
lay in its endorsement of the terminology of supply side integration and
integrated teams. Although these concepts remained ill-defined, they were
nevertheless seen to be of central importance. The list of supposed advantages of supply side integration was extensive, and included the following:

- increasing quality and productivity;
- reducing in project times;
- increasing cash-flow efficiency;
- reducing costs from getting it 'right first time';
- ensuring that people work within the process;
- delivering benefits during initial project delivery;
- securing best value throughout use of the completed project;
- maximising opportunities for sustainable solutions.

The benefits listed above are difficult to argue against, but it seems hugely
optimistic to accredit such a wide-ranging list of benefits to such a vague
set of concepts. 'Supply chain integration' and 'integrated' could perhaps
be read as rhetorical correctives to that old *bête noire* of the construction sector: 'fragmentation'. But, strangely, fragmentation is not mentioned in the
text of *Accelerating Change*. In contrast, the term 'integration' is cited 27
times, and 'integrated team' is mentioned an incredible 40 times. It seems
that 'integrated teams' had replaced 'lean thinking' as the most popular
term in the lexicon of industry improvement. Despite the original
Construction Task Force being impressed by the 'dramatic success' of lean
thinking, it seems that by 2002 the novelty had worn off - thereby creating
another problem for the Construction Best Practice Programme in its task
to maintain a checklist of approved best practice recipes.

Notwithstanding the long list of supposed benefits, the main emphasis
of the justification for supply side integration lay with creating 'value' – a
term that the *cognoscenti* found increasingly difficult to use without the
simultaneous raising of both eyebrows. The implication of this mannerism
was that the word was used to imply a much deeper set of philosophical
principles. The repeated emphasis on 'value' was increasingly central to
the espoused discourse of industry improvement, even if it remained
difficult to define and operationalise. But 'value' was clearly a good thing,
and clients understandably wanted more of it. And if integrated supply

chains secured greater value for the client then this must be a good thing too. A further difficulty lay in defining precisely what constituted an 'integrated supply chain'. *Accelerating Change* did make it clear that an integrated supply chain includes the client, as well as those who are 'pivotal' in providing solutions that meet the client's requirements:

> '... those involved in asset development, designing, manufacturing, assembling and constructing, proving, operating and maintaining, will have the opportunity to add maximum value by being integrated around common objectives, processes, culture/values, and reward and risk.' (Strategic Forum 2002; 24)

The implication is that membership of the integrated team is limited to those who are seen to be of 'pivotal' importance to the outcome. It was further emphasised that 'key' manufacturers must be part of the integrated team, although again the implication is that those manufacturers who are not 'key' should not be admitted to the integrated team. The essential argument of seeking to integrate key suppliers owes much to supply chain management (SCM), whereby key suppliers are embraced into partnering contracts and non-key suppliers are ruthlessly squeezed on the basis that they can easier be replaced. Such an approach undoubtedly makes good business sense, but it hardly constitutes a 'culture change'.

Notwithstanding the above, *Accelerating Change* gave particular encouragement to product manufacturers, suppliers and specialists who can develop solutions that involve less site processing. This was very much in line with the established mantra of modern methods of construction (MMC), whereby increased standardisation, pre-assembly and pre-fabrication are seen as means of improving quality and reliability. The engagement of such firms in the integrated team was further seen to be a way of 'unlocking their research expertise and deploying it to deliver value and enhance the finished product'. Once again, the overriding emphasis was on the advantages that can accrue to clients; there was much less emphasis on the benefits realised by the team members (other than the potential for relationship continuity). It could also perhaps be inferred that the concept of integrated teams was limited in its application to new construction; *Accelerating Change* as a whole had had very little to say about repair and maintenance, which accounts consistently for approximately 45% of construction sector output.

Accelerating Change also called for a significant investment in education and training to emphasise the importance of team working. The Construction Best Practice Programme was once again tasked with developing the necessary tools to support SMEs to become part of an integrated supply team. As previously, this presumably only applied to those SMEs who were judged to be 'pivotal' to meeting the client's requirements. Perhaps to alleviate the prevailing sense of vagueness, the Strategic Forum also promised to ensure that a toolkit would be developed by April 2003 to help assemble integrated

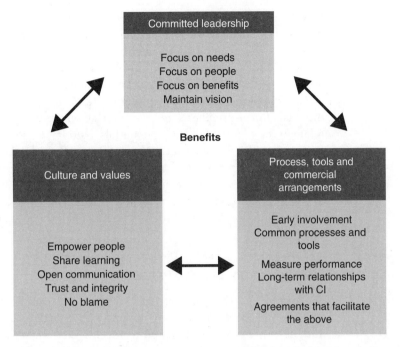

Figure 9.2   Principles of integration (*Source*: *Accelerating Change*, Strategic Forum, 2002)

teams, mobilise their value streams and promote effective team working. If only such a 'toolkit' for enhancing cooperation had been produced to follow through the recommendation of the Simon report:

> '... efficiency and success are dependent on an honest desire for co-operation.' (Simon, 1944; cited in Murray and Langford, 2003; 205)

The phenomenon at play here seems to be what Buchanan (2000) refers to as the 'continuous reinvention of teamworking as a management idea'. Each generation of construction professionals are exhorted to adopt team working as a new idea and to break with the industry's adversarial past. But *Accelerating Change* also provides an important thread of continuity in stating that integrated teams should ideally be based on strategic partnering. The relationship continuity inherent within strategic partnering is held to enable knowledge and expertise to be transferred more effectively from one project to the next. Especially notable within *Accelerating Change* is the stubborn refusal to define the meaning of an integrated team; a task which was delegated to the proposed 'toolkit'.

The promised toolkit was subsequently published on the Strategic Forum website, where it was advocated that the interdependent principles shown in Figure 9.2 should be adopted if effective and sustained integration were to be achieved.

Despite all the grandiose claims in favour of integrated project teams, it is ultimately difficult to see how the substance of the 'underlying principles' differ from those of strategic partnering. *Accelerating Change* further pledged its support for a greater focus on supply chain management and logistics as a means of facilitating integrated working and the elimination of waste. Supply chain management was seen to relate to the flow of goods and materials from supplier to the point of use, whereas logistics was seen to relate to the flow of goods and materials, equipment, services and people throughout the supply chain. Yet again the Construction Best Practice Programme, this time in conjunction with the Construction Products Association, was tasked with the responsibility of collating best practice in logistics and disseminating best practice. The chief executive of the Construction Best Practice Programme must have been starting to despair at this stage.

9.4.6 Accelerating culture change in people issues

Accelerating Change devotes an entire chapter to 'people issues', arguably in response to previous criticisms directed at *Rethinking Construction*. The authors express a determination to reverse the long-term decline in the industry's ability to attract and retain a quality workforce. The dominant instrumentalism is indicated by the way the workforce is described as the industry's most valuable 'asset'; as if it were something to be owned and exploited. Nevertheless, the overall tone is progressive and the arguments presented act as an important counter-balance to the top-down prescription which dominates elsewhere. Reference is made to the ten KPIs associated with the Respect for People initiative which include benchmarks on employee satisfaction, staff turnover rate, sickness, pay, safety and working hours. Further reference is made to no less than eight toolkits which were to be available in October 2002 to help managers evaluate their performance and focus on the following areas:

- workforce satisfaction;
- personal working environment;
- safety;
- health;
- work in occupied premises;
- training
- working environment; and
- equality and diversity.

Strong support was also expressed for the Construction Skills Certification Scheme (CSCS) as a means of improving the industry's health and safety performance. The CSCC was intended to enable employers to identify workers with a recognised level of competence in skills and in health and safety. The focus on supply chain integration was judged to have benefits *vis-à-vis* health and safety. For example, such integration was seen to aid pre-planning such that certain risks could be 'designed-out'. The emphasis on pre-assembly and pre-fabrication was also seen to have safety benefits. Particular emphasis was placed on the importance of consulting the workforce on health and safety matters. Practitioners were further exhorted to be more aware of their responsibilities under the CDM regulations. All of this deserves to be applauded.

In marked contrast to *Rethinking Construction*, specific mention was made of the need to offer attractive pay and conditions in comparison to other sectors. Criticism was directed at the 'long hours culture' within the construction industry, considered detrimental to health and safety and also costly for employers. Perhaps most remarkably, employers were further urged to ensure that all operatives were embraced within the industry's new stakeholder pension scheme. Pleas were also made in support of *Investors in People* and greater workforce diversity. However, it is noticeable that no KPIs were promoted to measure the extent to which firms were improving their performance against these criteria. Nevertheless, this was a much more progressive agenda than that advocated by *Rethinking Construction*.

In contrast to the trend within previous reports, supportive comments were made in respect of the role of professional institutions and the Construction Industry Training Board (CITB). But even here much of the discussion about training is directed towards the need to create the required integrated teams and supply chains. The involvement of the Major Contractors Groups (MCG) would seem to have neutralised any significant criticism of the main contractor. In consequence, SMEs were targeted as the fall guys and were criticised for not being as active in the area of 'people culture issues' as they might be. But even those SMEs which were 'paying attention' to people issues were apparently only doing so because their clients had been telling them to. Apparently, the reason for this lack of attention was that most SMEs fail to see the good business case in support of tackling people issues. SMEs were further criticised for being confused and struggling to decide what to do first. In light of their constant bombardment with ill-defined management techniques, it is perhaps not surprising that the construction sector's SMEs were in a perpetual state of confusion.

Universities were also considered a safe target for allocating blame for the perceived deficiencies in training and education. Support was expressed for the Construction Industry Council's 'Common Learning Outcomes' initiative which was aimed at achieving a greater degree of consistency across university degree curricula. *Accelerating Change* further recommended the requirement to include integrated project team-working within university courses as part of the accreditation process. Presumably the Strategic Forum was looking for such ideas to be accepted entirely uncritically on the basis of their recommendation.

Clients however were not let off the hook completely:

'[h]ow partners in the supply chain behave towards one another is important in developing the relationship of trust that underpins integrated teams.'

The implication was that clients had a responsibility to set an example in terms of trust. There is once again an obvious paradox between advocating trust, on the one hand, while advocating continuous performance measurement, on the other. But setting an example in terms of trust was not the challenge presented to clients. In contrast, they were tasked with the rather less onerous role of ensuring that the selected designers and consultants recognised working rule agreements, were in possession of excellent health and safety records and were committed to training their workforce. If these three points were satisfied, the implication was that clients would be doing their bit towards achieving the strategic vision of excellent performance and whole life value. Placing onerous requirements on private-sector clients was clearly not acceptable. Indeed, government was notably singled out as the industry's most important client, and as such was tasked with leading the way in terms of best practice. Here again, one can easily imagine how placing the emphasis on public sector clients met with the approval of Confederation of Construction Clients (CCC) and British Property Federation (BPF).

Within the Strategic Forum's meetings, it is easy to imagine the trade union representative struggling to get his points across and embedded into the text of *Accelerating Change*. The one issue that the unions would have lobbied consistently to have included would have been that of 'false' self-employment. There were at the time between 300–400 000 self-employed workers in the construction industry who were held to be falsely classified as self-employed when on the basis of any 'economic reality test' they should have been in direct employment (cf. Harvey, 2001). The large proportion of self-employed workers in the construction sector presents on obvious tension with the dominant focus on the 'integrated team'. However, the issue of false self-employment is conflated strangely with the informal construction economy.

Within *Accelerating Change*, the overriding inference is that the problem of bogus self-employment is confined to the grey economy. Firms who contravene tax and employment legislation were seen to provide unfair competition for 'respectable law abiding firms'. Strangely, there is no mention of government complicity in the systemic encouragement of self-employment, nor in the shift away from direct employment by the main contractors. The conflation of false self-employment with the informal construction economy served usefully to confine the problem to a different space from that occupied by the MCG and the CCC. *Accelerating Change* further maintained that it was this 'other' sector where there was most concern about health and safety and where 'people' issues were most consistently ignored.

Unfortunately, this was a gross misrepresentation of the picture painted by Harvey (2001). Harvey's research had demonstrated that workers were routinely falsely classified as self-employed on major construction sites, many of which involved members of the MCG as main contractors (see the discussion in Chapter 3). The association of false self-employment with the informal construction sector was at best disingenuous. Undoubtedly the informal sector is characterised by many of the problems described, but these very same problems also apply to the formal construction sector presided over by the membership of the Strategic Forum. But nothing could be said which went against the interests of the Strategic Forum's membership. It was much better to focus on vague exhortations in favour of 'integrated teams'. At least if the reader could not understand precisely what was being advocated, nobody could be offended.

9.4.7 Cross-cutting issues

The final chapter of *Accelerating Change* set out a list of 'cross-cutting' issues which could act as either enablers or barriers to change. The first such issue was sustainability, which the authors had seemingly belatedly realised had not received sufficient emphasis in the main body of the report. Care was taken to align *Accelerating Change* with the Sustainable Construction Task Group as then chaired by Sir Martin Laing. Approval was expressed for the advocated importance of whole life performance in terms of ensuring value through productivity in use. This may have been a very narrow interpretation of sustainability, but at least it was an improvement on *Rethinking Construction*.

However, the Strategic Forum notably offered its support to the then Secretary of State's call to Trade Associations to address environmental and social impacts. Despite having previously – and erroneously – claimed that 'radical improvement' in sustainability was a central objective of *Rethinking Construction*, the authors of *Accelerating Change* subsequently contradicted themselves by offering an explanation of why it had not previously featured as a core issue. Apparently, at the time *Rethinking Construction* was published, it was considered more important to focus on the 'fundamental flaws' in the construction procurement and delivery process. Precisely how these flaws had been eradicated in the intervening four years was left unexplained. Neither was it explained why lean thinking had received so much emphasis in *Rethinking Construction*, but was then seemingly abandoned by 2002. Nevertheless, *Accelerating Change* went on to emphasise that:

> '[e]very link of the supply chain has a critical contribution to make towards sustainable construction and development.' (Strategic Forum, 2002; 35)

However, there was very little coverage of the broader principles of sustainability. Despite having previously alluded to the broader concept of corporate social responsibility (CSR), the advice offered remained focused stubbornly on the elimination of waste in the cause of efficiency. In this respect *Accelerating Change* was entirely consistent with *Rethinking Construction*, and the linking of this with sustainability seemed somewhat convoluted. The 'waste' which was targeted primarily was not physical waste in the sense of environmental pollution, but inefficiency caused by a lack of alignment with the client's objectives. It is worthwhile recalling the original definition of waste promoted within *Rethinking Construction* relating to 'activities which do not add value from the customer's viewpoint' (see Chapter 5).

It is therefore difficult to hold up *Accelerating Change* for its emphasis on sustainability. The argument presented was broadly consistent with that mobilised previously by *Rethinking Construction*. And the original Egan report was no more representative of the principles of sustainability than was Frederick Taylor's scientific management (Taylor, 1911). The emphasis was not on the broader principles of sustainability, but on narrowly construed notions of economic efficiency. The attempt to weave sustainability into the argument while being consistent with the 'principles' of *Rethinking Construction* led to the authors of *Accelerating Change* almost tying themselves into rhetorical knots. The task was made especially difficult by the pre-determined position that the answer to all of the industry's perceived problems was the adoption of 'integrated teams':

> '[a] poorly specified brief perpetuates waste and increased costs; without integrated teams the ability to pre-plan is lost, thereby running the risk of even greater inefficiencies and potential accidents; an undervalued and undertrained workforce make mistakes which result in financial, environmental and, all too often human cost.' (Strategic Forum, 2002; 35)

All the construction sector seemingly had to do was to adopt integrated teams, coupled with a commitment to pre-planning. It was of course recognised that the required integration of the supply process would require a 'step change in the culture of the industry'. Every proposed improvement technique is seemingly dependent upon 'culture change'. According to the Strategic Forum, the envisaged change would be characterised by 'clients procuring and specifying sustainable projects, products and services'. And the supply side would in turn respond collaboratively in a way that enabled all of the parties involved in the integrated supply team to 'maximise the added value their expertise can deliver'. The construction sector was indeed fortunate to have the Strategic Forum to set such a clear strategic vision for the industry. Or perhaps clarity was not considered a core issue in 2002, just as sustainability had not been considered important in 1998.

Other so-called 'cross-cutting issues' included a further exhortation in favour of investment in high quality design. And of course, the investment had to be made by an integrated team, the existence of which was seen as an essential pre-condition. IT and E-business were held to be enablers and were seen to have transformed many operations in the construction sector, with much scope for further improvement. But of course *Accelerating Change* warned that the benefits accruing from E-business and virtual prototyping are not easy; their adoption was seen to require the construction industry:

> ... to transform its traditional methods of working and its business relationships. Key barriers to this transformation include organisational and cultural inertia, scale, awareness of the potential and knowledge of the benefits, skills, perceptions of cost and risk, legal issues and standards.'
> (Strategic Forum 2002; 36)

The advice on offer can once again be seen to fall someway short of being useful. Of particular note is the assumption that the construction sector is essentially static. Traditional methods are seen to dominate and the sector at large is seen to be characterised by an organisational and cultural inertia. Despite the lack of clarity, the essential mindset remains depressingly familiar. The quote above could equally have been lifted from the narrative of business process re-engineering (BPR) (see Chapter 7).

The final 'cross-cutting' issue deserving of mention relates to the need for the industry to invest in research and development (R&D), which was seen to be essential to underpin innovation and continuous improvement. The extent to which innovation is compatible with the advocated regime of performance measurement was, needless to say, not addressed. The support function provided previously by the Construction Research and Innovation Strategy Panel (CRISP) was dismissed as inadequate and a call was made for a 'new CRISP' with an expanded role and resources. Unsurprisingly, the Strategic Forum considered that an immediate priority was to focus research effort 'on filling the industry's knowledge gap in the development of integrated supply teams'. One can almost imagine the scenario throughout the industry's R&D departments:

> '... listen up everyone: ditch all that work on lean thinking, it's integrated teams from now on.'

But in truth by 2002 there were very few R&D departments left in the construction sector. Several large contractors had maintained R&D departments in the 1960s/70s, but these had all but disappeared by the late 1980s. Despite the alleged organisational and cultural inertia, the industry had already undergone extensive radical change prior to the propagation of the so-called Egan Agenda. However, the pre-existing dynamics of change were seemingly of little interest to the visionaries of the Strategic Forum.

Accelerating Change gave especially scant attention to topical issues such as PFI and the move towards public-private partnerships. Fleeting note is made of the OGC's recommendation that forms of contract should be adopted which 'encourage team integration' such as PFI, Prime Contracting and Design & Build. Whether or not PFI encourages integration is a moot point; but it was in any case a claim which was entirely peripheral to the host of criticisms which were at the time being directed at PFI. The Strategic Forum clearly felt no need to offer any qualifying observation; its members seemingly had no opinion on the efficacy of PFI. If the OGC said it was good for integration then this was good enough for them. Neither did they apparently have any opinions about the contracting out of public services or the move towards public-private partnerships. The same had also been true of the preceding *Modernising Construction* report. Reports about industry improvement kept well clear of commenting on PFI. In the case of *Accelerating Change*, all that seemed to matter were a lot of warm words about 'integration', together with the promise of yet more KPIs and toolkits. The link between those who were advocating change and the reality of what was happening on the ground was becoming evermore tenuous. This was after all the 'age of spin', and nobody seemed very worried about reality any more.

9.5   Summary

This chapter has described the context of construction improvement under the early years of the New Labour government. In the short term, there was no radical departure from the policies of enterprise enacted by the preceding Major government. The Blair administration was a very different from the Labour governments of the 1970s. Blair was fortunate to preside over a period of continuous economic growth throughout his term of office. The continuous growth allowed for significant new investment in health and education. It was only when Gordon Brown took over as Prime Minister in 2007 that dark clouds began to appear on the horizon. The preceding years under Blair represented very good times indeed for the UK construction sector, with many new opportunities around PFI/PPP and the contracting out of public services. Many construction firms diversified extensively in response to these emerging opportunities. At the same time, pre-existing notions of partnering merged into a new government-promoted discourse of social partnership.

Rethinking Construction (Egan, 1998) was initially important in cementing New Labour's business-friendly credentials. In many respects, the report represented the highpoint of the enterprise culture in construction. However, it must be recognised that this was not a form of enterprise which relied on the free market; this was a form of enterprise subject to continuous performance management in accordance with pre-determined objectives. But the early emphasis on implementing the Egan Agenda described in Chapter 5 was overtaken rapidly by events. Of particular importance was New Labour's espoused commitment to the 'Third Way', which promised a new combination of enterprise and social democracy. In the construction sector, this entailed a certain degree of new-age revisionism in terms of what the Egan Agenda was about. Public sector agencies were exhorted to work in partnership with the private sector for the purposes of securing best value. The machine metaphors of *Rethinking Construction* were supplemented progressively by the organic metaphors of corporate social responsibility. Of particular note was the way in which the battle lines which existed previously between the public and private sectors were re-drawn. Political ideology was claimed to no longer be important; what mattered now was what worked. Hence the notion of performance management became diffused across every policy arena, and the construction sector was no exception. The culture of enterprise had long since emphasised the private sector's capacity for innovation, but this could no longer be taken on trust. It now had to be measured alongside everything else in accordance with an endlessly expanding infrastructure of key performance indicators (KPIs).

The debate was also extended progressively beyond the domain of contracting, to include clients, professional advisors, subcontractors and the suppliers of materials. Integration became the new watchword, and supply chain integration became the essential rhetorical corrective to structurally-embedded fragmentation. And if the rhetoric of 'partnering' had by now become a little tarnished, there were always new buzzwords to be mobilised. Certainly by 2003 all the talk within the newly formed Construction Excellence was of 'integrated teams' and 'collaborative working'. Lean thinking had seemingly slipped out of fashion and was certainly not held to be the panacea it once was. But the model of the lean organisation remained central to way the contracting sector operated. And when recession returned in 2008, the emphasis of lean thinking on the elimination of waste became popular once again. The increasing policy focus on environmental sustainability also played a part in the rehabilitation of lean thinking, although this was a radically different interpretation of lean from that which was proposed in *Rethinking Construction*. But in the age of spin, reality no long seemed to matter. All that mattered was to be 'on message' with the advocated policy direction.

It must also be said that the vast majority of managers in the construction sector remained aloof from the details of such debates. If the client wanted an integrated team, they would provide an integrated team. And if the client wanted prime contracting, then prime contracting is what the contractor would provide. Given that none of these terms were clearly defined it was not difficult to comply with requirements. This was perhaps the logical endpoint of customer responsiveness in the age of spin. The important thing was to be 'Egan compliant', despite the fact that nobody bothered to remember precisely what it was that had been advocated. It was always possible to cite the demonstration projects as having provided clear evidence of the benefits of new ways of working, even if nobody was quite sure what the 'new ways of working' were. Few pointed out the inherent contradiction between advocating innovation, on the one hand, while seeking to measure performance against a pre-determined set of KPIs, on the other. However, the economy was growing and CEOs within the construction sector were broadly happy – barring lingering concerns about skills and health and safety. Few bothered to mention the absence of stable employment regimes as increasing numbers of migrant workers were sucked in from Eastern Europe. It was far easier to trot out the favoured buzzwords while advocating the need for an industry-wide culture change.

References

Adamson, D.M and Pollington, T. (2006) *Change in the Construction Industry*, Routledge, London.

Atkinson, J. (1984) Manpower strategies for flexible organisations, *Personnel Management*, **16**(8), 28–31.

Buchanan, D. (2000) An eager and enduring embrace: the ongoing rediscovery of teamworking as a management idea, in Proctor, S. and Mueller, F. (eds) *Teamworking*, MacMillan, Basingstoke, pp. 25–42.

Connolly, C., Martin, G. and Wall, A. (2008) Education, education, education: the third way and PFI, *Public Administration*, **86**(4), 951–968.

Collins, H. (2001) Regulating the employment relation for competitiveness, *Industrial Law Journal*, **30**, 17–46.

Collins, H. (2002). Is there a third way in labour law?, in J. Conaghan, R. Fischi and K. Klare (eds.), *Labour Law in an Era of Globalization*. Oxford University Press, Oxford, pp. 450–69.

Craig, D. and Brooks, R. (2006) *Plundering the Public Sector*, Constable, London.

DEFRA (2006) *Procuring the Future: Sustainable Procurement National Action Plan – Recommendations from the Sustainable Procurement Task Force*, Department for Environment, Food and Rural Affairs, London.

Driver, S. and Martell, L. (1999) New Labour: culture and economy, in L. Ray and A Sayer (eds) *Culture and Economy: After the Cultural Turn*, Sage, London, pp. 246–269.

Egan, Sir John. (1998) *Rethinking Construction*. Report of the Construction Task Force to the Deputy Prime Minister, John Prescott, on the scope for improving the quality and efficiency of UK construction. Department of the Environment, Transport and the Regions, London.

Evans, R., Carrell, S. and Carter, H. (2009) Man behind illegal blacklist snooped on workers for 30 years, *The Guardian*, Wednesday 27 May.

Faux, J. (1999) Lost on the Third Way, *Dissent*, **46**(2), 67–76.

Fisher, N. and Green, S. (2000) Partnering and the UK construction industry the first ten years - a review of the literature, in *Modernising Construction*, Report by the Comptrollor and Auditor General of the National Audit Office, The Stationery Office, London, pp. 58–66.

GCCP (2000) *Achieving Sustainability in Construction Procurement*, Government Construction Clients' Panel, London.

Giddens, A. (1998) *The Third Way: The Renewal of Social Democracy*, Polity Press, Cambridge.

Gray, C. and Davies, R.J. (2007) Perspectives on experiences of innovation: the development of an assessment methodology appropriate to construction project organizations, *Construction Management and Economics*, **25**(12), 1251–1268.

Hall, S. (2003) New Labour's double-shuffle, *Soundings*, **24**, 10–24.

HM Treasury (2003) PFI: Meeting the Investment Challenge, HMSO, Norwich.

Harvey, M. (2001) *Undermining Construction*, Institute of Employment Rights, London.

Ietto-Gillies, G. (2006) Is New Labour's 'Third Way' new or just hot air in old bottles?, *Post-autistic Economics Review*, **39**, 31–47.

Jenkins, S. (2006) *Thatcher and Sons: A Revolution in Three Acts*, Allen Lane, London.

Latham, Sir Michael (1994) *Constructing the Team*. Final report of the Government/industry review of procurement and contractual arrangements in the UK construction industry. HMSO, London.

Leiringer, R., Green, S.D. and Raja, J. (2009) Living up to the value agenda: the empirical realities of through-life value creation in construction, *Construction Management and Economics*, **27**(3), 271–285.

Marr, A. (2007) *A History of Modern Britain*, Macmillan, London.

Murray, M. and Langford, D. (2003) Conclusion, in M. Murray and D. Langford (eds) *Construction Reports 1944–89*, Blackwell, Oxford, pp. 196–217.

NAO (2001) *Modernising Construction*. Report by the Comptrollor and Auditor General of the National Audit Office, The Stationery Office, London.

NAO (2009) The Building Schools for the Future Programme: Renewing the Secondary School Estate, The Stationery Office, London.

OST (1995) *Technology Foresight Report: Progress through Partnership*, Office of Science and Technology, London.

Powell, M. (2000) New Labour and the Third Way in the British welfare state: A new and distinctive approach, *Critical Social Policy*, **20**(1), 39–60.

Roe, P. and Craig, A. (2004) *Reforming the PFI*, Centre for Policy Studies, London.

Simon, Sir Ernest (1944) *The Placing and Management of Contracts*. HMSO, London.

Smith, P. and Morton, G. (2006) Nine years of New Labour: Neoliberalism and workers' rights, *British Journal of Industrial Relations*, **44**(3), 401–420.

Strategic Forum (2002) *Accelerating Change*, Rethinking Construction, London.

Taylor, F.W. (1911) *Principles of Scientific Management*, Harper & Row, New York.

TUC (2008) *Hard Work, Hard Lives*. The full report of the Commission on Vulnerable Employment, Trades Union Congress, London.

Wolmar, C. (2007) *Fire and Steam: A New History of the Railways in Britain*, Atlantic, London.

10   A Legacy of Dilemmas

10.1   Introduction

The purpose of this final chapter is to bring the debate about construction improvement up-to-date. It also provides an opportunity to pull some of the threads together from the preceding chapters. It is appropriate first of all to re-visit the changing infrastructure of construction improvement. Many of the bodies advocating construction sector improvement which came into being during the 1990s were subsequently combined together to form Constructing Excellence in 2003. Despite criticisms of previous attempts to codify 'best practice' it is argued that inter-organisational improvement networks such as Constructing Excellence have an important role to play in promoting debate. Their value lies primarily in the provision of networks which bring different sectors of the industry together to promote localised innovation. The more recent decline of Constructing Excellence as an independent organisation dedicated to industry improvement is therefore an issue of concern. At the time of writing, the United Kingdom is entering a new age of austerity with dramatic cuts in public expenditure. In retrospect, it appears that the levels of public expenditure presided over by New Labour were unsustainable in the long term. The global financial crisis of 2008–10 must of course take much of the responsibility for the economic slowdown, which suddenly meant that Britain was once again living beyond its means.

Particular attention will be given to the post-Egan propensity for continuous target setting, which continued to be championed by the Strategic Forum. However, the current targets cannot sensibly be seen simply as a means of promoting the 'Egan Agenda'. Of the six currently espoused targets, four of them pre-date *Rethinking Construction* by several decades. And the other two bear little relation to the arguments promoted by Egan. Attention is also given to the recurring popularity of simplistic machine metaphors and the possibility that these reflect and reinforce pre-existing trends in the construction sector. From this perspective, mechanistic

---

Making Sense of Construction Improvement, First Edition. Stuart D. Green.
© 2011 Stuart D. Green. Published 2011 by Blackwell Publishing Ltd.

'improvement' recipes such as business process re-engineering (BPR) and lean thinking, rather than being part of the solution, can be seen to have been part of the problem. Small wonder that the sector possesses an 'adversarial culture' given the way it is continuously bombarded with over-simplistic improvement recipes. Consideration is further given to the 'disconnected agendas' which continue to characterise the construction sector improvement debate. For example, the recent focus on the value of good design does little to promote a stable employment context for the construction workforce. Health and safety in construction continues to be a huge concern. There is a strong argument that the industry's fragmented employment regime has a directly adverse effect on the level of on-site accidents. Despite all the talk of 'integrated teams', very few industry leaders seem interested in tackling the fragmentation caused by false self-employment. In this respect, there are two almost entirely disconnected agendas at work. The first is concerned with a 'value-orientated built environment'; the second is concerned with the needs of vulnerable workers. It seems these two agendas are destined to remain forever disconnected as the industry concentrates its attention on dealing with a new age of austerity.

The chapter is concluded by a brief review of *Never Waste a Good Crisis* (Wolstenholme, 2009), which offers a review of progress since *Rethinking Construction*. It will be argued that this ongoing fixation with Sir John Egan's report of 1998 continues to distract attention from the issues which need to be addressed.

10.2   The changing infrastructure of construction improvement

10.2.1   Constructing Excellence

The changing infrastructure of industry improvement has been referred to at several points throughout this book. Chapter 4 described how the Construction Industry Board (CIB) was created as a direct result of the Latham (1994) report. In the wake of *Rethinking Construction* (Egan, 1998) there was an explosion of different bodies and networks advocating the need for industry improvement. Some of these were publicly funded and others operated on the basis of private sector industry 'clubs'. *Modernising Construction* (NAO, 2001) concluded that such organisations had succeeded in raising awareness across different parts of the industry. The NAO report further commented on the extent of duplication of effort and a sense of confusion within the industry on the best source of assistance. It was hence recommended that there should be more coordination and better direction of their activities. In consequence, the various publicly funded agencies came together in 2003 under the brand of Constructing Excellence.

At the time of writing, Constructing Excellence is still in existence, although it operates as a much smaller organisation than it did when it was in its prime. It currently works in close collaboration with BRE Ltd. By means of a progression of mergers, Constructing Excellence lays claim to the legacy of many of the improvement bodies which have been referred to throughout this book, including:

- Reading Construction Forum (RCF);
- Design Build Foundation (DBF);
- Construction Best Practice Programme;
- Movement for Innovation (M4I);
- Local Government Task Force (LGTF);
- Rethinking Construction;
- Be (Collaborating for the Built Environment);
- Construction Clients' Group (CCG).

The overall trend is that as the various bodies have merged together, the smaller the overall entity has become. Bodies such as the Construction Best Practice Programme and the Movement for Innovation had originally been government funded. From the onset government took the view that the 'best practice' movement should ultimately become self-funding. This view prevailed throughout the years of the New Labour government. It was therefore inevitable that the gradual withdraw of publicly-funded support would present real challenges for the infrastructure of industry improvement. Bodies such as the Reading Construction Forum had previously operated very successfully as membership organisations funded through subscription. However, the onset of recession in 2008 made the funding of Constructing Excellence through membership fees progressively more difficult.

At the time of writing, Constructing Excellence continues to champion the cause of key performance indicators (KPIs) as a means of benchmarking performance. It also purports to conduct 'action research' as a means of piloting new ideas. It continues to support a programme of demonstration projects following on from the work of the Movement for Innovation (M4I). Perhaps in response to previous criticisms, the programme also includes a restricted number of in-depth 'innovation in practice' projects. It is to be hoped that these are subject to a more rigorous evaluation than was previously the case with the M4I demonstration projects. Construction Excellence continues to promote a regional structure of 'best practice' clubs, although some of these are more active than others. Many have become dormant. The regional 'best practice' clubs have in the past frequently been critical of the centralised London-based organisation, claiming that it is ill-placed to understand the nuances of their local markets. Construction practitioners possess a very human characteristic in that they do not like to be taught 'best practice' by others; they like to work it out for themselves.

10.2.2   *In favour of best practice networks*

Notwithstanding the above, the shrinking role of Constructing Excellence should be an issue of concern to all who are interested in innovation in the construction sector. The history of industry improvement bodies since the Latham report (1994) leaves much to be desired, but there is little doubt that they have an important role to play. The benefits do not lie in the advocated codified prescriptions, but in the processes through which such ideas are re-conceptualised by practitioners within localised contexts. In many cases, practitioners undoubtedly adopt some of the more fashionable sound bites while persevering with tried-and-tested routines. However, in other cases the advocated recipes may act as catalysts for localised innovation. The overriding difficulty is that practising managers too often look for management recipes which legitimise what they already do. Hence partnering and supply change management are adopted as best practice because they legitimise the reliance on subcontracting. In other words, such recipes are used as sense making mechanisms; they are used to make sense of the reality which practitioners experience taking place around them. They support practitioners' sense of self-identity by helping them ascribe to themselves an active role in the shaping of change, rather than seeing themselves as passive recipients.

But the difficulty lies in what goes unsaid. If the popularised improvement agenda is silent on the potentially adverse effects of accepted 'best practice', such issues do not become part of the discussion. Thus, managers focus on enacting 'supply chain management', while remaining seemingly blind to the corrosive effects of labour casualisation and false self-employment. In seeking out recipes which reinforce their sense of self-identify, managers act progressively to shape the language within which the debate is conducted. An active endorsement of supply chain management is consequently translated into a passive acceptance of the increased role of subcontracting. Hence the favoured discourse of industry improvement serves to legitimise and reinforce pre-existing trends. However, the best practice agenda is by no means homogeneous; there is plenty of scope for firms to innovate in different ways. Each new generation of managers has the opportunity to re-invent the improvement agenda; we are not necessarily destined to continuously re-cycle the debates of the past.

It must further be recognised that the best practice agenda is only one source of ideas; there are numerous other agendas against which practitioners continuously position themselves. But what is clear is that there is a much looser relationship between codified 'best practice' and what firms *do* than is commonly supposed. Different firms will have different path dependencies, and hence will respond to ideas in different ways. The benefits lie not in the implementation of best practice, but in the extent to which the advocated best practice initiates localised processes of innovation. Critiques of the accepted approaches thereby become equally important as prescriptive guides to implementation. It could even be argued that critiques are even more important in that they leave more scope for localised responses. It is further important to recognise that such processes do not take place solely within the boundaries of single firms, but span across dynamic networks which transcend organisational boundaries. Processes of innovation must therefore be understood in the broader context of sectoral change. They do not exist in isolation.

Organisations such as Constructing Excellence often claim to provide a unique bridge between industry, clients, government and the research community. This role is undoubtedly very important. The events and networks promoted by Constructing Excellence provide a quasi-neutral space where different parties can come together to debate issues of industry improvement. All of this is very healthy and arguably contributes directly to a more innovative construction sector. But what is less healthy is the requirement to align with pre-determined agendas such as that advocated by *Rethinking Construction*. Best practice organisations such as Constructing Excellence would serve the cause of innovation much more successfully if they were rather more open to the value of critique. They could also become slightly more orientated towards placing current debates within the context of long-term trends. The difficulty lies in the way in which participating individuals seek to align themselves with the established agenda because they think this important for the purposes of career progression. Hence individuals willing to promote arguments which go 'against the grain' are few-and-far between – it is simply perceived to be too much of a career risk to the individual, and too much of a commercial risk to their employing organisation. This was certainly the case in the wake of *Rethinking Construction*.

It would of course be a mistake to suggest that more critically-orientated discussions never take place, but more could undoubtedly be done positively to encourage them. For example, seminars on lean construction should not be limited to those advocating the supposed benefits of lean construction techniques; they should also include those who argue that the rhetorical adoption of such techniques legitimise and constitute a particular way of working. Of particular importance would be to position such ideas within the context of 'leanness' as an organisational form (see Chapter 3). This would be a relatively simple step towards reinvigorating current debates, and in the long term would serve to make bodies such as Constructing Excellence more relevant to the practical challenges faced by practitioners. But perhaps it has been the academic community which has been at fault rather than the advocates of best practice. Too many academics are content to promote abstract theoretical ideas in isolation of any engagement with the lived realities of practitioners. Others strive merely to echo the function of consultants by seeking only to reinforce the persuasiveness of popularised recipes. Academics are of course motivated increasingly by the need to secure research funding. In consequence, they often feel the need to echo the accepted improvement narrative in an attempt to make their research 'relevant' to the needs of industry. Some even complain that they are faced with some sort of censorship mechanism. Precisely who they think is operating the censorship mechanism is unclear. In truth, self-censorship is a much more invidious problem whereby academics limit themselves to research which they think will be received sympathetically. These are the processes through which researchers and practitioners become indoctrinated into the same discourse.

10.2.3 Targets and yet more targets

In addition to Constructing Excellence, the Strategic Forum also remains in existence, although it has become noticeably less assertive in recent times. The Strategic Forum still retains responsibility for 'target setting' in the spirit of *Rethinking Construction*. Its principal role is stated as being: 'to coordinate, monitor, measure and report on progress under the headline targets'. It is notable that the target areas have continued to evolve, although strangely nobody has yet admitted that the original targets were inappropriate. The target areas have further been reduced to six in number:

- procurement and integration;
- commitment to people;
- client leadership;
- sustainability;
- design quality;
- health and safety.

The areas listed above can all be linked back superficially to *Rethinking Construction*. But in truth the emphasis on sustainability and design quality developed primarily in response to notable omissions within the Egan report. Aesthetic notions of design quality were undoubtedly squeezed out by the report's obsession with narrowly defined efficiency. This reflected and reinforced the pre-existing culture of design-and-build architecture which continues to blight the United Kingdom's urban landscape. 'Lean architecture' has unfortunately been in fashion for some time, and the advocates of design quality indicators (DQIs) seem to be making limited headway in the face of fresh imperatives for cost-efficient construction. It has also been conveniently forgotten that sustainability received only the scantest of lip-service within *Rethinking Construction*. Subsequent attempts to present 'lean thinking' as a means of enhancing sustainability are guilty of conflating two diametrically opposed concepts.

It is further difficult to credit *Rethinking Construction* with any originality in respect to the remaining four areas of desired improvement. All four can be traced back at least as far as Emmerson (1962) and Banwell (1964). Murray and Langford (2003) have previously noted the recurring themes which have long since characterised the construction improvement agenda.

The Strategic Forum's propensity for setting endless performance targets echoed the policy orientation of the New Labour government. Although the 'New Labour' label was conspicuously down-played once Gordon Brown finally took over as Prime Minister in June 2007, the overriding commitment to performance targets remained firmly in place.

Rather than construing the current Strategic Forum targets as the direct continuation of *Rethinking Construction*, they could be rather more convincingly construed as representative of areas where the policy agenda of successive governments has failed. In many respects, the targets can be linked directly with the regressive side effects of the enterprise culture. This of course is not to say that the construction sector was perfect prior to the advent of the enterprise culture, because clearly it was not. Few remain nostalgic for the crisis years of the 1970s; and those who are nostalgic often have very warped memories. But collectively, the construction sector has chosen consistently to prioritise enterprise over other considerations. The cumulative effects of this prioritisation cannot be denied. The diverse needs of multiple stakeholders have been sacrificed on the altar of narrowly-construed competitiveness. This tendency was especially dominant throughout the 1980s and early 1990s, and resulted in structural flexibility becoming established as the dominant competitive model within the contracting sector. This bought significant advantages in terms of the ability of firms to expand and contract in accordance with fluctuations in demand, but it undoubtedly had adverse implications in terms of integrated working and skills development. The emergence of the hollowed-out firm, coupled with a widespread reliance on non-standard forms of employment, is also widely held to have had a detrimental effect on health and safely (Donaghy, 2009). The continued widespread reliance on casualised labour, coupled with an all too frequent conflation with vulnerable migrant workers, continues to sit ill-at-ease with the Strategic Forum's espoused 'commitment to people'.

The Strategic Forum's focus on 'procurement and integration' can similarly be positioned against a backcloth of extensive downsizing within private sector client organisations. This has often resulted in non-core capabilities – such as property management – being outsourced in accordance with the doctrine of business process re-engineering (BPR). It is this scenario which often provided the motivation to engage in 'partnering' (see the case studies in Chapter 7). Such changes have been even more pronounced within public sector client organisations characterised by the contracting-out of services. The trend is exemplified by the demise of the Property Services Agency (PSA) (see Chapter 3). In light of these long-term trends, it is perhaps not surprising that procurement policies within many client organisations remain under-developed. This is arguably especially true in terms of ethical sourcing and sustainability – two issues which were so conspicuously under-emphasised in *Rethinking Construction*. The outsourcing of property services further served to dissipate procurement expertise across a wide range of external private-sector consultancy firms, thereby contributing to ongoing concerns about a lack of 'integration'. Firms who have committed themselves to extensive programmes of outsourcing are also attracted to storylines about 'knowledge management'. As more and more activities are outsourced, organisations understandably become concerned with ways of retaining control over residual 'knowledge'.

According to the Strategic Forum, client leadership 'is vital to the success of any project and enables the construction industry to perform at its best'. This is undoubtedly true, and the Broadgate and Heathrow Terminal 5 developments continue to be cited as good examples of effective client leadership. The latter is especially deserving of mention in terms of its efforts to encourage direct employment. But the onset of customer responsiveness fatally undermined the 'mutuality of responsibility' for the construction sector's long-term development which prevailed during the 1960s. Clients have since become focused solely on what the industry can do for them on a project-by-project basis. The changing mindset was illustrated dramatically by the failure of the Construction Clients' Forum to support the aspirations of the Construction Industry Board (CIB) (see Chapter 5). It is also notable that BAA has withdrawn from the position of client leadership which they once occupied. This could in part be due to the take-over by *Grupo Ferrivial*, or to the absence of any long-term investment programme. Client leadership in the cause of self-interest is not a difficult concept to sell. Clients with long-term investment plans understandably adopt a longer-term perspective which is at least in part driven by enlightened self-interest. Perhaps even more striking is the retreat of public sector clients from any notion of 'mutually of responsibility'. Such aspirations have long since fallen victim to the culture of performance management initiated by market testing and compulsory competitive tendering (CCT). Hence construction sector development has become almost entirely dependent upon market forces. Many aspects of construction have improved immeasurably since the early 1980s, and many others have remained stubbornly unimproved. Others have become progressively worse. Training regimes and stable employment patterns fall firmly in the latter category. Client leadership in these areas remains especially lacking, with little immediate prospect of improvement.

Perhaps the biggest failure of the last 20 years lies in the area of sustainability. *Progress through Partnership* (OST, 1995) was undoubtedly ahead of its time in the emphasis given to the cause of sustainability. Yet, as already noted, sustainability received remarkably little emphasis in *Rethinking Construction*, which also downplayed corporate social responsibility (CSR) in favour of narrowly-defined customer responsiveness. Subsequent years have seen a plethora of reports urging public sector clients to encourage sustainability through procurement (GCCP, 2000; NAO, 2005; DEFRA, 2006). But these reports have never quite been embraced with the same enthusiasm as *Rethinking Construction*. Even Sir John Egan belatedly endorsed New Labour's vision of 'sustainable communities' in his foreword to *Skills for Sustainable Communities* (ODPM, 2004). Sustainable communities were defined as those which:

> '... meet the diverse needs of existing and future residents, their children and other users, contribute to a high quality of life and provide opportunity ad choice. They achieve this in ways that make effective use of natural resources, enhance the environment, promote social cohesion and inclusion and strengthen economic prosperity.' (ODPM, 2004; 7)

This was indeed a dramatic conversion on the road to Damascus. In common with the rhetoric espoused by Margaret Thatcher, the authors of *Rethinking Construction* were seemingly blind to the notion of externalities. The Egan report did not quite go as far as saying there is 'no much thing as society', but this was the overall implication. Yet six years later Sir John Egan was enthusiastically endorsing sustainable communities. Lean thinking had seemingly been well and truly ditched. No wonder so many in the construction sector were slow to 'change their culture'. They were understandably confused by so many vague and conflicting messages. Needless to say, the advocated approach to sustainable communities involved yet more targets and KPIs, with little guidance on how to make meaningful trade-offs in localised contexts.

In light of the above, it is pertinent to question the extent to which the Strategic Forum's reliance on voluntary regulation by means of an ever-evolving set of centrally-imposed targets is likely to be effective. It is certainly necessary once again to query whether generic top-down targets can ever be meaningful to the hugely heterogeneous construction sector. At least it can be said that 'something is being done'; but there is little evidence to suggest that private sector companies adjust their behaviour in response to centrally dictated targets. There is certainly no evidence that such targets are ever likely to induce culture change.

The above should not be taken to imply that practitioners are unable to take meaningful action in localised contexts. Although public-private partnerships have been much disparaged, they have at least tried to combine localised sustainability targets with the imperatives of efficiency (with admittedly mixed results). However, many construction firms have always been deeply embedded within their local communities, contributing widely to a range of local causes. This is equally true for small local building firms as it is for regional contractors. Such firms paid little attention to the Egan Agenda, and arguably many remain much more viable as a result. The construction sector is not, and never has been, populated with 'dinosaurs' which are somehow different from the rest of society. Construction practitioners are no less likely to be committed to sustainability aspirations than the public at large. But their behaviour is undoubtedly patterned and constrained by the institutional context within which they operate. Unfortunately, the various manifestations of the improvement agenda over the last 50 years are directly implicated in shaping the context within which they operate.

10.2.4 Enduring popularity of machine metaphors

A recurring theme throughout this book has been the way in which the problems of the construction industry are invariably conceptualised as impediments to machine efficiency. The overriding assumption is that complex organisations are subject to an 'engineering fix'. The rhetoric and imagery of re-engineering was persuasive amongst construction sector leaders because it resonated with the established favoured discourse (cf. Bresnen and Marshall, 2001). Construction leaders find it easy to endorse improvement recipes that exhort others to be more efficient. The rhetoric of BPR sought specifically to impose a management regime of command and control on the construction sector. Classical management theories such as Taylorism (Taylor, 1911) are characterised similarly by the same underlying machine metaphor. Organisations are perceived to be unitary entities with all parts working in harmony towards pre-determined objectives. The environment within which organisations operate is further assumed to be static. The primary task of management is to ensure that the machine operates efficiently.

The storyline of lean thinking is characterised similarly by a reliance on an underlying machine metaphor. Indeed, the same argument could also be made for the entire infrastructure of key performance indicators (KPIs). There is a further widespread tendency to interweave the machine metaphor with notions of 'teamwork' and 'leadership'. This was especially apparent in the case of partnering, which combines notions of efficiency with metaphors of teamwork and collaborative working. 'Good team players' are expected repeatedly to subjugate their own aspirations to those of the project. Leadership is concerned primarily with motivating team members towards pre-determined objectives. In many cases, the ultimate test for an effective project team is that it should 'work like a well-oiled machine'. The underlying model of organisations would have been readily recognised by Taylor (1911). Re-engineering is popular because it reflects and reinforces the way that construction industry leaders already think.

The danger of continuously advocating improvement recipes based on underlying machine metaphors is that it reinforces the reality that too many employees in the construction industry are required to act as mindless cogwheels in a remorseless machine. There is frequently little pretence that any efficiency gains will be shared equally amongst the diversity of stakeholders in the construction sector. Targets abound for reducing the cost of construction and enhancing profitability. The rhetoric is heavy in the machine metaphor whilst exhorting others to be more efficient. It is taken for granted that people are compliant, predicable and willing to be programmed in accordance with the requirements of a rationally designed system. There is no recognition that the continued imposition of simplistic machine metaphors may contribute to the 'bad attitudes' and 'adversarial culture' that industry leaders repeatedly decry.

The culture metaphor also features strongly in the rhetoric of construction improvement. The advocated storyline tends invariably to be hugely optimistic about the extent to which the construction sector's culture can be changed. The associated assumption that organisational performance depends upon an alignment between employee values and managerial strategy dates primarily from Peters and Waterman (1982). However, others have questioned the extent to which culture can be manipulated and controlled (e.g. Willmott, 1993; Antony, 1994; Legge, 1994). Of particular interest is the possibility that managerial action may promote unforeseen counter-cultures that are dysfunctional. Individuals are especially unlikely to be persuaded by the discourse of culture change when the espoused storyline is in direct conflict with experienced reality (Ogbonna and Harris, 2002). In the construction sector, culture change initiatives espousing integration and collaborative working are therefore unlikely to be persuasive when experienced reality accords with downsizing and outsourcing. Culture change is especially strongly emphasised within the discourse of partnering (Bennett and Jayes, 1998). However, observed discrepancies between rhetoric and reality render wholesale culture change a highly unlikely outcome. Memories of partnering are also likely to shape responses to subsequent calls for a culture change in favour of collaborative working. Construction practitioners are often cynical of new initiatives as a direct result of their memory of previous initiatives. Readers of this book will be perhaps even more cynical. But there is an important difference between informed cynicism and uninformed cynicism. The former provides a sensible basis for action, while the latter does not.

10.3   The disconnected agendas of construction improvement

10.3.1   The value of good design

Throughout the period reviewed by this book there have been numerous attempts to understand the meaning of 'value' and how it might be best achieved. Banwell (1964) was clear that good value depends upon clients allowing sufficient time during briefing and design. Too often clients were rushed into sending out the tender documents before they were fully complete. It was seen to be vital that the design was complete and tender documents 'watertight' if disputes leading to claims were to be avoided. The underlying message was that clients need to be more diligent in codifying their requirements in the tender documents. For those clients who continue to engage in lowest price lump sum tendering, this remains good advice.

But the above advice was based on a very narrow conceptualisation of value-for-money which related solely to the transaction between client and contractor. Several decades later, Latham (1994) was again to touch on value for money in the context of tender selection, but the emphasis here lay with taking costs-in-use into account:

> '... [v]alue for money and future cost-in-use should play an important part in the selection process.'

> '... those tenders which offer best value for money and show clear regard for cost-in-use should be accepted.'

This was progress of a kind, although scant attention was given to how cost-in-use might best be evaluated. Nevertheless, discussions about value for money remained framed in terms of meeting a defined specification at minimum cost. It is this interpretation of best value which leads directly to the industry's recurring obsession with efficiency. The dominant metaphor is that of the machine and the overriding challenge is that of operational efficiency. It was this predilection for machine metaphors which rendered business process re-engineering (BPR) so attractive to the authors of *Progress through Partnership* (OST, 1995). Popular myth would have it that Egan came along in 1998 with a radical new recipe for industry improvement. The rather more mundane truth is that *Rethinking Construction* served only to reinforce the construction sector's predilection for machine metaphors. This is illustrated throughout the Egan report by the repeated emphasis on the elimination of waste. Egan's obsession with waste elimination is exemplified by his interpretation of value management:

> 'Value management is a structured method of eliminating waste from the brief and from the design before binding commitments are made.' (Egan, 1998; 10)

Egan was apparently much less interested in how the brief might be better formulated in the first place, or how subsequent changes might be managed more effectively. As was argued at length in Chapter 6, those who are trapped into thinking in terms of the machine metaphor will inevitably deny themselves the benefits of alternative perspectives. This is not to say that efficiency is not important, because clearly it is. But, as previously argued, there is a danger in focusing on narrowly-defined efficiency alone. Egan was especially blind to the value of design quality, a failing which was only in part rectified by the formation of the Commission for Architecture in the Built Environment (CABE) and the subsequent development of the design quality indicators (DQIs). This was characteristic of a tendency for any alleged gap in the construction improvement agenda to be met by yet another quango supported by yet another set of KPIs. Unfortunately, each set of evermore detailed KPIs risks reinforcing the prevailing mechanistic mindset. Architects of course have always seen themselves as decidedly separate from the construction sector, and such attitudes are unlikely to be changed by the advent of the DQI. Indeed, it is a moot point whether such a change in attitude would even be desirable. Diversity in thought among the professions is an important catalyst of innovation, even if it does continuously render aspirations of 'integrated project teams' problematic.

Following rapidly on the heels of *Rethinking Construction*, the Royal Academy of Engineering published a report (Evans *et al.*, 1998) which emphasised the long-term costs of owning and using buildings. The stated intention was to shift the emphasis away of seeking capital cost reduction in isolation from any consideration of the long-term implications. Central to the adopted argument was the disputed claim that for every pound spent on the construction of a commercial office building, five pounds are spent on maintenance and operating costs and 200 pounds are spent on staff and business operations. At the time, this claim received relatively little attention, although it would subsequently become hotly contested and was eventually discredited as misleading (Hughes *et al.*, 2004; Ive, 2006). But the point of importance is that there was a counter-discourse to the prevailing narrow focus on building process efficiency. This was despite the continued insistence within the DETR that the industry's future depended on an ever-more enthusiastic application of the Egan 'principles'.

The Eganites were still demanding greater commitment to the cause of *Rethinking Construction* when the Department of Culture, Media and Sports (DCMS) published *Better Public Buildings* (DCMS, 2000). In contrast to Egan's emphasis on the elimination of waste in meeting client requirements, *Better Public Buildings* pleaded the case for good design:

> 'Good design is worth investing in. It is the key to giving the client maximum value for money throughout the whole life of a building.' (DCMS, 2000)

The view was further expressed that the achievement of value depends upon making sure that buildings better serve the needs of the organisations and people who use them. Good design was held to bring 'numerous benefits including: the sick healing faster, students performing better, workers more productive in their workplace'. This debate of course took place in an entirely separate space from that within which the Egan agenda was being promoted. But design and construction are not the separate entities they once were. They are situated increasingly within the same organisational framework in procurement approaches such as design-and-build and prime contracting. This is not quite the same thing as saying they are 'integrated'. Designers have long been criticised for poor buildability, a problem which was conveniently 'solved' by the subservience of designers to constructors within design and build projects.

In essence, one of the main limitations of *Rethinking Construction* (Egan, 1998) was the way it effectively reduced all construction to a design-and-build mindset whereby the reduction of waste is paramount. The failing was seemingly recognised subsequently given the more explicit promotion of design within *Accelerating Change* (Strategic Forum, 2002). But what was strange was the way that *Accelerating Change* later claimed that design quality had been one of the original objectives of *Rethinking Construction*. This is at best a highly partisan re-interpretation of what had been written originally. The leaders of the former Soviet Union had a propensity for re-writing history, although they were an unlikely source of inspiration for those involved in the construction improvement debate.

The design lobby continued to make itself heard, and the National Audit Office (2004) subsequently published a report entitled *Getting Value for Money from Construction Projects through Design*. The authors included Stuart Lipton, the then chairman of CABE and previously best known for his leading role on the Broadgate development with Stanhope. The NAO (2004) was admirably clear in its conceptualisation of value for money:

> 'A good project will continue to provide value for money and meet user needs throughout its lifetime.'

> 'The ultimate aim is to deliver construction projects that meet the requirements of the business and all stakeholders, particularly the end users.'

This was indeed a welcome corrective to the more myopic *Rethinking Construction*. But noticeably absent was any commitment to the construction sector which was required to build the designs which were being advocated. The 'mutuality of responsibility' between public sector clients and the construction sector had long since been forgotten. The crude customer focus of the 1980s had given way to the rhetoric of stakeholders much beloved by New Labour. But just as New Labour had seemingly forgotten about its Old Labour constituents, so had the design quality debate appeared to forget about the construction workforce. At least the Egan Agenda had its *Respect for People* initiative, despite its many limitations (see Ness, 2010). Here lies an essential dilemma at the heart of any construction improvement debate: on whose behalf are the improvements being made? Improvements directed solely at the needs of a commercial developer are unlikely to meet the long-term needs of the building occupiers. Neither will they necessarily meet societal aspirations for a quality public built environment. But even if we shift the emphasis from the former to the latter, there is still no guarantee that the improvements will contribute to the aspirations of the construction workforce. Indeed, it may well serve to marginalise them even further. Hence different agendas will always populate different spaces, and any unified vision of 'construction improvement' is destined to be forever elusive.

10.3.2 *Be valuable*

The design quality debate progressively became subsumed within the so-called 'value agenda', at least within Constructing Excellence circles. The pivotal contribution here was provided by Saxon's (2005) 'guide to creating value in the built environment', otherwise known as the *Be Valuable* report.

Richard Saxon had long been a champion of industry improvement having previously chaired the Reading Construction Forum (1999–2005) while at the same time acting as chairman of Building Design Partnership (1996–2002). The Reading Construction Forum merged with the Design Build Foundation in 2002 to form the bizarrely titled 'Be, Collaborating for the Built Environment'. 'Be' was subsequently absorbed into Constructing Excellence in 2005, the same year that Saxon's report was published. As has already been argued, the changing emphases of the construction improvement debate cannot be understood in isolation from an understanding of the ever-changing infrastructure through which such debates are conducted. Different networks are populated by different interest groups; and ideas are constituted differently within different networks. There is a need to understand all such activities as a seamless web whereby the improvement debate becomes constituted mutually with the particular social networks within which it is propagated. This in part is what is intended when we refer to the 'discourse of construction improvement', although many would extend this interpretation to include the material effects of the advocated storylines.

However, it is important to recognise that there is never only one discourse being mobilised at any particular time. The construction improvement terrain is invariably populated by multiple discourses which are propagated through different networks. Hence the co-existence of several disconnected agendas of construction improvement. Individual practitioners are invariably immersed within one particular discourse. But even this does not prevent them placing their own interpretations on the storylines on offer – and acting accordingly. Nor does it prevent them from drawing ideas from other discourses. It is therefore quite legitimate to argue that practice is shaped by discourse without falling into the trap of determinism.

Returning to the 'value agenda', the *Be Valuable* report did much to bring the 'value of design' theme into the remit of Constructing Excellence, and thereby into the mainstream of the construction improvement debate. Previously, the debate had been dominated by issues relating to contracting firms. Indeed, the construction industry had largely been seen to be synonymous with the contracting sector, as defined by Standard Industrial Classification (SIC) 45 used in the official statistics. Pearce (2003) was influential in arguing the case for a broader definition of the 'construction industry'. He referred to the narrow interpretation of construction as being predominantly limited to on-site assembly and the repair of existing buildings and infrastructure by contractors. In contrast, he pointed towards a broader definition of construction which includes the materials supply chain together with professional services such as architecture, engineering design and surveying. The Egan agenda clearly subscribed to the narrow interpretation of construction, seemingly with little interest in professional services. If consideration is limited to the narrow interpretation of construction, it is little wonder that the improvement debate tends to focus on delivering a defined product at minimum cost. Up until relatively recently this was the limit of the construction sector's responsibility. The corresponding advice directed at client organisations was to ensure that their requirements were fully defined in advance.

The changing focus of the improvement agenda is indicative of a combination of several factors. First, it comprises a broadening of the long-established focus on customer-responsiveness. The construction sector is now being asked to do much more than build efficiently; it is being asked to 'add value' to the client's business. For decades, the improvement agenda had focused on efficient construction. This emphasis is still there, but there is now an additional focus on the need to provide 'integrated solutions'. The focus on integrated solutions embraces the need to combine the provision of built facilities with aspects of through-life service provision. This is in no small way reflective of long-term trends within client organisations to outsource their property management capabilities. Clients had previously performed 'adding value' activities internally, but over a period of time these were increasingly outsourced. Not only are services now routinely being contracted-out, but so is the responsibility for 'adding value'. This in many ways is a direct continuation of the re-structuring of the industry's client base described in Chapter 3.

The 'integrated solutions' storyline provides a justifying narrative for the sharp escalation in the number of PFI projects initiated by New Labour. Whereas PFI had previously been a minority interest, by 2005 most of the large national contractors were directly involved. The desire to procure on the basis of 'value' was operationalised through the use of output specifications. Such an approach became central to the 'second generation' PPP initiatives relating to schools and hospitals. Prime contracting also placed significant emphasis on designing to performance specifications. Notwithstanding the significant practical difficulties associated with such aspirations, the accumulative experience contributed to a changing rhetoric of industry improvement focused upon outcomes. Hence the challenge of school design extended beyond the spatial layout of the building to the end goal of 'educating students'. Likewise, hospital design became focused on treating patients. The difficulty in both cases is that the design of the building does not determine the desired outcomes. The skills of the teachers (or surgeons) are always going to be of primary importance, and these thankfully remain beyond the control of the construction sector. Nevertheless, the focus on the 'up side' of construction's contribution provided a welcome break with the previous obsession with building efficiently. The focus on the interaction between the construction sector and broader societal goals is suggestive of an organic metaphor, and the imperative for firms to learn across projects is immediately suggestive of a cybernetic metaphor. These conversations should therefore be warmly welcomed, but as will be argued below, the perspectives on offer can hardly be construed as new.

Saxon (2005) positions the 'value agenda' rather differently, and contrasts it with the previous prevailing focus on cost:

> 'The era since World War II in UK property and construction matters has until recently been ruled by the word "cost". Buildings were designed down to a budget by consultants, then tendered for by contractors with the lowest cost tender winning.' (Saxon, 2005; 2)

The interpretation above is a fair characterisation as far as it goes. But what it fails to convey is any sense of the changing context within which the construction improvement agenda is situated. As described in Chapter 1, the priority in the immediate aftermath of World War Two was the provision of housing. The focus lay on building the maximum number of housing units for minimum cost. This led to an interest in prefabrication (later to be re-labelled 'modern method of construction'), which in turn led to concerns about quality. Much of the debate also focused on contractual procedures with a view to ensuring that client requirements were specified fully in advance. Such debates must be positioned against prevailing concerns about the capacity of the construction sector to deliver, and the damaging effects of the economic stop-go cycle on investment in training and skills. As has already been described, in the immediate post-war years there were real concerns that private sector contracting firms did not have the capacity – or the management skills – to deliver against the national agenda. Combined with fears about bid-rigging and corruption such concerns rendered nationalisation a real possibility. At the time, nationalisation was seen to be almost synonymous with modernisation.

Saxon (2005) notably omits any of the considerations above from his analysis, preferring to limit himself to a supposed dichotomy between value and cost:

> 'Good value in that era was defined as getting "it" for the lowest possible price. "It" was a generic, commoditised view of building, sufficient to get planning permission, pass building regulations and meet yardstick areas and specifications for its function.' (Saxon, 2005; 2)

As with the previous quote, this is a superficially appealing sound bite; but there is once again much which remains unsaid. Saxon fails to mention the extensive debate about 'value' which took place among building economists in the 1960s and 1970s (e.g. Stone, 1966; Turin, 1966; Hutton and Devonald, 1973). Such authors gave extensive attention to the 'use value' of buildings and the extent to which it could be optimised. Saxon's (2005) interpretation of value as the ratio of 'what you get' to 'what you give' is therefore at least 40 years out of date, and is arguably rooted in the idealism of a previous era.

10.3.3   Quality and value in building

Given the current popularity of the 'value agenda' it is worthwhile to rehearse briefly the previous debates about quality and value in building. In 1978, the Building Research Establishment published a report entitled a *Survey of Quality and Value in Building* (Burt, 1978). While not notable for its originality, the report was successful in encapsulating the understanding of both quality and value which prevailed at the time:

> 'Maximum value is….in theory obtained from a required level of quality at least cost, the highest level of quality for a given cost, or from an optimum compromise between the two.'

Burt (1978) also took care to define precisely what was meant by quality:

> '… quality is defined as the totality of the attributes of a building which enable it to satisfy needs, including the way in which individual attributes are related, balanced, and integrated in the whole building and its surroundings.'

Saxon's (2005) critique of the commodified view of building is probably a fair reflection of the frame of reference routinely adopted by contracting firms. But even here it says more about the institutional structures within which they work, rather than any underlying lack of imagination. Construction firms are still routinely contracted to build in accordance with a pre-determined specification. Even when engaged on a design-and build basis, they compete on the basis of meeting yardstick space requirements. But challenges to this frame of reference are by no means new – it was challenged repeatedly by the construction research community throughout the 1970s. Notions of optimising value at the time rested very much on what remained of the idealism of the 1960s. However, by the late 1970s researchers were already starting to give up on attempts at 'design optimisation'. Such efforts had been informed by the disciplines of operational research and systems engineering, which encapsulated prevailing assumptions about an unchanging environment.

Notions of design optimisation further rested on an entirely unrealistic reification of both 'quality' and 'value'. In other words, rather than recognise quality and value as abstract concepts building economists sought to treat them as if they were real, tangible and measurable. Such an approach now seems as dated as the planned economy within which such ideas were situated, but there is a danger that the same mistakes are repeated by every new generation. Similar processes of reification underpin many of the KPIs which are advocated routinely in response to aspirations which in reality do not lend themselves to simple measurement. The net effect is to reduce understanding to a limited number of quasi-objective criteria which too often stand as poor proxies for the desired outcomes. The Egan targets promoted in the wake of *Rethinking Construction* stand as good examples of how a process of reification can eventually become counter-productive. The endless cycle of target setting, backed up by ever more complex KPIs, serves ultimately to impoverish the improvement debate. The difficulty is that a mechanistic focus on 'performance' tends to distract attention away from the need to take meaningful action in an environment characterised by conflicting priorities.

Returning once again to the debates of the 1970s, the research community came to the view that terms such as 'quality' and 'value' are best read in terms of their metaphorical connotations, rather than in terms of their literal meaning. Different interpretations of value follow from different metaphorical perspectives. Attempts to optimise the 'totality of attributes' of a building were clearly shaped by an underlying allegiance to the machine metaphor. An interpretation of value in accordance with the organic metaphor would emphasise the importance of remaining responsive to changing circumstances (see Chapter 6). The culture metaphor would similarly lead to a different interpretation of value, likewise the cybernetic metaphor. But the metaphor which ultimately proved most persuasive was that of the marketplace. Research into value optimisation was outflanked by an axiomatic belief in the wealth generating capabilities of private sector enterprise. And the only requirements which needed to be satisfied were those required by the immediate paying customer. Activities which did not 'add value' to the customer were classified as waste, and were hence eliminated. This trend was legitimised by the doctrines of BPR, which reached their zenith in the construction sector with the publication of *Rethinking Construction*.

Most contracting firms realised in the early 1980s that clients were in the game of maximising their own financial returns, with little thought given to the long-term needs of the construction sector. Notions of shared responsibility were thus consigned to the past as the construction sector organised itself in accordance with the principles of the enterprise culture. In the context of a 'share-owning democracy', this was a much more preferable outcome than building nationalisation. The poor industrial relations climate was a further source of significant risk to both clients and contractors. Strikes and unofficial stoppages impinged directly upon the profits of main contractors as well as threatening the value which clients derived from construction. The path towards leanness and agility fitted the bill as a means of risk mitigation (Druker, 2007). The strategy was driven by an espoused vision of 'value creation', but the vision was shaped primarily by the discourse of the enterprise culture.

Under the New Labour government elected in 1997 the rhetoric of stake-holders complemented increasingly the rhetoric of enterprise. Value was not only achieved through satisfying customer requirements; it also became necessary to provide value to society – and even to stakeholders as yet unborn. Thus, the value agenda remains strong in terms of its persuasive rhetoric, but rather less strong as a guide to pragmatic action. Perhaps its most useful role is to provide discursive resources for firms seeking to legitimise their role within the domain of PFI/PPP. This remains true even if the underlying modus operandi of such firms remains much the same as it ever did (see Leiringer *et al.*, 2009).

10.3.4 *The industry formerly known as construction*

Returning to the script of the modern 'value agenda', Saxon (2005) unwit-tingly captures the essence of the shift which has taken place over a 30-year period:

> 'Customers in the great majority of cases do not seek to buy construction per se; they seek the use of facilities or the creation of assets. They find value in the availability of service space, developed and run to support their business or social service.'

In other words, the creation of value for customers continues to be priori-tised over any consideration of the sustainability of the construction supply chain. Stakeholder models of management which balance the needs of different interest groups may have come back into fashion elsewhere, but not seemingly for the major clients of construction. The tendency to pro-mote client value with little consideration to the long-term capacity of the construction sector is by no means limited to the United Kingdom. The following quote is especially telling in the context of the changes described throughout the preceding chapters:

> 'In many countries private clients are organised into groups and have used their enhanced power in recent years to force contractors to lower their costs and improve their delivery. Unfortunately, as we have seen, this has too often been at the expense of the workforce and of the invest-ment in human capital required to ensure the long-term capacity to deliver high-quality construction.' (ILO, 2001)

By the time *Be Valuable* was published in 2005, the adverse implications of the downward pressure on construction costs for the industry's workforce were becoming increasingly difficult to ignore. But Saxon (2005) and his col-leagues within Constructing Excellence were seemingly not interested in engaging with such deep-rooted and protracted problems. Issues relating to the construction workforce were sidestepped deftly by shifting the focus of the debate elsewhere. This was achieved by combining construction, property development and facility management into a continuum which was labelled the 'built environment'. The construction industry thereby became relegated to 'an industry formerly known as construction' (Saxon, 2002). To talk about the 'construction industry' was seen to perpetuate old ways of thinking. Hence the undesirable implications of the Egan Agenda and its subsequent variations were effectively barred from polite conversation.

The rhetoric of Saxon's (2005) concept of a value-orientated built environment industry was directly reflective of the New Labour ideology which prevailed at the time. This is most evident in the stated mission of:

> 'adding value for customers and society by shaping and delivering the built environment to their needs.'

The pre-existing doctrines of customer responsiveness and enterprise remain of central importance, but are sweetened with vague aspirations of social democracy. The explicit reference to 'society' would certainly not have occurred during the years of unconstrained enterprise. Prior to 1997 any reference to the needs of society would have had too many connotations with the discredited social consensus of the 1960s/1970s. But by 2005 'society' was back in fashion – if only in a rhetorical sense. Saxon (2005) further alludes to the 'triple bottom line' which must be satisfied if the built environment is to be sustainable. Reference is made to the need to satisfy aspirations of social equity as well as environmental and economic considerations. Social sustainability is further seen to require 'positive social and cultural value'. But no comment is offered on the need to seek social equity within the construction workforce, which had apparently slipped entirely off the agenda. The needs of society may well have been accepted as important, but this did not extend ostensibly to the two to three million operatives who worked within the construction sector. *Be Valuable* pointedly expressed no interest in how the supply side is organised in terms of the widespread reliance on subcontracting, false self-employment and the use of agency workers. Such issues had been rendered invisible.

10.3.5 Vulnerable workers in the 'built environment'

Following on from the above, Saxon (2002, 2005) undoubtedly deserves credit for re-energising the 'value agenda' after it had been neglected for the best part of 30 years. But Saxon's arguments enabled those in the 'industry formerly known as construction' to legitimise their long-standing lack of interest in issues relating to the construction workforce. Investment in the employment relationship has been progressively sidelined by fashionable management ideas such as supply chain management and collaborative working. Furthermore, this trend has been reinforced consistently by the construction improvement agenda ever since the advent of the enterprise culture in the late 1980s. Certainly the 'best practice' agenda promoted by Constructing Excellence and its predecessors has done little to promote an awareness of vulnerable workers in the construction sector.

The increased plight of vulnerable workers in the UK economy was highlighted by the report of the TUC Commission on *Vulnerable Workers* (2008), which estimates there are currently two million such workers in the United Kingdom. The report laments the way in which 'employees', 'workers' and the 'self-employed' possess different levels of employment protection. As described in Chapter 3, determination of employment status is by no means straightforward, and can often only be determined by a consideration of a range of factors derived from statute, contract and various test cases (TUC Commission, 2008*)*. The *Vulnerable Workers* report further describes how temporary agency workers continue to be paid less than directly employed workers doing the same job, with reduced rights to sick pay, holiday pay and pension contributions. Furthermore, operatives working in bogus self-employment often lack the autonomy of the genuinely self-employed. Yet at the same time they are denied the protections routinely provided to direct employees. Migrant workers are held to be particularly vulnerable, especially when forced to work for dodgy labour agencies by force of circumstance, or by immigration regulations which effectively bar them from direct employment. Thus, the employment terms and conditions of migrant workers are frequently dictated by the economic circumstances of the areas from where they are recruited, rather than by the 'going rate for the job' in the United Kingdom. Such extreme forms of fragmentation in the construction sector lie consistently beyond the construction improvement agenda. It appears that the favoured rhetoric of 'integrated teams' and 'collaborative working' does not apply to the construction workforce.

However, the problem of vulnerable workers is not limited to the construction sector. It also extends across a range of other low-paid sectors including care, cleaning, hospitality and security. Many of these sectors have followed similar trends to those experienced in construction. Outsourcing and the contracting out of services has led to a dramatic increase in the numbers of vulnerable workers, with similar discrepancies between the rights of direct employees and agency workers. Many such workers spend the majority of their working days within the context of the built environment, and would therefore fall within the definition of Saxon's (2002) 'industry formerly known as construction'. The operation of the new schools and hospitals procured under PPP arrangements are frequently dependent upon a low-paid and vulnerable workforce, despite the surrounding rhetoric of social sustainability. Vulnerable workers also provide cleaning and security services to the glitziest of modern office developments. Behind the espoused visions of 'educating students' and 'treating patients', the key source of competitive advantage is too often provided by low-paid workers who lack basic employment rights. This is the unacknowledged reality which too often lies behind Saxon's concept of a 'value-orientated built environment industry'.

10.4   Health and safety

10.4.1   Long-term trend in fatalities

Any consideration of the 'legacy of dilemmas' would not be complete without a discussion of health and safety in the construction sector. Construction professionals take health and safety extremely seriously, and its importance has in recent years been emphasised ever more strongly. It is of course much easier to advocate the importance of health and safety than it is to address the underlying reasons for the construction sector's continued relatively poor performance. Each year, without exception, dozens of construction sector workers are killed at work. The long-term trend in construction fatalities is thankfully inexorably downwards (see Figure 10.1). There were 53 fatalities in 2008–09, which represents a welcome reduction from 72 in the previous year. Both figures compare favourably with the corresponding statistics from the 1970s. For example, in 1974 there were 166 fatalities reported in the UK construction sector. There has undoubtedly been a significant increase in societal concern regarding health and safety since the 1970s. This shift in attitude has resulted in a welcome decrease in the number of accidents in the workplace across all sectors. Tabloid journalists are still often prone to refer in disparaging terms to a 'health-and-safety culture gone mad'; but the stark truth is the health-and-safety culture in the construction sector is not quite 'mad' enough. 53 fatalities in a year is still 53 too many, despite the overall downwards trend. It must also be said that the reduction in fatalities in construction has been less than that achieved in other industry sectors. Hence there is little justification for complacency. Any decline in the number of deaths is undoubtedly to be welcomed, but it nevertheless remains the case that much more could be done. Accidents on site are of course only part of the overall picture; consideration must also extend to issues of occupational health.

10.4.2   Statistical uncertainties

It must be conceded from the outset that there is little which is straightforward about health and safety statistics. Although it is the headline figure of number of fatalities per year which inevitably grabs the attention, a more meaningful statistic relates to the number of fatal accidents per worker (also indicated in Figure 10.1). However, as highlighted in Chapter 4, there are continuing concerns that the statistics on non-fatal accidents in the construction sector are rendered unreliable due to systemic under-reporting. Reporting rates are held to be especially low amongst the self-employed who often have little incentive to comply with statutory procedures. It is worth re-emphasising that many commentators have repeatedly further argued that customary subcontracting arrangements have negative implications for health and safety as a result of blurred responsibilities

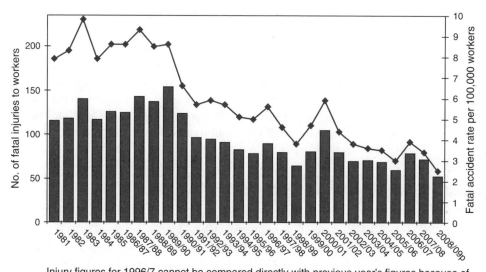

Injury figures for 1996/7 cannot be compared directly with previous year's figures because of introduction of RIDDOR 95

Figure 10.1   Fatal injuries to workers in construction January 1981 to March 2009 (*Source*: Department for Work and Pensions (2010)/Health and Safety Executive)

(Gyi *et al.*, 1999; Haslam *et al.*, 2005). Clarke (2003) is especially forthright in arguing that the possibility of a positive safety culture is seriously undermined by an increased percentage of contract and contingent workers. However, in the late 1990s such critical voices were easily contained by the *Rethinking Construction* target to reduce construction accidents by 10% per year. Provided that fatalities continued to fall nobody seemed to worry too much about the under-reporting of non-fatal accidents.

One particular source of statistical uncertainty relates to the amount of overtime being worked, which is again less likely to be recorded accurately among those who are engaged on non-standard forms of employment. Overtime is more likely to occur during periods of economic growth, thereby having an adverse impact on the health and safety statistics. The picture is further clouded by the grey economy of micro-firms which serve the domestic market. An increasingly important health and safety issue relates to the construction sector's high reliance on illegal migrant workers. Although this is frequently conflated with the grey economy, the presence of such workers is by no means confined to small projects. What is clear is that changing patterns of employment within the sector are likely to distort the available health and safety statistics.

A further issue relates to the number of accidents which subsequently prove to be fatal. The available statistics make no allowance for the significant improvement in ambulance response times and emergency medical treatment since the 1970s. An accident which resulted in a fatality in 1974 might not have had the same catastrophic consequence in 2010. The same

issue accounts in part for the lower recorded number of US military causalities in the Iraq conflict (2003–2010), in comparison with the period of
US involvement in the Vietnam War (1967–73).The recorded number of
causalities is not only an indication of the intensity of the conflict, it also
says something about the comparative standards of in-theatre medical
treatment. As an aside, there would seem to be even less interest in statistical data relating to the survival prospects of Vietnamese and Iraqi causalities. However, in the context of health-and-safety in the construction sector,
one issue is clear: statistics must always be treated with caution.

10.4.3   *Balancing voluntary regulation and legislation*

In charting the UK construction sector's progress in health and safety, it is
appropriate to begin in the 1970s during the years of the 'Social Contract'
(1974–79) (see Chapter 1). At the time, the government joined with the trade
union movement to place health and safety at the centre of the political
agenda. The aspiration was to strengthen the regulatory framework within
which issues of health and safety were managed. In contrast, business
organisations such as the Confederation of British Industry (CBI) consistently lobbied in favour of self-regulation as an alternative to 'restrictive'
government legislation (Beck and Woolfson, 2000). One of the first acts of
the incoming Thatcher administration in 1979 was to encourage the Health
and Safety Commission (HSC) to consider the 'economic implications' of
any proposed new regulations (Dawson *et al.*, 1988). Two subsequent White
Papers, *Lifting the Burden* (1988) and *Building Businesses Not Barriers* (1986),
further emphasised the need for a reduction in the 'regulatory burden' on
business (Green, 2009). The emphasis of government policy hence shifted
subtly away from legislation towards encouraging the Health and Safety
Executive (HSE) to be more sympathetic to the needs of business. During
the latter years of John Major's premiership, capacity within the Health
and Safety Executive (HSE) was further reduced by £15 million in budget
cuts, with direct implications for morale among inspectors (Beck and
Woolfson, 2000).

On the positive side, the Construction Design and Management
(CONDAM) regulations came into force in 1995 covering all sites where
five or more workers were employed. The regulations placed duties on all
those involved in the design and construction process. Clients were required
to appoint a designated planning supervisor and the principal contractor
was legally required to be competent in the management of health and
safety and to deploy sufficient resources. Designers were in turn required
to ensure that their designs carried minimum risks *vis-à-vis* health and
safety. The planning supervisor was awarded overall responsibility for
coordinating the health and safety aspects of the design. He (or she) was
further required to set up the initial stage of a health and safety plan which
would run throughout the project. Principal contractors were similarly

obliged to take account of health and safety when preparing tenders and to ensure the compliance of subcontractors. In addition, they were responsible for ensuring the provision of information and training for employees and for consulting with the self-employed on health and safety issues. The CONDAM regulations undoubtedly served to raise awareness of health and safety and to ensure that proper management regimes were in place. They were revised and updated in 2007 by the Construction (Design and Management) (CDM) Regulations. The revised Regulations replaced the planning supervisor with a CDM coordinator, whose job is to ensure that the Regulations are being observed and to act as an expert advisor to the client on health and safety procedures. The remit of the CDM coordinator extends throughout the design and planning phases of the construction work. Non-compliance with the regulations can result in unlimited fines, or imprisonment for serious breaches.

10.4.4   One Death is Too Many

Many of the issues above were bought into sharp focus in 2009 with the publication of Rita Donaghy's report to the Secretary of State for Work and Pensions entitled *One Death is Too Many* (Donaghy, 2009). Donaghy had previously been the chair of the Advisory, Conciliation and Arbitration Service (ACAS). When her appointment had been announced in 2008 it was roundly criticised by Alan Ritchie, the General Secretary of UCATT, on the basis that she had no background in construction. Yet, in sharp contrast to the prevailing trend within mainstream reports directed at construction improvement, the Donaghy report succeeds in capturing the very essence of construction:

> 'The Construction Industry generally is modelled to provide maximum flexibility. Consequently the majority of functions are contracted out and at least 40% of workers are self-employed or CIS[1]s. The advantages are obvious in that it reduces overheads. Some but not all argue that it improves profitability and productivity. The disadvantages are that it becomes more difficult for a safety culture to flourish, worker engagement is weak, employment security and continuity is minimal and skills training is at best patchy.' (Donaghy, 2009; 21)

One Death is Too Many certainly represents a sobering read for those who have followed the construction improvement debate over the preceding three decades. The Donaghy report reads in part like a catalogue of the

---

[1] The Construction Industry Scheme (CIS) requires subcontractors to register with the Inland Revenue to obtain a registration card which has to be presented to the main contractor before any payment can be made. The main contractor then deducts tax and national insurance from any due payments.

accumulated failings of the enterprise culture in construction. The report contains no less than 28 detailed recommendations, including an extension of the Gangmasters Licensing Regulations to cover construction. It was further recommended that the industry should renew its efforts to establish genuine consultative frameworks designed to encourage greater worker participation. Even more surprising is the plea that more should be done, particularly by the larger companies, to encourage joint working with the trade unions. Pleas of this nature were commonplace in the 1960s and 1970s, but such thoughts of collaboration with the trade unions were effectively displaced by the onset of the enterprise culture. It is notable that when organisations such as Construction Excellence champion the cause of 'collaborative working', it is invariably taken to apply primarily to clients and contractors. Professional consultants are undoubtedly embraced by the Strategic Forum's (2002) notion of the 'integrated team', which also extends to key suppliers. But collaborative working with trade unions consistently falls outside the accepted lexicon of industry improvement. Donaghy's advocacy for worker participation and constructive engagement with trade unions seems like an echo from a previous age. Donaghy's background with ACAS undoubtedly preconditioned her towards a process of consulting equally with employers and trade unions. This orientation led to a very different type of report from those which had dominated the industry improvement debate over the preceding couple of decades. One important distinction was the way in which the employment status of construction workers was given centre stage.

One Death is Too Many rehearses – and updates – many of the arguments about self-employment presented in Chapter 3. Donaghy certainly adopts a much broader interpretation of industry stakeholders than had prevailed during the Egan era:

> '[m]any stakeholders, particularly trade unions, some academics and bereaved families, feel strongly that self-employment, whether genuine or bogus, adds to the risk in the industry because self-employment is such a high proportion of the total. In London it is approaching 90%.'

The more usual interpretation of stakeholders is limited to those groups or individuals that the organisation depends upon for its continued survival. But Donaghy's interpretation includes not only the 'involved', but also the 'affected'. Such a view is in stark contrast to the guiding construct of lean thinking, whereby activities which do not add value to the client are classified as waste. Donaghy appeals to a much broader moral argument which is rooted in an expectation that everything which can be done should be done to keep construction workers alive. *One Death is Too Many* is particularly critical of the way in which construction fatalities are still seen to be socially acceptable on the basis that 'construction is a dangerous industry'. Prosecutions and sentencing were further observed to be ludicrously low.

The issue which lies at the heart of the Donaghy report is that construction is rendered more dangerous as a result of the sector's fragmented employment context.

One Death is Too Many offers a further damning indictment of the construction sector's record on occupational health. Although the brief of the inquiry was focused on fatal accidents, Donaghy recognises that far more workers die as a result of the chronic effects of ill health caused by long-term exposure to on-site hazards. The report points towards the high numbers of SMEs in the sector, and that for many any evaluation of occupational health risks consistently takes second place to 'making do' and getting the job done. Twenty tradesmen are estimated to die every week from asbestos related disease, with a further 12 construction workers dying every week from silica related lung cancer. However, statistical analysis of deaths resulting from occupational ill-health is even more problematic than that relating to accidents. Workers' health is often only affected 15–40 years after exposure, and it is invariably difficult to be sure precisely when exposure to hazardous substances took place. The fragmented employment context within construction also renders it extremely difficult retrospectively to identify who was responsible for any lack of preventative hazard control. There is of course no occupational ill-heath compensation scheme for construction workers.

An important point of reference is provided by the £4 billion British Coal Compensation Scheme. This was established in 1998 following the ruling of two High Court judges that the British Coal Corporation had been negligent in its hazard control procedures when the risks to miners were well known. Following the demise of British Coal in 1997, it was ruled that the government was responsible for the coal industry's historical failings. The first ruling of the Compensation Scheme related to vibration white finger, a debilitating illness that causes muscle spasms and can lead to gangrene. The second ruling related to chronic bronchitis and emphysema. These opened the flood gates to the largest-ever number of compensation claims against a single employer. A similar but much smaller scheme exists for radiation-linked diseases with the participation and support of major employers in the nuclear sector. UCATT have long campaigned for a government-sponsored occupational ill-health compensation scheme for the construction sector, with few imminent prospects of success.

10.4.5 *Government response*

The formal government response to *One Death is Too Many* (DWP, 2010), supported the majority of the report's findings. One important exception was the recommendation relating to the extension of the Gangmasters Licensing Regulations to construction. It was observed that this would involve the licensing of over 200 000 construction businesses, with significant capacity and cost implications. The view was expressed that the responsibility for ensuring the health and safety of construction workers was already clear under the CDM regulations. The principal contractor was seen to be responsible for all individuals on site, irrespective of their employment status. It was further emphasised that this applied equally to the directly employed, labour-only subcontractors and the self-employed. Each contractor working under the principal contractor was held to have a duty of care to those who are working under their control. Emphasis was therefore given to the effective enforcement of existing laws, rather the need for additional legislation. Subcontracting and self-employment were re-affirmed to be central to the way the construction sector operates, as had been clearly acknowledged in the Donaghy Report.

However, the government response did concede the issues associated with false self-employment and the adverse impact it has on levels of reporting relating to unsafe practices and serious accidents. In mitigation, attention was drawn to the government's support of various initiatives to improve awareness of vulnerable workers' rights and how to report abuses. It was further pledged that the government would continue to consider the possibility of applying the licensing regime to parts of the construction industry, so the possibility was at least left open for the future. At the time of writing, there is little evidence that gangmaster licensing in construction is being actively considered. The unfortunate truth is that it seems that there are a socially acceptable number of deaths in the construction sector, just as there are a socially acceptable number of deaths in road traffic accidents.

10.5   Never Waste a Good Crisis

10.5.1   Ten years on

The final report which deserves mention was published by Constructing Excellence in October 2009. *Never Waste a Good Crisis* (Wolstenholme, 2009) set out to review progress since *Rethinking Construction* and the extent to which the principles behind the Egan Agenda remain relevant. Particular attention was given to the extent to which the recommendations of *Rethinking Construction* have been implemented in practice. On the basis of the preceding chapters of this book it is clear that it is naïve to expect an instrumental relationship between codified best practice and what happens in practice. Nevertheless, it was this same instrumentalist way of thinking which characterised the Egan report and led to the subsequent policy-level obsession with key performance indicators (KPIs). For many, the justification for such an approach lay in the mantra of 'we can only manage what we can measure'. For others, the continued focus on KPIs served only to reinforce the industry's predisposition for simplistic machine metaphors.

Never Waste a Good Crisis was notable immediately for its reluctance to direct any criticism at *Rethinking Construction*. It seems that in Constructing Excellence circles the Egan report continues to enjoy an unchallengeable status. Constructing Excellence could have picked numerous other reports as the basis for such a review of industry progress, but other reports were seemingly not judged significant enough. Clearly nobody within Constructing Excellence had any interest in Egan's subsequent report on *Skills for Sustainable Communities* (ODPM, 2004). But the latter report comprised a very different kind of discourse which industry leaders had difficulty in relating to. Egan himself seems to forget about the latter report when talking to construction sector representatives.

The foreword to *Never Waste a Good Crisis* boasts contributions from Sir Michael Latham, Sir John Egan and Nick Raynsford MP, all of whom lend their enthusiastic support. Latham once again advocates the cause of partnering and close collaboration between client and the 'whole' construction team. Latham is nothing if not consistent. He further promotes the case for best practice, which he claims to mean that 'all have won and all must have prizes'. Unfortunately, as was described in Chapter 7, such an outcome does not necessarily follow from the instrumental implementation of partnering.

Egan predictably reflects back on 1998 and comments 'we could have had a revolution and what we've achieved so far is a bit of an improvement'. Most would agree with this statement, but the desire to instigate a 'revolution' was entirely unrealistic from the outset. Managerial regimes of target setting and performance measurement rarely produce dramatic improvements in performance. It did not work in the former Soviet Union, and neither does it seem likely to work in the UK construction sector. Even as late as October 2009, Egan was still seemingly obsessed with 'radical improvement', whereas most others had settled for continuous improvement. The difference in emphasis in part reflects the long-running battle between business process re-engineering (BPR) and total quality management (TQM). Egan consistently seems to opt for the more gung-ho storylines of BPR.

Nick Raynsford's contribution to the foreword looks back at the Movement for Innovation with no small hint of nostalgia, and recalls it as a time of hope and expectation. Tellingly, he also comments that it is not easy to reach a balanced judgement on what has been achieved, and this is certainly true. The difficulty is that change in the construction sector runs concurrent with changing expectations, thus rendering objective appraisal highly problematic. Even more problematic is the existence of multiple lobby groups representing starkly conflicting interests; hence the industry's fractured representative structure. The existence of multiple interest groups, which inevitably pull in different directions, presents an ongoing challenge to the espoused cause of 'integration'. This is no less true today than it was in the 1940s.

10.5.2 Re-writing history

The main body of *Never Waste a Good Crisis* takes an equally partisan view of the Egan improvement agenda. The Wolstenholme Report is perhaps best understood as a rallying call to arms rather than an objective appraisal of the merits of *Rethinking Construction*. The onset of recession in 2008 inevitably created a very different environment for any review of progress, and there were genuine fears that the construction sector would 'revert to type' by abandoning even its rhetorical commitment to collaborative working. The construction sector had experienced continuous growth since 1994 and firms had grown used to an ever-expanding market. Many industry practitioners had no memory of the previous recession which followed the 'Lawson boom' of the late 1980s. Certainly by October 2009 the electorate was waking up to the need for stringent cuts to address the accumulated budgetary deficit. And cuts in public expenditure are bad news for the construction sector. During the credit crunch of 2008–09, billions of pounds of public money had been diverted to bail out the banks, thereby preventing the economic recession turning into a full-blown depression. Strangely there was little talk of the banks being lame ducks that had to fend for themselves in the marketplace. Simply put, their role in the economy was too important for the government to allow them to fail. They did of course continue to pay their senior managers handsome performance bonuses, much to annoyance of the man (or woman) on the Clapham Omnibus. The crisis was of course global in nature and the UK economy was by no means alone in catching a severe cold. Some accorded Gordon Brown significant credit for steering the nation through the crisis. Others blamed him for creating the mess in the first place. It will be some time before such recent events can be judged dispassionately.

To say that *Never Waste a Good Crisis* was entirely uncritical of *Rethinking Construction* is of course not the same as saying that it offered an identical storyline. The Wolstenholme Report was equally enthusiastic about Saxon's vision of an industry based on the built environment, and the necessity to understand how value is created over the whole life cycle. The following quote is particularly powerful in communicating the vision adopted by Wolstenholme's review team:

> '[w]e want our industry to embrace the whole, complex picture of how people can interact sustainably with the environment to maximise health, wealth and happiness.' (Wolstenholme, 2009; 5)

While the vision stated above may have been compatible with Saxon's (2005) concept of a value-orientated built environment industry, it was far removed from the arguments which were promoted originally in *Rethinking Construction*. The Egan (1998) report had emphasised primarily the need to add value through efficiency improvement; the dominant metaphor was that of the machine. In contrast, the dominant metaphor behind Wolstenholme's vision of the industry is that of an organism striving to survive in complex interaction with its environment. In comparison with *Rethinking Construction*, the vision quoted above reads like it was written in a 1960s hippie commune. This was a very different message from that of 1998 Egan Report, and yet there was no acknowledgement that there had been a change in philosophy. Criticising the Egan report was clearly off the agenda, but at the same time it was seemingly important to be equally positive about subsequent reports which had received Construction Excellence's endorsement. Here lies a further notable characteristic of the construction improvement debate; previous reports are never cited to have been wrong. They are either supported entirely uncritically, or they are forgotten. The difficulty is the continuity of people which are involved in each progressive report. They are drawn essentially from the same clique of politicians and industry leaders. During the years of New Labour, individuals moved increasingly seamlessly between the two categories. Only occasionally is a report written by an outsider who adopts a very different viewpoint. One of the most striking examples is Rita Donaghy's report *One Death is Too Many*.

Never Waste a Good Crisis was noticeably bold in declaring the era of client-led change to be over. The gauntlet was thrown down to the supply side to demonstrate how it could improve. But ever since *Accelerating Change* (Strategic Forum 2002), the improvement agenda had been much more about 'integrated teams' striving to improve performance collectively. Client-led change had of course always been the exception rather than the norm. Wolstenholme had previously been Capital Projects Director at BAA where he was responsible for Terminal 5 as well as the Heathrow Express rail link. He was therefore one of relatively few individuals who could speak of client-led change with any degree of conviction. BAA had undoubtedly been an excellent client which had led the industry in a number of innovative approaches. But, as has already been observed, it did have the important advantage of being a privatised quasi-monopoly, and was hence relatively sheltered from the harsh winds of competition which prevail throughout open markets. It is unlikely that BAA would have been able to exercise the same degree of leadership had they been in a more competitive situation.

The Wolstenholme Report exhorts government to make greater efforts to understand the benefits which could be achieved through more intelligent design. It undoubtedly remains the case that many government departments continue to undervalue the importance of good design. But it should also be noted that in *Better Public Buildings* the DCMS (2000) was advocating good design and through-life value creation when the Eganites were still insisting upon compliance with simplistic machine metaphors about waste elimination. Wolstenholme might therefore have more meaningfully directed his criticism at *Rethinking Construction*. This recurring failure to point out the limitations of previous reports does little to encourage open debate, and ultimately constrains innovation. In truth, *Rethinking Construction* contributed as little to the design quality debate as it did to sustainability.

Rather than investigating progress since *Rethinking Construction* it might have been more useful to assess the extent to which the principles of the Egan report merely acted to legitimise changes which had already been initiated by the onset of the enterprise culture. Such a starting point would have led to a very different diagnosis.

Never Waste a Good Crisis further perpetuates the myth that Egan's Construction Task Force set out carefully thought-through targets for industry performance improvement. What is repeatedly forgotten is that the authors of *Rethinking Construction* were seeking primarily 'to illustrate the kind of targets which the Task Force wants to see construction adopt'. The suggested set of targets was readily admitted to be impressionistic and partial. It is therefore odd that the same targets should still be quoted some 11 years later. The continued adherence to the Egan targets risks reinforcing the mechanistic thinking which was so dominant in the immediate aftermath of *Rethinking Construction*. So much had changed since 1998, both within the construction sector and beyond, and yet in 2009 the industry improvement debate was still unhealthily obsessed with Egan's original targets. In the years following *Rethinking Construction* the Department of the Environment, Transport and the Regions (DETR) was undoubtedly under pressure from the Treasury to justify its investment in bodies such as the Construction Best Practice Programme and the Movement for Innovation. The targets therefore suited the Treasury's agenda, but it is doubtful if they ever suited the diverse needs of the heterogeneous construction sector.

There is of course a broader irony. It has become axiomatic that the private sector is more competitive, innovative and efficient than the public sector. This was the repeated justification provided over the course of several decades for initiatives such as compulsory competitive tendering (CCT) and 'Best Value'. It was also the justification for PFI. For many years the debate between public and private sectors had been essentially ideological, but progressively such arguments became outdated. Tony Blair famously declared himself to be in favour of 'whatever works'. The irony of course lies in unleashing private sector innovation, on the one hand, while continuing to insist on monitoring the construction sector against a set of generic targets, on the other. This was of course a classic New Labour characteristic. But even more bizarre is the way in which the original Egan targets related to issues which would surely have been best left to the marketplace. At the same time, there seemed to be little interest in setting targets for the issues which the market was patently failing to deliver, such as stable employment and training regimes.

10.5.3 Progress so far

Wolstenholme (2009) cites the 500 demonstration projects as the strongest body of evidence in favour of the principles of *Rethinking Construction*. Clearly, his Review Team did not share the reservations about the demonstration projects which had been expressed previously by *Modernising Construction* (NAO, 2001). Furthermore, the Egan 'principles' had been vague from the outset, and had been reinterpreted subsequently to such an extent that it was decidedly unclear which particular principles the demonstration projects were supposed to demonstrate. Wolstenholme certainly never succeeded in engaging with the paradoxes which lie behind supposed best practice recipes such as partnering and lean construction.

Nevertheless, the *Never Waste a Good Crisis* team did conduct a survey which revealed that the commitment to the 'principles' of *Rethinking Construction* often remain only skin deep, even in those cases where they have allegedly been adopted. There is a direct correlation here with the 'responses to lean production' discussed in Chapter 8. Wolstenholme's Review Team clearly had an expectation that the Egan principles would be adopted in accordance with the 'universal applicability model'. In other words, their expectation was that firms would stop operating in the way they had previously, and begin operating in an entirely different way. However, the survey results are indicative of the 'lip-service' model, where by adoption tends to be limited to cherry picking isolated techniques together with a few supporting sound bites. In truth, this should not have been too much of a surprise.

Questionnaire surveys of this nature are of course of limited value in determining the effectiveness of planned organisational change. The primary problem is that both the respondents and the evaluators are too fully immersed in the particular change discourse to be able to offer any degree of detached objectivity. A further difficulty is that only those who are broadly sympathetic to the questions being asked tend to complete such questionnaires, thus leading to a very biased sample. In light of this inbuilt bias it was therefore no surprise that 90% of the respondents reported a positive impact from *Rethinking Construction*. But it was interesting that many at the same time pointed towards its limited uptake. Especially interesting is the reported finding that the most widely perceived benefit was a greater emphasis on integration, collaboration or partnering. But many respondents added the caveat that the benefit was patchy and did not extend into the supply chain. Such responses would be much more meaningful if there were commonly agreed definitions for these supposed 'new ways of working', all of which have been advocated repeatedly at least since the Second World War. The point is well illustrated by the surprise expressed that only 56% thought that 'integrating the process and the team around the product' remains important. The question was raised that this could mean that the respondents felt that sufficient progress had already been made, or alternatively that there was still a large section of the industry that has yet to understand the supposed benefits of an integrated process. In other words, the responses did not mean very much at all. Integration and collaboration are undoubtedly words that will be trotted out for many years still to come, but ultimately they are only words. To understand how such words can be translated into meaningful action is a different challenge altogether. And any discussion of this challenge needs to rest on experience and an understanding of the ever-changing context within which such words are enacted. It is to such a broader understanding that this book has sought to contribute.

An associated problem is the difficulty of ascertaining which innovations are inspired by the Egan Agenda and which are not. Many attempts to implement partnering may well have been inspired by the Latham report (1994), or even the notoriously hackneyed *Partnering the Team* (CIB, 1997). In truth all such reports merge into a continuously fluxing discourse of industry improvement, with many internal contradictions and competing sub-narratives. The changing nature of this discourse has been plotted throughout the course of this book; as has its relationship with the changing structural characteristics of the sector and the broader socio-political context. Such issues cannot be understood as unrelated constituent parts, but need to be understood as a single complex entity.

Notwithstanding the above, *Never Waste a Good Crisis* clearly acknowledges the difficulty of knowing whether the industry's alleged improved profitability is due primarily to a decade of favourable economic conditions, or to the effects of process efficiency and waste elimination. The authors might equally have expressed caution about the difficulties of separating the efficiency gains due to lean thinking from, for example, those which were due to BPR. The overriding problem lies in the expectation of a linear relationship between cause and effect. Yet such linear models have become increasingly discredited in favour of other models which emphasise the non-linear, iterative and multi-agent character of the innovation process (Perkman and Walsh, 2007). This is a debate about innovation with which the mainstream advocates of construction improvement have yet to engage. Too many advocates of industry change remain obsessed with measuring the impact of ideas in isolation of any understanding of the way in which ideas are diffused and re-configured in practice.

Like many other harbingers of construction improvement, the authors of *Never Waste a Good Crisis* seem blind to the extensive re-structuring which took place in the construction sector in the 20 years preceding the Egan report. It is insufficient to advocate innovative ways of working in isolation from an understanding of the context in which they must be applied. Advocated innovations will inevitably be re-configured in interaction with the pre-existing dynamics of industry change. However, the extensive changes which took place prior to 1998 were seemingly of no interest to the Wolstenholme Review Team. This reflects a recurring tendency; in the assessment of industry improvement it has become commonplace to accept 1998 as the base date for analysis. Nothing that happened prior to *Rethinking Construction* is judged relevant. Notwithstanding the occasional acknowledgement of Latham's role in 'paving the way' for Egan, 1998 has become the equivalent of Cambodia's Year Zero. It is this fallacy above all others which it is hoped this book has helped overcome.

10.6   A final word

In seeking to pull the threads of this book together it is appropriate to point towards the continued popularity of collaborative working, integration and lean. Despite being advocated repeatedly for many decades, these ideas are still seemingly persuasive among those who promote the cause of industry improvement. The continued currency of such ideas is evidenced by the comments of Construction Excellence Chief Executive Don Ward in the April 2010 newsletter:

> 'Collaborative working, integration and lean will be essential to deliver low carbon buildings with much lower budgets, and our mission continues apace to save the industry from the damage of a reversion to lump sum lowest price tendering.'

Collaborative working, integration and lean would of course be even more essential if they could be defined in any meaningful way. But in truth, the longevity of such concepts is related directly to their essential vagueness. For those who have not yet slipped into cynicism, these over-lapping storylines may indeed inspire localised innovation. Others are more likely to appropriate such fashionable sound bites for the purposes of legitimising the structural changes of the last 30 years. As has been argued at length, fashionable management storylines can also play an important role in bolstering practitioners' sense of self-identity in the face of their day-to-day lived realities.

However, what is clear is that the repeated emphasis on collaborative working cannot be understood in isolation from long-term trends in outsourcing and subcontracting. Integration is similarly attractive as a mantra of industry improvement because it offers a rhetorical corrective to recurring problems of industry fragmentation. It is much easier to issue calls repeatedly for more integration than it is to do anything about the industry's hard-wired reliance on numerical flexibility. Yet 'integration' is rarely advocated as an alternative to the construction sector's fragmented employment patterns – such conversations seemingly take place in entirely disconnected spaces. The overriding importance attached to 'leanness and agility' has led directly to the dominance of the hollowed-out firm, with widely acknowledged adverse implications for health and safety, employment conditions and training. Construction firms became experts in 'leanness' even before the terminology became fashionable. The present book of course has not offered any solutions to the problem of 'construction improvement', but it has offered a more contextualised understanding of the way the construction industry improvement debate has ebbed and flowed over time.

The most recent challenge for construction improvement relates to the need to move towards a low carbon built environment. The HM Government

(2010) Innovation and Growth Team (IGT) report on *Low Carbon Construction* points towards the need for the biggest change management programme since Victorian times if the commitments of the Climate Change Act are to be met. The suggestion is that if 10–30% cost savings can be secured from integration and modernisation then low carbon buildings could feasibly be delivered for the same price. Even greater problems lie with the existing building stock, where the problem has more to do with user behaviour than with technological solutions. Specific barriers to progress identified by the IGT (HM Government, 2010) include:

- the structure of the industry, particularly in the lack of collaborative integration of the supply chain;
- a growing need for a general up-skilling of people in all parts of the supply chain to address the design, construction and operation of low carbon, energy efficient buildings;
- a continuing preoccupation, on the part of many public and private clients, with initial capital cost, instead of appraising projects on a whole life basis;
- the lack of drivers for a change in customer demand, without which the supply side lacks the confidence to invest in new products and services for which there may be no market at a profitable price.

These barriers to progress are not very different from those identified previously in *Building Britain 2001* (University of Reading, 1988) and *Progress through Partnership* (OST, 1995). Much within the construction sector has changed beyond all recognition, and yet the cited barriers to progress remain essentially the same. The underlying problem has been that the incentives for sustained improved performance have not been sufficient to overcome the institutionalised barriers to change. If the barriers identified by the IGT are to be overcome, it will require a very different approach to industry reform than the one that has prevailed over the last 30 years. In assessing the challenges of the future, it is hoped that much can be learned from the failures of the past.

Postscript

On 6 May 2010 the years of New Labour came to end. Gordon Brown received insufficient electoral support to form a government. David Cameron's rebranded Conservative Party secured the greatest number of seats, but still fell 20 seats short of an overall majority. For the first time since 1974 there was a hung parliament with no single party in command of an overall majority. After several days of political negotiation, a coalition government involving the Conservative Party and the Liberal

Democrats was announced on 11 May 2010. Gordon Brown exited 10 Downing Street, thus bringing to an end 13 years of Labour government. David Cameron, leader of the Conservatives was named as Prime Minister, with Nick Clegg of the Liberal Democrats taking on the role of Deputy Prime Minister. The electorate had apparently not been convinced by Gordon Brown's track record of economic management. But in an age when personalities routinely take precedence over policies, he undoubtedly suffered as a result of his inability to smile as effectively as David Cameron (or indeed Tony Blair). For the voters of Middle England, Gordon Brown also had the distinct disadvantage of being Scottish. But in truth the electorate had become tired of the Blair/Brown dynasty, in much the same way that they had previously tired of the Thatcher/Major period. What is clear is that the United Kingdom's economic situation and societal priorities inherited by David Cameron are very different from those inherited by Tony Blair.

The new coalition government wasted little time before announcing extensive cuts in public expenditure, with severe consequences for the construction sector. Public investment in roads was severely curtailed and the *Building Schools for the Future* programme was abruptly cancelled, with the loss of 700 planned new schools. The new coalition government also embarked upon yet another major reorganisation of the National Health Service (NHS). The expressed intention was to move hospitals outside the NHS to create a 'vibrant' industry of social enterprises. This represents a further re-structuring of one of the construction sector's major clients, with a corresponding dissipation of procurement expertise.

The *Building Schools for the Future* (BSF) programme had previously been heralded as an exemplar of New Labour's commitment to public-private partnerships. It was further held up as a highly innovative procurement context whereby the construction sector could at last compete on the basis of 'adding value'. BSF projects had also been at the forefront of the sustainability agenda, with many projects winning significant awards for their innovation. But suddenly all this changed. BSF was suddenly denounced by government as a bureaucratic monster which provided the taxpayer with poor value for money. It may well have been have been an important context for best practice under New Labour, but this was very clearly not the case under the incoming coalition government.

At the time of writing, the extent to which performance targets will characterise the management style of the coalition government led by David Cameron remains to be seen. Early signs are that the bureaucratic processes of target-setting may be ditched in favour of a greater reliance on unregulated enterprise. But the challenges of the future will not be the same as the challenges of the past. The coalition government remains committed to moving towards a low-carbon built environment, while implementing simultaneously severe cuts in public expenditure. It remains to be seen how this circle can be squared.

References

Antony, P.D. (1994) *Managing Culture*, Open University Press, Milton Keynes.

Banwell, Sir Harold (1964) *The Placing and Management of Contracts for Building and Civil Engineering Work*. HMSO, London.

Beck, M. and Woolfson, C. (2000) The regulation of health and safety in Britain: from old Labour to new Labour, *Industrial Relations Journal*, **31**(1), 35–49.

Bennett, J. and Jayes, S. (1998) *The Seven Pillars of Partnering*, Thomas Telford, London.

Bresnen, M. and Marshall, N. (2001) Understanding the diffusion and application of new management ideas in construction, *Engineering, Construction and Architectural Management*, **8**(5/6), 335–345.

Building Business – Not Barriers (1986), Cm 9794, HMSO, London.

Burt, M. (1978) *A Survey of Quality and Value in Building*, Building Research Establishment, Garston, Watford.

CIB (1997) *Partnering in the Team*. A report by Working Group 12. Construction Industry Board, London.

Clarke S. (2003) The contemporary workforce: Implications for organisational safety culture, *Personnel Review*, **32**(1), 40–57.

Dawson, S., Willman, P., Bamford, M. and Clinton, A. (1988) *Safety at Work: the Limits of Self Regulation*, Cambridge University Press, Cambridge.

DCMS (2000) *Better Public Buildings: A Proud Legacy for the Future*, Department of Culture, Media and Sport, London.

DEFRA (2006) *Procuring the Future: Sustainable Procurement National Action Plan – Recommendations from the Sustainable Procurement Task Force*, Department for Environment, Food and Rural Affairs, London.

Department for Work and Pensions (DWP) (2010) *The Government Response to the Rita Donaghy Report*, Cm 7828, TSO, London.

Donaghy, R. (2009) *One Death is Too Many*. Report to the Secretary of State for Work and Pensions. Cm 7657, TSO, London.

Druker, J. (2007) Industrial relations and the management of risk in the construction industry, in Dainty, A., Green, S, and Bagilhole, B. (eds) *People and Culture in Construction*, Spon, London, pp. 70–84.

Egan, Sir John. (1998) *Rethinking Construction*. Report of the Construction Task Force to the Deputy Prime Minister, John Prescott, on the scope for improving the quality and efficiency of UK construction. Department of the Environment, Transport and the Regions, London.

Emmerson, Sir Harold (1962) *Survey of Problems Before the Construction Industries*. HMSO, London.

Evans, R., Haryott, R., Haste, N. and Jones, A. (1998) *The Long Term Costs of Owning and Using Buildings* Royal Academy of Engineering, London.

GCCP (2000) *Achieving Sustainability in Construction Procurement*, Government Construction Clients' Panel, London.

Green, S.D. (2009) The evolution of corporate social responsibility in construction: defining the parameters, in *Corporate Social Responsibility in Construction*, (eds. M. Murray and A. Dainty), Taylor & Francis, Abingdon, pp. 24–53.

Gyi, D.E., Gibb, A.G.F., Haslam, R.A. (1999) The quality of accident and health data in the construction industry: interviews with senior managers, *Construction Management and Economics*, **17**, 197–204.

Haslam, R.A., Hide, S.A., Gibb, A.G.F., Gyi, D.E., Pavitt, T., Atkinson, S., Duff, A.R. (2005) Contributing factors in construction accidents, *Applied Ergonomics*, **36**, 401–415.

HM Government (2010) *Low Carbon Construction IGT: Emerging Findings*, Department for Business, Innovation and Skills, London.

Hughes, W., Ancell, D., Gruneberg, S. and Hirst, L. (2004) Exposing the myth of the 1:5:200 ratio relating initial cost, maintenance and staffing costs of office buildings. *In:* Khosrowshahi, F (Ed.), *20th Annual ARCOM Conference*, 1–3 September 2004, Heriot-Watt University, Vol. 1, 373–81.

Hutton, G.H. and Devonald, A.D.G. (1973) The value of value, in *Value in Building*, Hutton, G.H. and Devonald, A.D.G. (eds), Applied Science Publishers, London, pp 1–12.

ILO (2001) *The Construction Industry in the Twenty-First Century: Its Image, Employment Prospects and Skill Requirements*, International Labour Office, Geneva.

Ive, G. (2006) Re-examining the costs and value ratios of owning and occupying buildings, *Building Research & Information*, **34**(3), 230–245.

Latham, Sir Michael (1994) *Constructing the Team*. Final report of the Government/industry review of procurement and contractual arrangements in the UK construction industry. HMSO, London.

Legge, K. (1994) Managing culture: fact or fiction, in Sisson, K. (ed.) *Personnel Management: A Comprehensive Guide to Theory and Practice in Britain*, Oxford, Backwell, pp. 397–433.

Leiringer, R., Green, S.D. and Raja, J. (2009) Living up to the value agenda: the empirical realities of through-life value creation in construction, *Construction Management and Economics*, **27**(3), 271–285.

Lifting the Burden (1988) Cm 9577, HMSO, London.

Murray, M. and Langford, D. (2003) Conclusion, in M. Murray and D. Langford (eds.) *Construction Reports 1944–89*, Blackwell, Oxford, pp. 196–217.

NAO (2001) *Modernising Construction*. Report by the Comptrollor and Auditor General of the National Audit Office, The Stationery Office, London.

NAO (2004) *Getting Value For Money from Construction Projects through Design*. Report prepared by Davis, Langdon & Everest, National Audit Office, London.

NAO (2005) *Sustainable Procurement in Central Government*, National Audit Office, London.

Ness, K. (2010) The discourse of 'Respect for People' in UK construction, *Construction Management and Economics*, **28**(5), 481–493.

ODPM (2004) *The Egan Review: Skills for Sustainable Communities*, Office of the Deputy Prime Minister, London.

Ogbonna, E. and Harris, L.C. (2002) Organizational culture: a ten year, two-phase study of change in the UK food retailing sector, *Journal of Management Studies*, **39**(5), 673–706.

OST (1995) *Technology Foresight: Progress Through Partnership*, Office of Science and Technology, London.

Pearce, D. (2003) *The social and economic value of construction: the construction industry's contribution to sustainable development*. nCRISP, Davis Langdon Consultancy, London.

Perkman, M. and Walsh, K. (2007) University-industry relationships and open innovation: towards a research agenda, *International Journal of Management Reviews*, **9**(4), 259–280.

Peters, T.J. and Waterman, R.H. (1982) *In Search of Excellence: Lessons from America's Best Run Companies*, Harper & Row, New York.

Saxon, R. (2002) The industry 'formerly known as construction': an industry view of the Fairclough Review, *Building Research & Information*, **30**(5), 334–337.

Saxon, R. (2005) *Be Valuable: A Guide to Creating Value in the Built Environment* Constructing Excellence in the Built Environment, London.

Strategic Forum (2002) *Accelerating Change*, Rethinking Construction, London.

Stone, P.A. (1966) *Building Economy: Design, Production and Organisation: A Synoptic View*, Pergamon Press, Oxford.

Taylor, F.W. (1911) *Principles of Scientific Management*, Harper & Row, New York.

TUC (2008) *Hard Work, Hard Lives*. The full report of the Commission on Vulnerable Employment, Trades Union Congress, London.

Turin, D.A. (1966) *What Do We Mean By Building?* An inaugural lecture delivered at University College London on 14 February 1966, H. K. Lewis, London.

University of Reading (1988) *Building Britain 2001*, Centre for Strategic Studies in Construction, Reading.

Willmott, H. (1993) "Strength is ignorance: slavery is freedom": managing culture in modern organisations, *Journal of Management Studies*, **30**(4), 515–552.

Wolstenholme, A. (2009) *Never Waste a Good Crisis*, Constructing Excellence in the Built Environment, London.

Index